Contents

KW-325-647

List of figures   vii
Notes on contributors   ix
Acknowledgments   xiii

Introduction: gender and landscape: renegotiating morality
and space   1
LORRAINE DOWLER, JOSEPHINE CARUBIA AND
BONJ SZCZYGIEL

PART I
A man's home is his empire   17

1   Home alone? Masculinity, discipline and erasure in
mid-nineteenth-century Ceylon   19
JAMES DUNCAN

2   The labourer's welcome: border crossings in the English
country garden   34
KAREN SAYER

3   Transplantation of the Picturesque: Emma Hamilton,
English landscape, and redeeming the Picturesque   55
SHIN-ICHI ANZAI

PART II
Mobile homes   75

4   At home aboard: the American railroad and the changing
ideal of public domesticity   77
AMY G. RICHTER

5   "The salt water washes away all impropriety": mass culture
and the middle-class body on the beach in turn-of-the-
century Atlantic City   94
DEBBIE ANN DOYLE

6   How to travel with a male   109
    THOMAS M. HEANEY

7   A wilderness for men: the Adirondacks in the photographs
    of Seneca Ray Stoddard   124
    FRANK H. GOODYEAR, III

PART III
Memories of home   143

8   Mapping the Amazon's salon: symbolic landscapes and
    topographies of identity in Natalie Clifford Barney's
    literary salon   145
    SHEILA CRANE

9   Pincushions, dormitory kitchens, and seed gardens:
    gender identity and spiritual place at the West Union
    Shaker village   162
    CHRISTINE GORBY

10 Cleaning house: or one nation, indivisible   182
    JESSICA BLAUSTEIN

11 "Virgin land," the settler-invader subject, and cultural
    nationalism: gendered landscape in the cultural
    construction of Canadian national identity   203
    PAUL HJARTARSON

PART IV
Writing home   221

12 The manly map: the English construction of gender in
    early modern cartography   223
    DALIA VARANKA

13 The importance of being provincial: nineteenth-century
    Russian women writers and the country   240
    HILDE HOOGENBOOM

14 "My garden, my sister, my bride": the garden of "The Song
    of Songs"   254
    KENNETH I. HELPHAND

15 Gendering ghetto and gallery in the graffiti art movement,
    1977–1986   269
    KRISTINA MILNOR

Index   282

Long Loan

This book is due for return on or before the last date shown below

St Martins Services Ltd

Gender and Landscape

Gender and Landscape is a feminist inquiry into a long-ignored area of study, the landscape. Although there has been an exhaustive investigation into issues of gender as they intersect with space and place, very little has been written about the gendering of the landscape.

This volume provides a bridge between feminist discussions of space and place as something "lived" and landscape interpretations as something "viewed". The first of its kind, this book demonstrates that feminist critiques of landscape and place are not exclusive ontological realms; rather, they are mutually constitutive. The chapters in this book argue that the gendering of space and place create moral codes within societies that can be read in the landscape.

A truly unique collection, this book will be of interest to all those academics whose research crosses into gender studies and geography.

Lorraine Dowler is Associate Professor of Geography and Women's Studies at Pennsylvania State University, USA.

Josephine Carubia is Chief Academic Liaison Officer at Pennsylvania State University, USA.

Bonj Szczygiel is Associate Professor of Landscape Architecture at Pennsylvania State University, USA.

Routledge International Studies of Women and Place

Series editors: Janet Henshall Momsen and Janice Monk

1 Gender, Migration and Domestic Service
Edited by Janet Henshall Momsen

2 Gender Politics in the Asia-Pacific Region
Edited by Brenda S.A. Yeoh, Peggy Teo and Shirlena Huang

3 Geographies of Women's Health
Place, diversity and difference
Edited by Isabel Dyck, Nancy Lewis and Sara McLafferty

4 Gender, Migration and the Dual Career Household
Irene Hardill

5 Female Sex Trafficking
Vidya Samarasinghe

6 Gender and Landscape: Renegotiating Morality and Space
Edited by Lorraine Dowler, Josephine Carubia and Bonj Szczygiel

Also available in this series

Full Circles: Geographies of Women over the Life Course
Edited by Cindi Katz and Janet Monk

'Viva': Women and Popular Protest in Latin America
Edited by Sarah A. Radcliffe and Sallie Westwood

Different Places, Different Voices: Gender and Development in Africa, Asia and Latin America
Edited by Janet Momsen and Vivian Kinnaird

Servicing the Middle Classes: Class, Gender and Waged Domestic Labour in Contemporary Britain
Nicky Gregson and Michelle Lowe

Women's Voices from the Rainforest
Janet Gabriel Townsend

Gender, Work and Space
Susan Hanson and Geraldine Pratt

Women and the Israeli Occupation
Edited by Tamar Mayer

Feminism/Postmodernism/Development
Edited by Marianne H. Marchand and Jane L. Parpart

Women of the European Union: The Politics of Work and Daily Life
Edited by Janice Monk and Maria Dolors Garcia-Raomon

Who Will Mind the Baby? Geographies of Childcare and Working Mothers
Edited by Kim England

Feminist Political Ecology: Global Issues and Local Experience
Edited by Dianne Rocheleau, Esther Wangari and Barbara Thomas-Slayter

Women Divided: Gender, Religion and Politics in Northern Ireland
Rosemary Sales

Women's Lifeworlds: Women's Narratives on Shaping Their Realities
Edited by Edith Sizoo

Gender, Planning and Human Rights
Edited by Tovi Fenster

Gender, Ethnicity and Place: Women and Identity in Guyana
Linda Peake and D. Alissa Trotz

Gender and Landscape

Renegotiating morality and space

**Edited by
Lorraine Dowler, Josephine
Carubia and Bonj Szczygiel**

Routledge
Taylor & Francis Group

LONDON AND NEW YORK

First published 2005
by Routledge
2 Park Square, Milton Park, Abingdon, Oxon OX14 4RN

Simultaneously published in the USA and Canada
by Routledge
270 Madison Ave, New York, NY 10016

Reprinted 2006

Routledge is an imprint of the Taylor & Francis Group, an informa business

Typeset in Baskerville by Florence Production Ltd,
Stoodleigh, Devon
Printed and bound in Great Britain by
MPG Books Ltd, Bodmin

British Library Cataloguing in Publication Data
A catalogue record for this book is available from the
British Library

Library of Congress Cataloging in Publication Data
A catalog record for this book has been requested

ISBN 10: 0–415–33949–9
ISBN 13: 978–0–415–33949–0

Figures

3.1 *Emma, Lady Hamilton*, Tommaso Piroli, after Friedrich Rheberg, 1794   57

3.2 *View of the English Garden at Caserta*, Jakob Philipp Hackert, 1793   60

3.3 *Encampment in the Great Namaqua Country*, François Le Vaillant, *Travels into the Interior Parts of Africa*, frontispiece, 1790   61

3.4 *View of Fusiyama from Yosiwara*, Alcock, 1863   63

3.5 *Lady H******* Attitudes*, Thomas Rowlandson (?), *c.*1800   67

5.1 "Watching the Bathers from the Boardwalk," *c.*1900–1908   101

5.2 "$5.00 Reward if you find me in this crowd of bathers, Atlantic City, NJ," *c.*1916   103

6.1 The "divided couple"   111

6.2 The "pampered woman"   114

6.3 "The call of the open road"   119

7.1 *Indian Head, Ausable Pond*, Seneca Ray Stoddard, *c.*1880   132

7.2 *Game in the Adirondacks*, Seneca Ray Stoddard, 1889   133

7.3 *Hotel Ampersand, Saranac Lake*, Seneca Ray Stoddard, 1891   135

7.4 *Stoddard's Photographic Platform*, Seneca Ray Stoddard, *c.*1880   136

7.5 *Drowned Lands of the Lower Racquette, Adirondacks*, Seneca Ray Stoddard, 1888   139

8.1 "Le Salon de l'Amazone," Natalie Clifford Barney, frontispiece to Barney, 1929   147

8.2 Barney's *pavillon* at 20 rue Jacob in Paris, *c.*1920   148

8.3 Temple de l'Amitié (Temple of Friendship) in Barney's rear garden at 20 rue Jacob in Paris, *c.*1920   150

8.4 *Portrait of the Amazon*, or *Portrait of Natalie Clifford Barney*, Romaine Brooks, 1929   155

9.1 Reconstructed drawing of the entire West Union Shaker village, based on the historic surveyor's map dated between 1824 and 1827   162

9.2   Drawing overlay of the original West Union Shaker village
        with the present day landscape, showing the extent of the
        Shakers' holdings in 1827   166
9.3   Composite drawing of the archaeological findings of the
        "Center Family" area   167
9.4   Drawing reconstruction of the Center Family area of the
        West Union site   170
10.1 Directors, artists, and sculptors of the 1893 Chicago
        World's Columbian Exposition   185
10.2 The White City's Court of Honor: Grand Basin looking
        westward from the Statue of the Republic   192
10.3 Souvenir map of the World's Columbian Exposition   194
10.4 The Broom Brigade, Children's Building   196
11.1 Aerial view, National Gallery of Canada   205
11.2 *Terre Sauvage*, A.Y. Jackson, 1913   206
11.3 *Stormy Weather, Georgian Bay*, F.H. Varley, 1921   215
12.1 "Novissima totius terrarium orbis tabula," John Seller,
        *c.*1676   227
12.2 "The World in Planisphere" [the "cosmography"],
        Herman Moll, *c.*1700   228
12.3 "A mapp of Virginia discovered to ye Hills, and in its Latt.
        From 35.deg: & 1/2 neer Florida, to 41.deg: bounds of
        New England," Virginia Farrar, 1975   234
14.1 Mary with garden symbols and other attributes, 1578   257
14.2 "Solomon had a Vineyard," Wharton Esherick, 1927   260
14.3 Ze-ev Raban, 1923   261
14.4 Arthur Wragg, *c.*1930   262
14.5 Granville Fell, 1897   266

Contributors

Shin-ichi Anzai is Associate Professor at the University of Tokyo. He is the author of *Paradoxes of the Hortus Apertus: Aesthetics of the English Landscape Garden* ((in Japanese) The University of Tokyo Press, 2000). His papers in Western languages include: "Illusionismus und Enttäuschung in englischen Gartentheorien um 1800," in G.-H. Vogel and B. Baumüller (eds), *Carl Blechen* (Greifswald, 2000).

Jessica Blaustein is a Mellon Postdoctoral Fellow in English at the University of Pennsylvania, where she teaches and writes at the intersections of American literature, architecture, visual culture, and critical theory. She has published articles on past and contemporary architectural design and suburbanization, and she is currently working on a book project entitled *Counterprivates: Household Subjects Gone Awry*, about the spatial politics of privacy in the early twentieth century. She holds a PhD from the Program in Literature at Duke University.

Josephine Carubia is the Chief Academic Liaison Officer for the Pennsylvania State College of Medicine. She holds academic appointments in the Department of Medical Humanities, the Department of English, and the Women's Studies program at Penn State University. Her research and writing have evolved from a focus on literature, particularly on the work of Virginia Woolf, to an interest in organizational leadership. Exploration of theories of knowledge from a feminist perspective continues as a major thread in her work.

Sheila Crane is a Lecturer in the History of Art and Visual Culture at the University of California, Santa Cruz. Her research focuses on the history and theory of modern architecture, with particular interests in issues of gender and sexuality, memory, place, habitation, and violence in the built environment. Her work on the landscapes of Natalie Barney's literary salon is part of a book manuscript in progress entitled *Constructing Sapphic Subjectivities: Natalie Barney, Romaine Brooks, and the Architectural Topographies of Female Homoeroticism.*

Lorraine Dowler is an Associate Professor of Geography and Women's Studies at Pennsylvania State University. Her interests focus in the intersection of gender with violent nationalism. Her previous research has focused on issues of identity politics in Northern Ireland. She recently started an investigation of the role of women as firefighters and civic nationalism in the United States.

Debbie Ann Doyle received her PhD from American University in August 2003. Research on her dissertation, "'The World's Playground' Tourism and Mass Culture in Atlantic City", was supported by grants from the National Museum of American History, Smithsonian Institution and the New Jersey State Historical Commission. She works for the American Historical Association.

James Duncan is a geographer at Cambridge University and author of numerous books that focus on issues of landscape representation. Some of his titles include *The City as Text: The Politics of Landscape Interpretation in the Kandyan Kingdom* (Cambridge University Press, 1990), a co-edited volume, *Writing Worlds: Discourse, Text and Metaphor in the Representation of Landscape* (Routledge, 1992), and a co-authored book, *Landscapes of Privilege: The Politics of the Aesthetic in the American Suburb* (Routledge, 2003).

Frank H. Goodyear, III is Assistant Curator of Photographs at the National Portrait Gallery, Smithsonian Institution, and an affiliated faculty member in the Department of American Studies at The George Washington University. He is the author of *Red Cloud: Photographs of a Lakota Chief* (University of Nebraska Press, 2003).

Christine Gorby is an architect (RIBA) and Associate Professor of Architecture at Pennsylvania State University. Her research focuses on the design and interpretation of spiritual environments including landscape and built form. She is currently working on a book *The Spiritual Dimensions of Experience: Women, Gender and Environment* that considers sacred American places outside the realm of the "church."

Thomas M. Heaney holds a PhD in history from the University of California, Irvine, where he completed his dissertation *"The Call of the Open Road": Automobile Travel and Vacations in American Popular Culture, 1935–1960*. He presently teaches at Feather River College where he is in the history department.

Kenneth I. Helphand is Professor of Landscape Architecture at the University of Oregon. He is a recipient of the University's Distinguished Teaching Award, a Fellow of the ASLA and an honorary member of the Israel Association of Landscape Architects. Helphand is the author of *Colorado: Visions of an American Landscape* (Roberts Rinehart Press, 1991), co-author of *Yard Street Park: The Design of Suburban Open Space* (John Wiley & Sons, 1994), *Dreaming Gardens: Landscape Architecture and the Making of Modern Israel* (University of Virginia Press, 2002), and former editor of *Landscape Journal*.

Paul Hjartarson is a Professor of English at the University of Alberta. His primary scholarly and critical research interests focus on twentieth-century Canadian literature and culture. In support of his research on the role of 1920s painting and literature in the cultural construction of Canadian national identity, he was awarded a grant by the Social Sciences and Humanities Research Council of Canada. He and Tracy Kulba have recently published *The Politics of Cultural Mediation: Baroness Elsa von Freytag-Loringhoven and Felix Paul Greve* (University of Alberta Press, 2003).

Hilde Hoogenboom is Visiting Assistant Professor of Russian at Macalester College. She is publishing a new translation from French, with introduction and commentary, of Catherine the Great's memoirs with Modern Library at Random House (2004). She has been awarded fellowships by the National Humanities Center and the Social Sciences Research Council for her book project *Identity and Realism: Russian Women Writers in the Nineteenth Century.*

Kristina Milnor is Assistant Professor of Classics at Barnard College in New York City. Her interest in the graffiti art movement springs out of work that she has done on women's voices in graffiti from the ancient Roman city of Pompeii. She is currently at work on a book entitled *Domestic Politics: Gender, Place, and Ideology in the Age of Augustus.*

Amy G. Richter is an Assistant Professor of History at Clark University. Her dissertation, *Tracking Public Culture: Women, the Railroad, and the End of the Victorian Public,* was awarded the Organization of American Historian's Lerner-Scott Prize (2001). Her book based on this project is forthcoming from the University of North Carolina Press.

Karen Sayer is Head of History at Trinity and All Saints College, University of Leeds, UK. She is the author of *Women of the Fields: Representations of Rural Women in the Nineteenth Century* (Minnesota University Press, 1995) and *Country Cottages: A Cultural History* (Minnesota University Press, 2000).

Bonj Szczygiel is Associate Director of The Center for Studies in Landscape History and Associate Professor of Landscape Architecture at Pennsylvania State University. Her recent research has been on women's voluntary organizations as they impacted on the nineteenth-century built environment and as they influenced the City Beautiful Movement. She has also written about the intersection of design and medical theory in urban planning and development of the same period.

Dalia Varanka holds a PhD in Geography from the University of Wisconsin–Milwaukee. Her research has been supported by grants from the National Science Foundation and fellowships from the John Carter Brown Library of Brown University, The Newberry Library, and the Graduate School of University of Wisconsin–Milwaukee. As a Research Geographer of the US Geological Survey, Varanka's research addresses the study of urban and temporal landscape change.

Acknowledgments

First and foremost we would like to thank the contributors for their ideas, patience and cooperation; it is obvious that this book would not exist without them. We are grateful to have worked with Janice Monk, Janet Momsen, Joe Whiting, and Terry Claque at Routledge, who walked us through this process with great kindness. We would also like to thank Kat Kleman Davis for her editorial support, which proved to be invaluable. We would also like to thank Colin Flint for his support and guidance throughout this process. Much thanks to Pennsylvania State University's College of Arts and Architecture, Continuing and Distance Education Program, The Institute of Arts and Humanistic Studies, the College of Liberal Arts; and the Department of Geography for their financial support.

Introduction

Gender and landscape: renegotiating morality and space

Lorraine Dowler, Josephine Carubia and Bonj Szczygiel

Several years ago, *The New Yorker*, in one of their humorous cartoons, illustrated a couple of offbeat hikers, a man and a woman, gazing from a hilltop over a forested landscape with the accompanying caption: "What, may I ask, does landscape have to do with gender?" *(The New Yorker* 2000). Although more pithily stated than most academics would prefer, the caption cuts to the heart of this book and asks a question that ironically has been overlooked by many scholars. In other words, the study of landscape, as it relates to gender, has been somewhat ignored. Although feminist scholars have actively engaged in analysis of spatial units such as, the body, public and private space, place, the homefront, the city and the built environment, we have, to some extent, disregarded the examination of landscapes as a system of power relations which are vital to the production of gendered identities.

To this point in time, landscapes have been the object of study by both artists and scholars. However, the critical examination of the gendering of these landscapes is a relatively recent event (see Monk 1992; Bondi 1992; Rose 1993; Domosh 1996; Johnson 1995; Currie and Rothenberg 2001; Jones *et al.* 1997; Deutsche 1996; Kolodny 1984; Westling 1996). We could argue the lack of a feminist critique of landscape is simply a hangover from the patriarchal history of the academy; however, although true in part, it does not explain why a feminist critique of landscape waned in favor of other spatial units. We argue it was to be expected, for the reason that despite the initial impression of universality, the scholarly study of landscape has not only been disciplinary but also elitist in nature. Still, as the result of the emergence of critical interdisciplinary paradigms, stemming from cultural, post colonial, subaltern, sexuality and gender studies, feminist scholars are now moving past disciplinary boundaries and engaging in discussions of both the locations and perspectives of groups that at one time had been roundly ignored in traditional scholarship. As a result, feminists now argue that the landscape not only reflects certain moral codes but performs as a medium to perpetuate socially constructed gender stereotyping.

The gendered nature of the landscape was rarely acknowledged until the development of feminist research in the 1970s, which not coincidentally mirrors the increase in the number of women in the academy. Historically the landscape was assumed to be masculine in that it represented the universal rather than the specific in most scholarly writing. Monk points to the difficulty in "seeing" gender in the landscape:

> It is not difficult to recognize the more obvious expression of class, race or ethnicity in the material landscape. The quality of residences and their decoration, the signs on shop windows, the graffiti on walls, the manicured lawns or the jumble of weeds and rubble convey to us impressions of affluence or poverty, diversity or homogeneity, and feeling of familiarity or strangeness, comfort or anxiety. *But gender?*
>
> (Monk 1992: 123)

Monk details the urgency for a language and critical methodology to situate gender in the landscape. It is the aim of this book to contribute to such a language and methodology. This volume is the product of a conference – Gendered Landscapes: An Interdisciplinary Exploration of Past Place and Space – which was hosted in 1999 by the Center for Studies in Landscape History, The Women's Studies Program and the Department of Geography at Pennsylvania State University. The aim of the conference was to assemble scholars who approach landscape studies from specific disciplinary perspectives, but who are interested in creating a new language that is not only interdisciplinary but grounded in both women's and men's everyday experiences. This book, as a result, is an interdisciplinary examination of the establishment of moral codes that render gendered landscapes.

Creating a new language

Westling is one of a group of scholars who have surveyed the history of literature, myth, drama and other texts in the traditions of the West to recover the stories of how landscape came to be gendered. Drawing upon work by DuBois (1988) and Lerner (1986), among others, Westling (1996) traces the establishment of moral codes through landscape metaphors that are inescapably gendered. She describes the societal standard of associating indigenous femaleness with "howling" and a disordered wilderness, as opposed to a benign and abundant pastoral paradise. Most importantly, Westling argues that it is one thing to observe how the metaphors of gender extend to the landscape; it is another to make explicit the link between what we think of as literary strategies and the enacted policies of governments. Westling encourages this by introducing the question of morality. She contends that if we come to accept indigenous peoples, women and the land as viewed through these pervasive cultural metaphors disseminated by the full weight of myth and literature, then perhaps we can trace the power negotiations behind immoral policies and actions with renewed perceptions. Westling suggests inquiries of this nature will lead us to narratives of accountability and moral responsibility.

Historically, landscapes have been exempted from moral responsibility due to their imagined nature. "Places" are made meaningful through lived experience while landscapes are viewed as deliberate scenes (Dennis 1999), or as Kolodny argues, simple fantasy, thereby rendering places as experienced and landscapes as contemplated. In this sense, landscape requires a deliberate act of looking and this act engages all sorts of interpretive strategies. In the past "Landscape" was situated in the realm of the visual, alongside other choreographed representations such as photographs, painting and film (Dennis 1999). Rose refines our understanding of the "visual" by exploring the difference between vision and visuality. For Rose "vision is what a human eye is capable of seeing" (6) while visuality refers to how vision is constructed in different ways, "how we see, how we are able, allowed, or made to see, and how we see this seeing and the unseeing therein" (Foster in Rose 1988a: ix).

The study of landscape as something "viewed" has a problematic past, due to a lack of understanding of the inherent power of "visuality." Historically the subjects of the vision, as in many of the cases in this volume, women, were ignored in terms of the visuality of landscapes. This could be attributed to disciplinary isolation, which in some cases limited the study of landscape to an elite terrain of great works such as art and literature. However this is not to imply that the "everyday" was ignored in past explorations of landscape. Scholars such as J.B. Jackson, Peirce Lewis, David Lowentahl and Donald Meinig focused their interest on the "ordinary landscape" (see Meinig 1979) that certainly recognized different ways of reading the landscape. However these earlier studies did not take into consideration issues of reflexivity and positionality in terms of the author's relationship to the landscape, in other words how other individuals might experience the same landscape differently. Once issues of gender (see Monk 1992 or Rose 1993), class (see Mitchell 1996), religion (see Withers 1996) or race (see Ley 1983; Schein 2003; Dwyer 2002) were introduced to the study of landscape, discussions, as Monk points out, became grounded into the exclusionary practices of everyday landscapes. For example, it is evident in the early studies of landscape that there was literally a "love" for the landscape. As the cultural turn has proven, this was certainly a blinded love, which was "unseeing" of the landscape as an active system of oppression.

During the 1980s and early 1990s new directions in landscape studies critiqued previous observations as biased through the positionality of the academic, who, in most cases, would be a white, heterosexual male. More recent studies have redefined landscape as a 'way of seeing' (rather than an image or an object), which incorporates socio-cultural and political processes (Duncan 1990; Schein 1997; Cosgrove 1984). However, despite the inclusion of such questions of power, feminist interpretations of landscape waned until the critiques of Monk (1992) and Rose (1993), who argued that although landscape studies had evolved, discussions of gender were once again ignored. The reason for this brings us to our next point, which is that very few women are represented in the study of landscape. However, as Karen Till points out, there has been some significant growth:

> Feminists explore the exclusionary process by which groups create, produce, and represent landscape to legitimize gendered ideologies. Some authors examine literature and imagery produced by or of women (Kolodny 1975, 1984; Norwood and Monk 1987), whereas other scholars examine how cityscapes are spatially and symbolically gendered and sexed (Domosh 1996; Hayden 1981; Wilson 1991). Recent work on official national landscape explores how public institutions, statuary, performance and myth reify and embody gendered notions of citizenship and private/public spheres in the built environment (Dowler 1998; Johnson 1995; Warner 1985). Feminists also advocate landscape-based citizen activism by arguing that marginalized groups can empower themselves by transforming taken-for-granted material landscapes to make their perspective and voices materially visible.
>
> (Hayden 1995 as quoted in Till 1999: 148)

As stated earlier, most of the past feminist critiques of the material environment have focused on discussions of space and place. It is important to state that these discussions should not be ignored in the study of landscape; rather, the study of the gendering of the landscape can be informed by this body of literature. Most importantly, an understanding of the relationship between public and private space is critical to the establishment of an interdisciplinary language located at the intersection of gender and landscape. In other words, feminist critiques of landscape and space are not exclusive ontological realms; rather, they are mutually constitutive.

The public and private language of landscape

As in the study of landscape, the exploration of the relationship between public and private space has been explored by many scholars across several disciplines. Traditionally, private spaces have been associated with the home and designated as feminine, whereas public spaces (or spaces outside the home) have been determined as masculine. Feminist theorists have explored alternative definitions of the public and private by analyzing the public in relation to the private (Staeheli 1996). Moreover, other scholars address the social interaction and power relationships that exist in and across public and private space such that "those who have the greatest power over space have both the greatest power of access and greatest exclusion" (Kilian 1998: 27). The power to allow access or to exclude individuals from a space is important to consider when analyzing the interaction between the public and private spheres. However, it is equally important to understand the productive and reproductive acts that occur across public and private space which can disrupt these traditional designations.

In feminist scholarship there has been a great deal of attention to the relationship of power to the construction of masculine and feminine identities in space and place (Spain 2001; Nash 1996; Walter 1995; Rose 1993; Spain 1992; Valentine 1989; McDowell 1983). Feminist scholars determined that gender divisions of space serve to confine women's access to knowledge, thereby reinforcing men's power advantage (Szczygiel 2003; Nash 1994; Spain 1992; Valentine 1992; Ardner 1981; McDowell 1983). In this vein, Rose argues that one of the most "oppressive aspects of everyday space" is the division between public and private spaces (Rose in Dowler 1998). As Benhabib contends, when examining the inter-dependent relationship of public and private space, we need to dismantle Arendt's notion of the 'shadowy interior of the household' (Arendt in Benhabib 1998). In other words, feminists now challenge conventional notions of family and personal life. As Landes argues:

> Feminists did not invent the vocabulary of public and private, which in ordinary language and political tradition have been intimately linked. The term "public" suggests the opposite of "private": that which pertains to the people as a whole, the community, the common good, things open to sight, and those things that are accessible and shared by all. Conversely, "the private" signifies something closed and exclusive, as in the admonition "private property – no trespassing".
>
> (Landes 1998: 1)

This book, through a discussion of landscape, challenges the historical legitimacy of the descriptions of space as private and public. Most specifically, this collection will transgress traditional views of the domestic landscape, such as the home, as a static immobile place of oppression. Rather, the authors examine the home as a dynamic site of re-evaluation and mediation of power relationships. As bell hooks argues, "the home is a place to regain lost perspective, give life new meaning ... that space where we return for renewal and self-recovery, where we can heal our wounds and become whole" (hooks 1990: 49).

Currie and Rothenberg (2001) gather concurring views on the so-called public/private dichotomy in their recent volume. McGibbon, in particular, argues that the "boundary between the public and private is largely 'fictional'" (McGibbon in Currie and Rothenberg 2001: 101). Many scholars in the Currie and Rothenberg volume, as in this one, seem to feel that the most fruitful research will examine the political and economic resources that are allocated around the public/private dichotomy and thus attempt to ascribe motivations based on power and resource differentials which call into question the establishment of a moral landscape. This methodological approach to public and private space also allows for regional variations such as differences in how gender intersects public and private space in different places in the world.

Deutsche proposes a revealing correlation between public/private distinctions and morality. She states "How we define public space is intimately connected to what it means to be human, the nature of society, and the kind of political community we want" (Deutsche 1996: 296). In a worldview that prizes harmony and surface unity, conflict and dissention are likely to be swept out of view, necessitating a distinction between what is allowed in the public view and what must be hidden out of view. "A rigid public/private dichotomy . . . consigns differences to the private realm and sets up the public as a universal or consensual sphere" (Deutsche 1996: 326). The position that Deutsche conveys is significant in placing spatial practice at the center of human consciousness and thus of identity and value formation. She holds that conflict is the structuring activity of social space (Deutsche 1996: xxiv), and that this is the guarantee of democratic practice. Morality is forged in these same negotiations that structure social space as democratic, autocratic, totalitarian, or otherwise; therefore, the concept of a public sphere mirrored by a private sphere conveys along with it a particular set of beliefs and habits that would be called morality. The concepts are inextricably intertwined.

The moral landscape

Duncan has argued that by encoding within a landscape various conventional signs of such things as group membership and social status, "individuals tell morally charged stories about themselves and the social structure of the society in which they live" (Duncan 1992: 39). Similarly, Walker argues morality as a set of "practices of certain social orders and individually varied lives" (Walker 1998: 15). In common language, she contends that morality is "locate[d] . . . in practices of responsibility that implement commonly shared understandings about who gets to do what to whom and who is supposed to do what for whom" (Walker 1998: 16). Just as we know that concepts and practices associated with landscape are related to identity, so also with morality, according to Walker:

> In the ways we assign, accept, or deflect responsibilities, we express our understandings of our own and others' identities, relationships, and values . . . Moral practices, in fact, cannot be extricated from other social practices, nor moral identities from social roles and institutions.
>
> (Walker 1998: 16, 17)

Similarly, as we have learned to understand different accessibility and agency in relation to landscape, so Walker contends a non-unitary conception of morality: "Because social segmentation and hierarchical power-relations are the rule, rather than the exception, in human societies, the commonplace reality is different moral identities in differentiated moral-social worlds" (Walker 1998: 17). Illustrative of the construction of such social worlds is Anderson's study of Vancouver's Chinatown in the late nineteenth century, which demonstrates how through racist discourse Chinatown was constructed as an immoral landscape which threatened Vancouver's white majority (Anderson 1991). Similarly, Driver highlights the "moral geography" of the mid-nineteenth-century English city, which viewed urban landscapes as "morally" diseased (Driver 1998), while Gruffudd's study of English nationalism demonstrates how the English countryside represented a place of "spiritual wholeness" (Gruffudd 1994: 61). To this end Sörlin argues that landscapes are "used and abused" for the purposes of power and control (Sörlin 1999: 103). Therefore landscapes are actively produced, programmed and scheduled. In other words they are not innocent; rather, they are the palette of a specific moral agenda.

While there has been a great deal of focus on the moral landscape, very little has been devoted to the gendering of that morality. Work has focused on such processes in the geography of institutions, in the use of architecture and landscape design to promote particular moral principles, and in the production of consciously "alternative social spaces" (Matless 1997, 2000). In this way, landscapes which are an embodiment of moral codes can point to heretical landscapes which, in turn, point to the production of alternative social spaces (Cresswell in Matless 1997). In other contexts, morality has been defined as an act of transgression of or obedience to moral codes which render individuals as ethical or deviant subjects (Foucault in Matless). To this end, the present volume addresses how individual women and men were viewed as ethical or deviant as their actions were interpreted by way of adaptation or transgression of the moral landscape. Most importantly, the book examines how certain transgressions, which at first glance one might think were deemed immoral, actually were considered appropriate in certain landscapes. In other words, landscapes often dictated morality rather than the other way around.

Having said that, it becomes clear that a discussion of moral landscape is more than an examination of architectural styles which juxtapose changes to the exterior of a home with the interior organization of space. More to the point, even here, the perseverance of the domestic landscape has to be viewed in a wider framework of moral messages about respectability, consumption and female domesticity (Pratt in Ley 1993). The domestic landscape, for example, can be seen to reinforce identities as well as the subordination of women or the mobility of men. For the landscape is not only an expression of dominant values but also tends to reproduce them as part of the natural order (Ley 1993).

About the essays

In the first part of the book, "A man's home is his empire," the authors question our ontological construction of the domestic landscape. These chapters focus on questions of scale at the level of the house, garden, nation and ultimately empire, whereby the vision of the English home was transplanted abroad as part of a nationalist vision of England. This transplantation of the English sense of home is illustrative of the power and morality inherent in human acts of looking and seeing in conjunction with one's positionality. In these chapters the act of viewing is dislocated, placed out of position, whereby the viewer eludes agency in constructing or even transfiguring the viewed scene. The authors demonstrate how ideologies of gender intersect with the wide variety of habits and operations that we collectively label "vision."

Illustrative of this is Duncan's argument in Chapter 1. Duncan examines the grafting of the English vision of home on to the colonial plantation in Ceylon. He argues that in a historical context, something different emerges from the colonial vision of the home: not England, not Ceylon, but a hybrid understanding of the private, a "morality of climate" which also pervaded the discourse of the public, i.e. moral masculinity. Duncan argues that in the case of Ceylon, the tropical highlands were represented as an adversary that presented a moral test of a planter's manhood, race and class.

Sayer further explores this dichotomy when she focuses her investigation of public and private spaces on the English country garden. She argues that the garden, as an extension of the home, was constructed as a parallel landscape: both feminine and masculine. Most importantly, the garden was a landscape of moral existence for both men and women, whereby the home, as an extension of nature, was associated with virtuousness.

The gendering of nature was inherent to the colonial experience. In examining the hybridity of nature, Anzai examines the role of the garden as a trope of the English home transplanted to the non-western world. Most specifically he examines the relocation of the garden as a metaphor for home to South Africa and Japan. To this end he examines the notion of the Picturesque, a mode of vision which emerged in the eighteenth century, which views the world as a series of established pictures. This mode of vision has been criticized for violently assimilating an abbreviated world (western and male) into the "other" (non-European, female body).

This transplantation of the home – whether it be at the scale of the house or of a nation – which disrupted notions of strict interpretations of public and private spheres, was not limited to the colonial period. In the next part of the book, "Mobile homes," we draw on the expansion of social science interest in mobility – the mobility of peoples, cultures and objects. The authors not only examine how people tour cultures but how cultures and objects themselves travel (Rojek 1997). Most specifically we examine how our understanding of public and private space constructs travel landscapes.

Illustrative of Americans' "attachment to home" in a public setting is Richter's examination of "public domesticity" in which she contends that the distinctions between the home and the public arena are mutable. She points to the second half of the nineteenth century when Americans learned to feel at home in public settings. Many scholars have demonstrated that the domestication of public spaces occurred repeatedly throughout the century as respectable women increasingly entered into this realm. The domestication of public spaces, however, reflects more than women's entry into the public realm. As both men and women made themselves at home in public, the meaning of home and public were themselves changing. Richter focuses on the experiences of railroad travel in order to trace the multiple and changing meanings of home as they were applied to specific, but representative, nineteenth-century public spaces.

The process of transplanting the home into a public setting changes our understanding of acceptable behavior in the public sphere. For example, Doyle examines how the very "public" landscape of the beach becomes a place where women can adopt new rules of conduct, displaying their bodies and engaging in playful, undisciplined motions that would be prohibited in any other environment. Actions that previously would have only been appropriate in the home were now presented in the form of a spectacle and consumed for a mass audience of tourists.

As a point of entry to the marketing of images to a mass audience of tourists Heaney examines how magazines such as *Holiday* and *The Saturday Evening Post*, while promoting family travel, constructed a gendered conception of the public and the private. These advertisements constructed a bounded world where a woman who traveled outside of the home needed a man to accompany her. These illustrations depict a limited travel landscape in which women traveled in planes and trains because they had schedules and destinations. Heaney argues that it was important to steer women away from places such as the wilderness, and that even though the wilderness still remained feminine, it was a place for "men only."

In this vein Goodyear examines how male identity is shaped and changed by the wilderness by way of examining the construction of the sportsmen in the landscape photography of Seneca Ray Stoddard. During the last thirty years of the nineteenth century, he was instrumental in transforming the region in upstate New York known as the Adirondacks into a marketable tourist product. However, through the process of constructing a consumable "wilderness," Stoddard also reconfigured it as a space that purposefully excluded all but a select segment of the population.

The chapters in Parts I and II of this volume highlight women's access to nature, whether it be the frontier, the beach or the wilderness, in very controlled settings. At first glance this type of surveillance could seem to be prohibitive; however, in the historical context women were transgressing the boundaries of the home and engaging with the public sphere in ways that would have been at one time thought impossible. Interestingly, when we shift discussion from the gendered gaze over the natural environment to that of women's transgression of the public spaces in the human-made built landscape, we see the blurring of the home with the public arena. In the third part of this book, "Memories of home," we examine the moral implications of how women redefined public spaces as "the home." The authors examine how we represent and remember the home from the smallest scale of the literary salon to the larger and larger scales of the Shaker village, White City, and the nation.

The built environment – both private and public – has long been structured as a means to implement control across lines of class, race and gender in a variety of ways. There is now a growing body of work that addresses past place and space as products of a complex society in which women also worked, lived and played. Most importantly this section of the book addresses how the identities of groups that were mediated by way of the built environment can now be repositioned and relocated in the landscape.

Starting from a drawing by Natalie Clifford Barney, Crane examines the ways in which representations of spaces, such as Barney's literary salon in Paris, transcended contemporary meanings of public and private space. Crane details how Barney focused on a small structure in her garden that took the form of a Greek temple. This place, although located in a private garden, comprised a site within and against which Barney publicly framed both herself and the literary community she gathered there.

Similarly, through the reconstruction of a historic Shaker site, Gorby explores how a community is represented through its public activities and how activities that took place in the home were erased. When considering the cultural geography of this Shaker site as a social process, the negotiations between public and private become important considerations. Shaker women, in particular, are examined within their own unique spatial structure that could be perceived as entirely public. However, given their chosen isolation from outside society and the fact that unlike many male members they had no business dealings or contact with the outside world, their activities were often considered (by the very same outside world) as homebound and thereby unimportant to preserve.

On the same scale, but with a different purpose the "White City" of the 1893 Columbian Exhibition articulated the fundamental relationship between space and representations of sexual difference. In this chapter Blaustein moves beyond traditional scholarship and criticism concerning the White City. Instead she examines a romance between two major characters – sex and architecture – in a bizarre fairy tale of national architecture. The first part of this chapter explores the 1893 World Fair's architectural construction of a national identity. Through close and critical reading of the architects' rhetoric as well as the built space itself, Blaustein understands the "White City" as a model of home, a national domestic landscape built by a self-declared family of men. Part Two of her chapter situates the proper "woman" in this national household, linking the architects' concern for aesthetic coherence with the disciplined display of a certain kind of female body. In short, Blaustein critically investigates the processes by which the Columbian Exposition "put on" a public face of the nation.

As Chicago's White City represented both the domesticated and eroticized female body, the National Gallery played a central role in representing Canada as a virgin body. Hjartarson contends that in forging the equation of nation with wilderness landscape the English-Canadian elite both redefined the field of cultural production in national terms and asserted patriarchal ownership over the means of production and consumption. The National Gallery of Canada played a central role in that work by representing Canada as an unpeopled "virgin" territory. This produced an interdependent private and public narrative of a national culture whereby it was written as a feminine body of nature.

Whether written, painted or built, a landscape's meaning draws on cultural codes of the society for which it was made, codes which are embedded in social power structures. In the fourth part of the book, "Writing home," the authors demonstrate how various textual mappings subvert traditional understandings of masculine (public) and feminine (private) spaces.

In the first essay Varanka examines the historical conventions and options upon which current cartographic design principles are based in order to disrupt what historically seemed to be neutral or objective science. Illustrative of this was the adoption of "Plain Style" which eliminated from cartographic design all that was thought to be feminine, such as emotion and subjectivity. Varanka moves past arguing that the feminine was written out of history; rather, she delves into the political motivations of why feminine style was abandoned. In this way she rewrites the feminine, which was at one time very much part of the public sphere of professional cartography, back into the profession.

Hoogenboom examines another type of textual mapping investigating how, in the mid-nineteenth century, some important Russian women writers represented themselves as provincial by positioning the rural Russian landscape prominently in their work. Hoogenboom argues that the reason for this provincial positioning was the restrictive nature of a public urban female identity. By placing their characters in the rural landscape these authors were not limited by preconceived notions of public behavior for women. In the countryside the writers and their characters fell beyond the watchful eye of an urban patriarchy, and could transcend the gender codes of public and private space.

Similarly, Helphand takes on the sacred by examining the illustrative history of the biblical poem "The Song of Songs." He suggests that love and marriage (to God and church) are two themes that pervade in both Hebrew and Christian illustrations that accompany the sacred text. But in extending beyond those well-accepted interpretations he introduces the notion of a private and public moral function of the garden. Connecting the imagery to the medieval *hortus conclusus*, Helphand introduces an expanded interpretation of secular imagery that speaks to both (private) earthly human lust as well as (public) sanctified love associated with religious devotion.

In the more contemporary setting of New York's commercial art galleries Milnor examines how graffiti art – during the mid-1970s in the streets of the south Bronx – gave rise within the mainstream cultural establishment to a heated debate which juxtaposed the "feminine" interiors of the professional art world to the "masculine" public spaces of the urban jungle. This chapter suggests that such a debate offers both a window on to the process by which ideologies of gender intersect with race and class to create a discursive map of the city, and, in turn, a means of exploring how that map becomes implicated in definitions of "real" cultural production.

Each contributor to this book points to one or more facets of the cultural frameworks and conceptual schema structuring our social lives and moral landscapes. Moreover, they suggest how historically established distinctions, hierarchies and correlations can be reversed, attenuated, erased entirely or dramatically reimagined. One of the most crucial contributions of scholars and activists working at this historical moment is bound up with the seemingly simple prefix "re-". The landscapes into which we are born and in which we make our livings and live our lives are ones in which relations of power are always already negotiated: we have not had a voice in the negotiations resulting in just these distributions of power, prestige and privilege. They are configured in certain decisive ways and, in accord with these inherited configurations, they are interpreted in certain conventional ways. The multifaceted task before us is, hence, to *renegotiate* the relationships enacted by material artifacts, from landscapes to buildings to our own bodies.

Bibliography

Anderson, K. (1991) *Vancouver's Chinatown: Racial Discourse In Canada, 1875–1980*, Montreal and Kingston: McGill-Queen's University Press.

Ardner, S. (1981) *Women and Space*, London: Croom Helm.

Benhabib, S. (1998) "Models of Public Space: Hannah Arendt, the Liberal Tradition, and Jurgen Habermas," in Landes, J. (ed.) *Feminism, The Public and the Private*, Oxford: Oxford University Press.

Bondi, L. (1992) "Gender Symbols and Urban Landscapes," *Progress in Human Geography*, 16 (2): 157–170.

Cosgrove, D (1984) *Social Formation and Symbolic Landscape*, London and Cambridge: Cambridge University Press.

Currie, G. and Rothenberg, C. (eds) (2001) *Feminist (Re)visions of the Subject: Landscapes, Ethnoscapes, and Theoryscapes*, Lanham, MD: Lexington Books.

Dennis, S.F. (1999) *Seeing the Lowcountry Landscape: Race, Gender and Nature in Lowcountry South Carolina and Georgia, 1750–1950*, Unpublished Dissertation, Geography Department, Pennsylvania State University.

Deutsche, R. (1996) *Evictions: Art and Spatial Politics*, Cambridge, MA and London, England: MIT Press.

Domosh, M. (1996) "A 'Feminine' Building? Relations Between Gender Ideology and Aesthetic Ideology in Turn-Of-The-Century America," *Ecumene*, 3 (3): 305–324.

Dowler, L. (1998) "'And They think I'm Just a Nice Old Lady', Women and War in Belfast, Northern Ireland," *Gender, Place and Culture*, 5 (2): 159–176.

Driver, F. (1998) "Moral Geographies: Social Science and the Urban Environment in Mid-nineteenth Century England," *Transactions of the British Institute of Geographers, New Series*, 13: 275–287.

Duncan, J. (1990) *The City As Text: The Politics of Landscape Interpretation in the Kandyan Kingdom*, Cambridge: Cambridge University Press.

—— (1992) "Elite Landscapes as Cultural (Re)production: The Case of Shaughnessy Heights," in Anderson, K. and Gale, F. (eds) *Inventing Places: Studies in Cultural Geography*, Melbourne: Longman Cheshire, 37–51.

Dwyer, O. (2002) "Location, Politics and the Production of Civil Rights Memorial Landscapes," *Urban Geography*, 23(1): 31–56.

Foster, H. (1988) *Vision and Visuality*, Seattle, WA: Bay Press.

Gruffudd, P. (1994) "Back to the Land: Historiography, Rurality and the Nation in Interwar Wales," *Transactions of the Institute of British Geographers, New Series*, 19: 61–77.

Hayden, D. (1981) *The Grand Domestic Revolution: A History of Feminist Designs for American Homes, Neighbourhoods, and Cities*, Cambridge, MA: MIT Press.

hooks, b. (1990) *Yearning: Race, Gender and Cultural Politics*, Boston, MA: South End Press.

Johnson, N, (1995) "Cast in Stone: Monuments, Geography, and Nationalism," *Environment and Planning D: Society and Space*, 13: 51–65.

Jones III, J.P., Nast, H. and Roberts, S. (eds) (1997) *Thresholds in Feminist Geography: Difference, Methodology, Representation*, New York: Rowman & Littlefield.

Kilian, T. (1998) "Public and Private, Power and Space," in Light, A. and Smith, J. (eds) *Philosophy and Geography II: The Production of Public Space*, Lanham: Rowman & Littlefield.

Kolodny, A. (1975) *The Lay of the Land: Metaphor as Experience and History in American Life and Letters*, Chapel Hill, NC: University of North Carolina Press.

—— (1984) *The Land Before Her: Fantasy and Experience of the American Frontiers, 1630–1860*, Chapel Hill, NC and London: University of North Carolina Press, 1984.

Landes, J. (1998) *Feminism: The Public and the Private*, New York: Oxford University Press.

Ley, D. (1983) *The Social Geography of the City*, New York: Harper & Row.

—— (1993) "Co-operative Housing as a Moral Landscape: Re-examining 'the Postmodern City'," in Duncan, J. and Ley, D. (eds) *Place/Culture/Representation*, London and New York: Routledge.

McDowell, L. (1983) "Towards an Understanding of the Gender Division of Urban Space," *Environment and Planning D: Society and Space*, 1: 59–72.

McGibbon, J. (2001) "Family Business: The Household, Gender, and Generational Relations," in Currie, G. and Rothenberg, C. (eds) *Feminist (Re)visions of the*

Subject: Landscapes, Ethnoscapes, and Theoryscapes, Lanham, MD: Lexington Books, 87–106.

Matless, D. (1997) "Moral Geography in Broadland," *Ecumene*, 1: 127–155.

—— (2000) "Definition of 'Moral Landscape'," in Johnston, R., Pratt, G. and Watts, M. (eds) *The Dictionary of Human Geography*, Blackwell (4).

Meinig, D.W (1979) *The Interpretation of Ordinary Landscapes*, Oxford: Oxford University Press.

Mitchell, D. (1996) *The Lie of the Land*, Minneapolis, MN: University of Minnesota Press.

Monk, J. (1992), "Gender in the Landscape: Expressions of Power and Meaning," in Anderson, K. and Gale, F. (eds) *Inventing Places: Studies in Cultural Geography*, Melbourne: Longman Cheshire.

Nash, C. (1994) "Remapping the Body/Land: New Cartographies of Identity, Gender, and Landscape in Ireland," in Blunt, A. and Rose, G. (eds) *Writing Women and Space, Colonial and Postcolonial Geographies*, New York: Guilford Press.

—— (1996) "Men Again: Irish Masculinity, Nature, and Nationhood in the Early Twentieth Century," *Ecumene*, 3(4): 427–453.

The New Yorker (2000) Cartoon by Edward Koren, October 9.

Rojek, C. (1997) "Indexing, Dragging and The Social Construction of Tourist Sights," in Rojek, C. and Urry, J. (eds) *Tourism Culture, Transformation of Travel and Theory*, New York: Routledge.

Rose, G. (1993) *Feminism and Geography, The Limits of Geographical Knowledge*, Minneapolis, MN: University of Minnesota Press.

—— (2001) *Visual Methodologies*, London: Sage.

Rothenberg, C. (2001) in Currie, G. and Rothenberg, C. (eds) *Feminist (Re)visions of the Subject: Landscapes, Ethnoscapes, and Theoryscapes*, Lanham, MD: Lexington Books.

Schein, R.H. (1997) "The Place of Landscape: A Conceptual Framework for Interpreting an American Scene," *Annals of the Association of American Geographers*, 87: 660–680.

—— (2003) "The Normative Dimensions of Landscape," in Wilson, C. and Grath, P. (eds) *Everyday America*, Berkeley, CA: The University of California Press, 199–218.

Sörlin, S, (1999) "The Articulation of Territory: Landscape and the Constitution of Regional and National Identity," *Norsk Geografisk Tidsskrift*, 53: 103–112.

Spain, D. (1992) *Gendered Spaces*, Chapel Hill, NC: University of North Carolina Press.

—— (2001) *How Women Saved the City*, Minneapolis, MN: University of Minnesota Press.

Staeheli, L. (1996) "Publicity, Privacy and Women's Political Action," *Environment and Planning D, Society and Space*, 14 (5): 601–619.

Szczygiel, B. (2003) "'City Beautiful' Revisited, An Analysis of Nineteenth-Century Civic Improvement Efforts," *Journal of Urban History*, 29(2): 107–132.

Till, K. (1999) "Definition of Landscape," in McDowell, L. and Sharp, J. (eds) *A Feminist Glossary of Human Geography*, London and New York: Arnold.

Valentine, G. (1989) "The Geography of Women's Fear," *Area*, 21: 385–390.

—— (1992) "Images of Danger: Women's Sources of Information About the Spatial Distribution of Male Violence," *Area*, 24: 22–29.

—— (1993) "(Hetero)sexing Space: Lesbian Perceptions and Experiences of Everyday Spaces," *Environment and Planning D: Society and Space*, 11: 395–413.

Walker, M. (1998) *Moral Understandings: A Feminist Study in Ethics*, New York and London: Routledge.

Walter, B. (1995) "Irishness, Gender and Place," *Environment and Planning D: Society and Space*, 13: 35–50.

Warner, M. (1985) *Monuments and Maidens: The Allegory of the Female Form*, London: Picador.

Westling, L.H. (1996) *The Green Breast of the New World: Landscape, Gender, and American Fiction*, Athens, GA: University of Georgia Press.

Wilson, E. (1991) *The Sphinx in the City: Urban Life, the Control of Disorder and Women*, London: Virago.

Withers, C.W.J. (1996) "Place, Memory, Monument: Memoralizing the Past in Contemporary Highland Scotland," *Ecumene*, 3(3): 325–344.

Part I

A man's home is his empire

1 Home alone?

Masculinity, discipline and erasure in mid-nineteenth-century Ceylon

James Duncan

From the 1830s until the 1880s, young English and Scottish men went to Ceylon to try to make their fortunes as coffee planters. British women were not to follow them into the hills in any numbers until the end of this period. For many young men it was not only the first time they had left Britain, but also the first time they had left their parental homes. The move represented a first home of their own where the discipline of the British family was to be replaced with Christian, manly, self-discipline.[1] Within the colonies, self-discipline was racialised and gendered; it helped constitute what it meant to become a European man.[2] As Stoler put it, the self-disciplining of individual colonial male Europeans was seen as tied to the survival of all Europeans in the tropics and hence the bio-politics of racial rule (Stoler 1995: 45). The home, the domestic milieu, was considered a prime site where such bourgeois values and consciences could be cultivated, where impulsive passions could be tamed and controlled. But a home on the frontier was also seen as "away from civilisation" and therefore a site where metropolitan norms of respectability could be somewhat relaxed while being simultaneously reaffirmed as a primary mark of whiteness. The resulting tension between the colonial and metropolitan homes produced an ambivalence about what constituted appropriate behaviour. Such a tension was manifested both in the self-censoring of written accounts of planter life and in a more acute form as a feeling of loss of Europeanness, a cultural anxiety about racial and gendered identities and differences.

Creating imperial homes

Visions of the economic and cultural transformation of tropical nature by English capitalism were common in mid-nineteenth-century Ceylon as elsewhere.[3] For example, the planter William Boyd wrote:

> [a] new era is dawning on Ceylon . . . The steam engine will be heard in every hollow, the steam horse will course every valley; English homes will crown every hillock, and English civilisation will bless and enrich the whole country, causing the wilderness to blossom as a rose, and

> making Ceylon, as it was in former times, a garden of the world and
> the granary of India.
>
> (Boyd 1888a: 410)

It is important to note that in the imagination, this place is not transformed
into England. Rather, through England, which is to say through English
capitalism, English civilisation and English homes, it is transformed into
a proper colony. But in the hills of mid-nineteenth-century Ceylon such a
transformation was still a pipe dream. In fact, the impact of the tropical
climate and native culture was seen as possibly more transformative of the
British than vice versa. The reverse impact destabilised notions of progress, development and the civilising mission, contributing to a crisis of
British, masculine identity (Gikandi 1996; Williams 1963), or at the very
least, a state of limnality that Phillips (1997: 13) identifies as inverted,
contradictory and reconstitutive.[4] A pioneer planter's first home in the hills
in the 1850s might be more like P.D. Millie's (1878) first "bungalow."[5]
He describes a mud and wattle hut, approximately 20 feet by 12 feet, par-
titioned in half by a straw mat. One half consisted of a combined servant's
room and kitchen, the other was the planter's bedroom and living room.
There was a hole cut in the mat for the servant to come back and forth.
Each room had a window and door, each with a plank on a leather hinge
to shut it. The simple furniture consisted of a bed, a table and chair. After
the first year on a new plantation, the planter would construct a more
substantial wooden dwelling consisting of a central living room, a kitchen
and bedrooms off the living room. But such bungalows were, more often
than not, ill-furnished and ill-kept.[6] Planters thought of such bungalows as
transitional, a first step on the way to constructing proper homes for themselves and eventually for British women. A proper home was thought to be
a key site in the transformation of the country. It represented a locus of the
"home country" in the foreign. Below I will explore the tension in the dual
meaning of home: as dwelling in the hills and as nation – Britain.

Moral hygiene and the tropical home

At the heart of this project of transforming a tropical place into a proper
colony was the theory of environmental determinism, a theory that, as
Livingstone (1991, 1999) argues, attributes a morality to climate. The
theory of environmental determinism, which can be traced back to Hippocrates assumes that human behaviour is controlled by the environment
(Glacken 1967). The effect of the heat and humidity of the tropics was
thought to be injurious to people of all races, although particularly to
Europeans who were unused to it. Such a theory was combined with
nineteenth-century racism to explain why peoples in some parts of the
world were less technologically advanced than Europeans. It was argued
that certain races had degenerated because they had dwelt too long in

tropical regions. There was also a concern that peoples from cooler regions, such as north-western Europe, would suffer if they remained in the tropics for too long. Anxieties about the possibility of successful white settlement in the tropics weighed heavily on the minds of imperialists. Arnold argues that, in spite of such concerns, environmental determinism played an important function in imperialism (Arnold 1996: 174). For according to prevailing determinist views, South Asia "was subject to nature to a far greater degree than Europe: hence, too, the need for the British rule over India, to bring about 'improvement,' 'order' and 'progress' and so liberate Indians from their servitude to nature." (Arnold 1996: 174). While the relative coolness of the hills in South Asia was sought out by some Europeans as a healthy respite from the heat of the lowlands (King 1976; Kenny 1995, 1997; Kennedy 1990, 1996), others questioned whether tropical coolness was the same as temperate coolness. Unlike in the temperate zone, coolness in the tropics was thought to be produced by humidity and great fluctuations of temperature, both of which were thought to have ill effects on the health of Europeans (Arnold 1996: 152).

Women, as the "weaker sex," and children were thought to succumb even sooner than men to the tropical climate (Arnold 1996: 37). The planter Millie (1878) was certainly of the opinion that European women could not survive for long in highland Ceylon. The climate took its toll on Europeans in a number of different ways. Again, Millie writes that "the climate of Ceylon has a strong action on the nervous system: after a long residence the nerves get unstrung, one easily gets irritated" (Millie 1878). Such ideas of climatically produced anxiety, which were thought to have a more pronounced effect on women than men, were later formalised by Woodruff in the late nineteenth century as the theory of tropical neurasthenia and was used to attempt to explain why whites could not acclimatise to the tropics (Kennedy 1990; Livingstone 1992: 232–41). Devastating to men and women alike was the cholera epidemic of 1845 where one-fifth of the population of Kandy died within a few weeks (Boyd 1888b: 281). Boyd, looking back on the early days of planting wrote:

> So great has been the mortality amongst young men whom I knew, that out of seventeen who lived twenty seven years ago within a few miles of me in one district, only two, myself and another, now survive; and this I fear is also the experience of the few survivors of the old days of which I write.
>
> (Boyd 1888b: 275).

It was tropical nature, the material reality as well as the imaginary geography of the tropics, its climate, its jungles and its poorly understood human and plant diseases that posed the greatest threat to the British planting community. Because there were so many more British men than women in the hills, the narrative structure of the writing by and for the planting community was of whiteness and masculinity. There developed a notion of moral masculinity struggling against tropical nature. Diaries, reminiscences and recruitment literatures were scripted through narratives of heroic masculine adventures. The ideal was challenged, however, by the material realities, often including utter failure, which produced insecure and strained imperial masculinity. While earlier models of moral masculinity were explicitly Evangelical, later constructions became more secularised into a cult of character, the gentleman and manliness. Due to the physical dangers of the tropics, the actual practices of masculinities deviated from British ideals in that it produced a highly embodied male. Extreme feelings of vulnerability produced a sense of embodiedness associated more often with women, the lower classes and racial others.

During the nineteenth century there was a feeling that through proper behaviour, or what was termed "moral hygiene", the impact of the tropics on the male European body could be somewhat alleviated.[7] For Ceylon to become a proper colony, not only had local people to be civilised, but equally important, the British colonists had to guard against losing civilisation in themselves. As Stoler argues, "European was not a fixed attribute, but one shaped by environment and class. It was contingent rather than being secured by birth" (Stoler 1995: 104). In other words, "civilisation" was something that was performative and embodied, in the gesture, the word, the act, the dress. The notion of European civilisation was explicitly bourgeois. Poor Europeans, like Europeans who had dwelt in the tropics for too long, were seen as lacking the degree of self-control necessary to perform a proper European identity. As such, they occupied a position somewhere between "real" Europeans who were recently arrived from the home country and mixed race locals.[8]

A key aspect of moral hygiene was proper housing. As one planter put it, "[b]ad housing accommodation, after a time begins to tell on a man's general character . . . When one gets dirty and careless in personal accommodation and appearance, worse is not far off . . ." (Millie 1878: n.p.). Absolutely central to this struggle was European, masculine, self-control. The visible, outer condition of slovenliness was thought to be both a product of and a signifier for the onset of tropical degeneration. Planters, especially in the early days, were widely dispersed and as they were unlikely to be accompanied by European wives, they had to battle against weakness produced by loneliness and the temptations of the tropics. Such conditions were thought to lead the planter away from Europeanness towards degeneracy and miscegenation. It was believed that a man's moral fibre once weakened by loneliness led him into such temptations as drink, sloth and uncontrolled sexuality. The planter Brown, in a locally published book aimed at the planter community, describes the planter as a "modern day Robinson Crusoe" (Brown 1845). Such a trope resonated with readers at that time, because the first plantations were thought of as widely interspersed islands of civilisation in a vast sea of tropical nature and natives. Like servants in Britain, the natives were present physically, but not socially. Millie, in his reminiscences of planting in the 1840s, writes:

As far as the eye could see the horizon was bounded by this perpetual jungle ... A sense of utter loneliness came over me; it was worse than being at sea: the ship moves and we get out of the waste of water in time, but here is a fixture in the settled gloom of never-ending forest. Frequently, as evening approached, enveloped in thick mist, not a sound was to be heard but the sharp bark of the red elk, the scream of the night hawk, varied by the crashing of the elephants in the forest. The birds have no song during the day, insect sound is mute, and the silence can almost be felt. The moaning sound of the wind passing over the forests serves also to increase the feeling of gloom.

(Millie 1878: n.p.)

Millie (1878) later describes the early plantations as being like prisons cut out of the forest. The images of the deserted island, the ship at sea and the prison are all images of the solitary white man.

And yet, there were many Tamil workers on the plantations and Sinhalese in the villages nearby. But the early planters spoke little Tamil or Sinhalese and furthermore fraternising with male field hands was considered unacceptable. Household servants, who were nearly always male, occupied an intermediate status, however (Carpenter 1910: 23, 75–80). The planter might chat with them, but would not admit them to be his friends. During the period of which I write, it was common to have a Sinhalese concubine in the bungalow, often with a room of her own (Hyam 1990: 118). But little or nothing was written of her. Such mistresses were thoroughly erased from the literature of the bungalow, appearing nowhere in the planters' accounts. They appear only in such documents as police accounts as a victim of planter violence or as the cause of violence between planter and house boy. Hence, bourgeois propriety portrayed planting life as even more lonely than in fact it was.

But even where there were other planters, loneliness could be a problem, especially for the newly arrived. For example, the seventeen-year-old James Taylor expressed his home sickness in letters home written shortly after he arrived from Scotland to assume his post as an assistant superintendent on a coffee plantation in 1852 (23 June 1852). He often asked for more letters from home "which would be a great comfort to me in this lonely wilderness" (Taylor 1851–92). And he missed the homely landscape of his native Scotland. He asked about the potatoes at home:

I wish we had them here though they were diseased. I would give something for a peck of oatmeal too or a pint of milk – and have the cows been eating the corn? ... As yet I mind everything [at home] as distinctly as though I had left yesterday – every cut in the road and every large stone and all the blue hills and knolls.

(Taylor 1851–92)

Temptations

While loneliness for isolated planters and the newly arrived was seen as a potential source of weakness leading to drink and the unhealthy associa-tion with native men and women, the association with other male planters was also seen by some as potentially dangerous. The planter life was characterised by a sense of fraternity, of masculine camaraderie and homo-eroticism among the small number of British men living in the hills. William Boyd wrote that the planters in the 1830s and 1840s were almost all bachelors who "kept open house" which was "extended to all and sundry who possessed a white face, who would conform to the habits of good society and possessed the rank of gentleman" (Boyd 1888b: 266). He went on to add that "it was even extended to many who had, perhaps at one period possessed the rank of gentleman, but whose habits had excluded them from all the privileges of their order" (Boyd 1888b: 266). Anderson refers to the social condition of colonialism as a "middle-class aristocracy" where issues of character rather than class background were paramount (Anderson 1983: 137). Stoler, however, argues that this varied from colony to colony and that previously more fluid class distinctions rigid-ified toward the end of the nineteenth century (Stoler 1991). What Boyd seems to be suggesting, however, is that the behaviour of some of these men did not conform to proper metropolitan norms of respectability and manliness. Such lapses were thought to be caused by a number of factors. The first was the tropical climate itself that weakened character. The second was isolation from positive civilising influences such as the church. And the third was the absence of clean and orderly homes cared for by proper middle-class women who could create an atmosphere of respectability, defined as sexual and other forms of bodily repression.[9]

The main forms of recreation for the planters were eating, drinking and hunting with neighbouring planters (Weatherstone 1986: 176). For some, especially those from privileged backgrounds, the colonies were a place where British country sports could be played. Some kept hounds and used native beaters to hunt elk (A Planter 1886: 29). Some local commentators worried that too many young planters spent their time pursuing such enter-tainments rather than working. This again was seen to be caused by a tropical environment that undermined the desire in Europeans to work, as it had done for millennia to native peoples. The frivolousness of such young planters was contrasted unfavourably with what Stoler refers to as the "pioneering Protestant ethic" in which ultimate success results from hard work and perseverance (Stoler 1991: 125). Such a Protestant ethic, it was thought, was greatly encouraged by having the supportive environment of a bourgeois home.

These bachelor homes were sites of ambivalent masculinity. On the one hand, they could not be proper homes, for they lacked a key element, a white bourgeois woman. On the other hand, as Tosh notes, around the mid-nineteenth century there was the shifting of the meaning of masculinity so that young middle-class and upper middle-class men increasingly defined their masculinity against the domesticity of the metropolitan bourgeois home (Tosh 1991: 68). In part to escape this home space, which they saw as a realm of femininity, they left for a bachelor life in the colonies. There was a cult of single male company linked to the experience of public schools as a homo-social environment (Mangan 1985; Tosh 1991; Rutherford 1997). Rutherford argues that such an environment produced a state of "perpetual adolescence" among some upper middle-class and middle-class Englishmen whereby the home and club were thought of as an extension of boarding school (Rutherford 1997: 23–24). The ambivalence came from the fact that they knew that such a minority lifestyle was contradicted by the Victorian ideal of the woman-centred home. Furthermore, the bachelor home was also the site of illicit homosexual and heterosexual relations and as such was a site of social shame, as evidenced by the erasure of such relations from their detailed accounts of planter life.

Drink and sexual relations with local people were activities about which planters as a whole were decidedly ambivalent, at least in print. While it was admitted that planters drank, and often to excess, the latter was bemoaned. Lewis wrote of "scenes of dissipation and extravagance" and "absence of restraint" which "seduced ... [some planters] from their work" (Lewis 1855: 14–15). Brown wrote of the weakness of some planters whose:

> [m]oral principles have not been strong enough to enable [them] to resist temptations to which a solitary life, distant from social amenities and religious restraints and privileges, has added force. Comfort is found in stimulants; the man takes to drink; that leads to habits and associations which deprive the victim of his own self respect and the respect of even his coolies it is his business to command.
>
> (Brown 1880: 56)

These habits of the "solitary life" were not detailed. Unquestionably they entailed association with Tamils and Sinhalese – sexual liaisons with male or female workers or village women.

The subject of women was much more delicate than that of drink, for while there were plenty of condemnations of the evils of drink, especially from those of a religious persuasion, it was extremely rare to find any mention of sex at all. By the mid-nineteenth century, the impact of the Evangelical revival and the social purity movement made British men even less open about liaisons with native women than earlier, and consequently references to sexual relations were guarded and often disapproving (Ballhatchet 1980: 5). Having said this, it was common before the 1880s, as I have already pointed out, for planters to take Sinhalese women as concubines. Until the 1860s, they commonly lived in the planter's bungalow, but later, as sexual relations with native women became more disapproved of, they were displaced from the home and kept in their villages off the estate (Weatherstone 1986: 180). It was an unwritten rule not to take Tamil concubines from the plantation workforce, as this was thought to produce conflict with the male Tamil workers. However, as one might expect, that rule was not infrequently broken. A Tamil gang boss by the name of Carpen wrote of the frustration male workers felt at the sexual harassment of their women by planters (Moldrich 1989: 86). It appears that planters had a preference for girls in their early teens. This was justified in terms of tropicality. Tropical peoples were thought to develop more rapidly than Europeans, especially sexually. Jenkins writes "[i]n the east a girl is a woman at thirteen, a mother at fourteen, and a grandmother before thirty" (A Planter 1886: 12). Boyd echoes this when he writes that native women were "models of feminine beauty when in the first blush of their opening charms", but "by the time they reached the age when English women are at their best, their beauty has faded" (Boyd 1888a: 413).

The Victorian model of the family home, which was seen by many as a model for colonial development, was rarely put into practice in the hills prior to the 1880s. In fact as I described, the bachelor dwelling was the antithesis of the proper home. After a number of years some planters began to feel that they were no longer able to participate in the Victorian model of the home, for the colonial home had become the primary site of miscegenation and what Stoler has termed "unproductive eroticism" (Stoler 1995: 135). The spectre of racial degeneration lay heavily on some of these men. For example, on 17 September 1857 the formerly homesick James Taylor wrote to his father after five years in Ceylon:

> now I am a confirmed bachelor and have not spoken to a European woman except once or twice for the last four years . . . Faith I shall become as afraid of them as [Mr.] Pride [his supervisor] was. He never went near them and would even shut his eyes when passing a native woman.
>
> (Taylor 1851–92)

In another letter home (3 October 1857) he wrote:

> [t]hough I would like well to get hold of a suitable wife, I don't know how. I can scarcely bring one from home as I have changed and been changed since I left and have so much to do with black people.
>
> (Taylor 1851–92)

Taylor felt he had been degraded by the tropics and his association with native women. He could no longer fit into the model of the Victorian metropolitan home, even though he wished he could. He was no longer British in the full sense of the term for that key definer of Victorian identity, the domestic realm, was no longer available to him. He continued in his letter of 3 October 1857: "I'll certainly never marry one [a native woman] and I don't know if a white woman would suit and a half-a-half is worse than either . . . striped."

It is worth considering Taylor's letter in rather more detail. First, he expresses great ambivalence about white women, on the one hand expressing a desire to have a white wife and yet on the other expressing fear of white women and suggesting at the end of his letter that any sort of wife, white, native or mixed race, would be incompatible, it being only a question of which would be worse. And yet, while his relation to white women is one of fear, his relation to women of colour is one of disavowal and aversion. Put slightly differently, on the one hand he fears that he is no longer British enough for a British wife, while on the other he fears that marrying a native woman would cast him altogether beyond the pale of Britishness. Taylor in the mid-nineteenth century was clearly unprepared to take the step which British men in South Asia during the eighteenth century had routinely taken, to marry a native woman.[10] One can see here Taylor's fear of miscegenation. The native woman, he says, is out of the question, for the offspring of such a union would be miscegenation. "Striped" he contemptuously calls the offspring of racial mixing. As Young points out, there was a great deal of debate in the nineteenth century over the issue of racial mixing, or hybridity as it was often termed (Young 1995). The degenerative view was propounded by racial ideologues such as Gobineau. In this view, racial mixing would weaken the white race and strengthen the dark race. But, as Young points out, many whites held an ambivalent attitude towards sexual relations with other races. While white men were attracted to native women, they feared that attraction. Perhaps even more so, the hybrid was simultaneously celebrated as an object of beauty and demonised as a mutation of humanity. But one wonders, as one reflects further on Charles Taylor's letters home, to what extent his rejection of non-white women is a projection of his feelings that his Europeanness was slipping away from him under the impress of living in the tropics. Kristeva's concept of the abject is useful in understanding Taylor's state of mind. Abjection is expressed in feelings of disgust as a subject responds to that which he finds difficult to secure a firm boundary against. Iris Marion Young argues that racism, sexism and homophobia are structured by abjection. As she says, "the abject provokes fear and loathing because it exposes the border between the self and other as constituted and fragile, and threatens to dissolve the subject by dissolving the border" (Young 1990: 144). She continues that "the subject reacts to this abject with loathing as the means of restoring the border separating self and other" (Young 1990: 145). Ambivalent feelings of attraction and repulsion are experienced at a deep visceral level of bodily reaction, including aesthetic judgements. Beautiful Tamil girls in their early teens are seen as "models of feminine beauty" but by their late twenties they have become "disgusting objects to look upon and repulsive beyond all imagination" (Boyd 1888a: 413).

Men like Taylor felt even less at home in Britain, the place they still referred to as "home". For "going home" meant to leave the island, and return to Britain. But, as one long-time planter (Millie 1878: ch. 27) who tried retiring to Britain put it, Britain has "a harsh, cold uncongenial climate, . . . and an entirely new mode of life, . . . [which are] uncongenial". He adds that, when one returns to Britain, one is made to feel like a "peculiar fellow, a sort of oddity, 'one who, you know, has lived all his life amongst black people, who is ignorant of [British] customs'". He concludes with a word of advice. If you go back, it is best to try and live in a hotel so that you can keep your freedom to be yourself. It is telling that it is *in the home* in Britain where the old hand will feel least *at home*.

Bringing women out: "angels for the home"

In 1871 the barrister Richard Morgan travelled up to the hills and observed the following:

> It was a fine sight to see stalwart men assembled from miles around to join in divine worship. There was a harmonium to aid the singing which was very fine. There was one wont . . . there were no ladies. The climate is fine and will suit Europeans. It would be a happy thing if each bungalow had a lady to adorn it, and planters would adopt the place as their home, and have each his family smiling around him.
>
> (Digby 1879: 97)

Within a decade, Morgan's wish was beginning to be granted. Health conditions in the hills were improving and the death rates of the early decades were greatly reduced. The 1869 opening of the Suez Canal reduced the length of the trip to Ceylon and allowed white women, with what was thought to be their more delicate constitutions, to return "home" more often. This led to increasing numbers of British women going out to South Asia. Whitehall encouraged this as women were thought to have a civilising influence on European men in the tropics, acting as moral guardians of the home realm (Hyam 1990: 119). Metcalf argues that with the arrival of white women in South Asia the distinctions between the home and the world were reinforced (Metcalf 1995: 94). The sharp distinction between the private and public, he says, lies at the heart of Victorian domestic ideology.[11] In fact, he says that this colonial distinction reinvigorated the ideology of separate spheres back in Britain. Indian women were set up as a constitutive other of British women. Progress was measured in part by the nobility and morality of the women. The higher the civilisation, the more private, pure and domestic the women. British women would, it was felt, help create a more self-contained white society by "rooting out" the native woman from the colonial home. Their duty would be to create an atmosphere of rules, manners and restraint conducive to Christian manliness as defined by self-mastery, the ability, as Mosse defines it, to restrain sexuality and other passions and impulses (Mosse 1985). Such a manliness, of course, was necessary to the colonial project of legitimating the white master as a role model for non-whites. Consequently, white bourgeois women were recruited to the colonies as evidenced by the following letter in a London paper in 1884:

> There is just one word of advice I should like to give to fathers and brothers. To the latter, if you go to Ceylon arrange after you have a house of your own to get your sister out with you. England is overstocked with women, who are clamouring for votes and husbands too. Now England is sending out some of her best blood to its distant possessions. Why should the young men go and not the young women? I am convinced that the presence of his sister would have saved many a young fellow, in the pioneering days in the tropics, from drink and ruin, if she had been there to look after his bungalow and minister to his wants. Fellows used to come in from a hard days' work on the mountain slopes, fagged and weary, to their bungalow. There was food prepared for them by native servants but it was often not fit to eat. So some went to the beer or brandy for consolation. Things are better now and ladies are more numerous; but still, in colonising, whether to tropical or temperate climes, sister and brother may well go out together.
>
> (Ferguson 1884)

But while the presence of white women could morally regenerate the colonial home, making it more respectable and bourgeois, it could not completely stop what was thought to be the environmentally induced racial degeneration of those who remained too long in the tropical colony. There would always be a great gulf between the colonial home and Britain as home.

The ambivalence of the colonial home and Britain-as-home can be seen in the reminiscences of a second-generation planter born in Kandy. He writes: "All this while my hopes lay on the prospects of going to England – 'Home' – as I regarded England, though I had never seen the place!" (Lewis 1926: 29). Upon his arrival, he discovered that Britain was not really his home, and his Britishness was called into question:

> Although I was born in an atmosphere of rigid British ideas; ideas that made me feel as thorough a Briton as if I had never been out of England, yet I lived in a climate that was tropical; surrounded with all the contracting influences, that unconsciously or subconsciously become Colonial; uneducated, untamed and with nothing to back me,

pecuniary or otherwise – these influences, in the aggregate, were bound to produce peculiar ideas of life, and angles of observation, completely different from those of our home-born countrymen.

(Lewis 1926: preface)

The feelings of inferiority to "real" Englishmen well up even more strongly later in his biography:

All who knew me, knew that I was born in the country and utterly ignorant of English home life: and experience has often showed that the locally born was a far inferior creature; mostly despised, generally distrusted, and invariably classed as "country bottled," a particularly offensive appellation.

(Lewis 1926: 32)

Many of those who had either never been to Britain, or left it years before, found that their idea of a proper British home had become outdated. In Britain, the Victorian ideal of domesticity and the gender relations it prescribes, had been changing over the decades into a more fully developed and progressively more restrictive ideology. The separation of the home realm as feminine and the public as masculine had become more rigid (Tosh 1991).

Conclusion

As Stoler argues, "If we accept that 'whiteness' was part of the moral rearmament of bourgeois society, then we need to investigate the nature of the contingent relationship between European racial and class anxieties in the colonies and bourgeois cultivations of the self" (Stoler 1995: 100). The colonial home was idealised as a site of whiteness where the British domestic ideology of privacy, security and respectability could be nourished. But Hepworth (1999) has argued that the middle-class Victorian home in Britain should be conceptualised less as a haven and more as a battleground where these values struggled against the pressures of a dangerous world, a world that not incidentally included the ideology of colonialism already in crisis (Gikandi 1996). But if the metropolitan home was a battleground unable to defend its integrity against a constitutive outside imperial world in crisis, the bachelor home in the tropics was ever so much more embattled. Even with the coming of British women, it never provided the planter with either the sense of physical security or certainty of moral superiority necessary to fully legitimate his manly, Christian, imperial identity.

Notes

1 Self-discipline was one of the most important values of evangelicalism in Victorian England. On this in relation to the home, see Hall (1992).

2 On the history of evangelical, Christian manliness and the enthusiasm for the ideal among early Victorian middle classes, see Rutherford (1997). Manliness included physical courage and a taste for adventure, as well as being a gentleman.
3 For a discussion of this trope more generally, see Pratt (1992) and Spurr (1993).
4 I wish to make it clear from the outset that the ambivalent transcultural impacts, insecurities and anxieties found on the margins of the British empire should not in any way be conflated with a fragility in the political economy of empire. The British rule over this part of empire was secure during this period and consequently our understanding of its exploitative impact should not be diminished through attention to British psyches. The latter is, however, the focus of this paper. On this point, see Parry (1998).
5 The term bungalow was used even if the dwelling in question was a mud hut, for the term bungalow distinguished a European dwelling in the hills, no matter what its actual form, from a native dwelling. For the definitive work on the bungalow as a colonial house type see King (1995).
6 On similar bachelor homes in India, see King (1995: 43).
7 Gay argues that it was the idea of regulating the will and channelling energy that separated the Victorian bourgeoisie from other classes in Britain (Gay 1998: 19).
8 For a discussion of the complexities of whiteness in colonial Barbados, see Lambert (2001).
9 On this point see Young (1990: 136)
10 It is estimated that in the mid-eighteenth century in India, 90 per cent of British residents married Indian women. By the late eighteenth century increasingly it was the lower classes of British residents who married Indian women. There were still old high-ranking officers who were married to Indians, however, but they were looked down upon as evidenced by the following letter from Lord North, the Governor of Ceylon to his sister in 1801: "The ships are just going to sail, which I am very glad of, as they are to carry off such a set of whores and rogues as I suppose never saw collected in so small a space ... General Fullerton, an old twaddler from Bengal, ... walks about with a round straw hat and a green gauze veil on account of his eyes. He has a Hindoo wife, whom he carried once to Scotland, where she shocked the neighbours by plastering the walls of her drawing room with cow dung." (Ludowyk 1966: 102). By 1835, however, the British East India Company forbade its personnel from marrying Indians and throughout the nineteenth century attitudes towards racial mixing hardened in Britain.
11 Hall (1992) traces the history of a "recodification" of British ideas about women as domestic beings to evangelicalism in Britain during the early nineteenth century.

Bibliography

Anderson, B. (1983) *Imagined Communities*, London: Verso.
Arnold, D. (1996) *The Problem of Nature: Environment, Culture and European Expansion*, Oxford: Blackwell.
Ballhatchet, K. (1980) *Race, Sex and Class Under the Raj*, London: Weidenfeld and Nicolson.
Boyd, W. (1888a) "Autobiography of a periya durai", *Ceylon Literary Register*, 3.
Boyd, W. (1888b) "Ceylon and its pioneers", *Ceylon Literary Register*, 2.
Brown, A. (1880) *The Coffee Planter's Manual*, Colombo: Ceylon Observer Press.
Brown, S. (1845) *Life in the Jungle, or Letters From a Planter to his Cousin in London*, Colombo: Herald Press.
Carpenter, E. (1910) *From Adam's Peak to Elephanta*, London.

Digby, W. (1879) *Forty Years in a Crown Colony*, Volume 2, London.

Ferguson, J. (1884) "The prospects of England's chief colony. An interview with a Ceylon journalist (Mr. J. Ferguson)", *The Pall Mall Gazette*, August 29. Reprinted in Ferguson, A.M. and J. (1885) *Ceylon and Her Planting Enterprise: in Tea, Cacao, Cardomom, Chinchona, Coconut and Areca Palms. A Field for the Investment of British Capital and Energy*, Colombo: A.M. and J. Ferguson.

Gay, P. (1998) *Pleasure Wars: The Bourgeois Experience. Victoria to Freud*, Volume 5, London: Fontana.

Gikandi, S. (1996) *Maps of Englishness: Writing Identity in the Culture of Colonialism*, New York: Columbia University Press.

Glacken, C. (1967) *Traces on the Rhodian Shore*, Berkeley CA: University of California.

Hall, C. (1992) *White, Male and Middle Class: Explorations in Feminism and History*, Cambridge: Polity Press.

Hepworth, M. (1999) "Privacy, security and respectability: the ideal Victorian home", in T. Chapman and J. Hockey (eds) *Ideal Homes?*, London: Routledge, pp. 17–29.

Hyam, R. (1990) *Empire and Sexuality: The British Experience*, Manchester: Manchester University.

Kennedy, D. (1990) "The perils of the mid day sun: climatic anxieties in the colonial tropics", in J.M. MacKenzie (ed.) *Imperialism and the Natural World*, Manchester: Manchester University, pp. 118–40.

Kennedy, D. (1996) *The Magic Mountains: Hill Stations and the British Raj*, Berkeley CA: University of California Press.

Kenny, J.T. (1995) "Climate, race, and imperial authority: the symbolic landscape of the British hill station in India", *Annals*, Association of American Geographers, 85: 694–714.

Kenny J.T. (1997) "Claiming the high ground: theories of imperial authority and the British hill stations in India", *Political Geography*, 16.

King, A.D. (1976) *Urban Colonial Development*, London: Routledge.

King, A.D. (1995) *The Bungalow: The Production of a Global Culture*, Oxford: Oxford University Press.

Lambert, D. (2001) "Limnal Figures: poor white, freedman, and racial re-inscription in colonial Barbados", *Environment and Planning D: Society and Space*, 19: 335–50.

Lewis, F. (1926) *Sixty Four Years in Ceylon*, Colombo: Colombo Apothecaries.

Lewis, R.E. (1855) *Coffee Planting in Ceylon, Past and Present*, Colombo: Examiners Office.

Livingstone, D.N. (1991) "The moral discourse of climate: historical considerations on race, place and virtue", *Journal of Historical Geography*, 17: 413–34.

Livingstone, D.N. (1992) *The Geographical Tradition*, Oxford: Blackwell.

Livingstone, D.N. (1999) "Tropical climate and moral hygiene: the anatomy of a Victorian debate", *British Journal of the History of Science*, 32: 93–110.

Ludowyk, E.F.C. (1966) *The Modern History of Ceylon*, London: Weidenfeld and Nicolson.

Mangan, J.R. (1985) *The Games Ethic and Imperialism*, Middlesex.

Metcalf, T.R. (1995) *Ideologies of the Raj*, Cambridge: Cambridge University Press.

Millie, P.D. (1878) *Thirty Years Ago: or Reminiscences of the Early Days of Coffee Planting in Ceylon*, reprinted from the *Ceylon Observer*, Colombo: A.M. and J. Ferguson.

Moldrich, D. (1989) *Bitter Berry Bondage*, Colombo: Ceylon Printers.

Mosse, G. (1985) *Nationalism and Sexuality*, New York: Fertig.

Parry, B. (1998) *Delusions and Discoveries*, London: Verso.

Phillips, R. (1997) *Mapping Men and Empire: A Geography of Adventure*, London: Routledge.

Planter, A (Richard Wade Jenkins) (1886) *Ceylon in the Fifties and the Eighties, a Retrospect and Contrast of the Vicissitudes of the Planting Enterprise During a Period of Thirty Years and of Life and Work in Ceylon*, Colombo: A.M. and J. Ferguson.

Pratt, M.L. (1992) *Imperial Eyes: Travel Writing and Transculturation*, London: Routledge.

Rutherford, J. (1997) *Forever England: Reflections on Masculinity and Empire*, London: Lawrence and Wishart.

Spurr, (1993) *The Rhetoric of Empire: Colonial Discourse in Journalism, Travel Writing and Imperial Administration*, Durham NC: Duke University.

Stoler, L.A. (1991) "Rethinking colonial categories: European communities and the boundaries of rule", in N.B. Dirks (ed.) *Colonialism and Culture*, Ann Arbor MI: University of Michigan, pp. 119–52.

Stoler, L.A. (1995) *Race and the Education of Desire: Foucault's History of Sexuality and the Colonial Order of Things*, Durham NC: Duke University Press.

Taylor, J. (1851–92) *Papers of James Taylor (1835–92)*, (MSS. 15908–10), National Library of Scotland, Department of Manuscripts.

Tosh, J. (1991) "Domesticity and manliness", in M. Roper and J. Tosh (eds) *Manful Assertions: Masculinities in Britain since 1800*, London: Routledge.

Weatherstone, J. (1986) *The Pioneers: The Early British Tea and Coffee Planters and Their Way of Life, 1825–1900*, London: Quiller Press.

Williams, R. (1963) *Culture and Society, 1780–1950*, Harmondsworth: Penguin.

Young, I.M. (1990) *Justice and the Politics of Difference*, Princeton NJ: Princeton University Press.

Young, R. (1995) *Colonial Desire: Hybridity in Theory, Culture and Race*, London: Routledge.

2 The labourer's welcome

Border crossings in the English country garden[1]

Karen Sayer

> When we return from visiting other lands, we notice with gratified eyes these homely wayside gardens, which are peculiarly English. Englishmen have always loved their gardens, and all classes share in this affection.
>
> (Ditchfield 1994: 84)[2]

In the nineteenth century, gardens, gardens large and small, be they in town or country, carried with them the most pleasing aspects of the rural; they represented a Nature that was homely, not savagely appetitive, but pastoral gentle and green. Situated between the sublime landscapes of the Romantic past and the disordered-but-cultured city of the Victorian present, gardens functioned at the level of "humanized landscape", as knowable havens that could soothe and delight (Waters 1988: 149–151). Quintessentially English, evocative of embeddedness, of rootedness and community, the *cottage* garden in particular was synonymous with Nature tamed. Always already old-fashioned and representative of earthy vitality (Waters 1988: 48–58), cottage gardens with their resident flowers, cultivated or self-seeded, vegetables and livestock, conjured up powerful and nostalgic associations of dwelling at a time of heightened colonial territorialism, as the import and export of goods and capital connected people world-wide. As a component of the idealised country cottage, the cottage garden was encompassed by the domestic sphere, and, like the cottage itself, was therefore seen as a private space or retreat from capital, not as a place of visible work, or of wider social relations.[3] But the cottage garden also gained its own connotations of abundance, permanence and thrift. Working at the level of the small scale and the communal, it became associated with the useful and the pleasurable (Hoyles 1991: 228). The image of the cottage garden has consequently come to symbolise those values we should mourn the loss of and remains suggestive of their resurrection.[4]

In 1981 Stephen Constantine remarked:

> Historians of the garden have been dazzled by the large and the beautiful. Whole forests have been felled to feed the appetite for books

describing the history and appearance of the handful of prestigious gardens attached . . . to the nation's [i.e. England's] stately homes, [at the cost of the history of the ordinary and everyday].

(Constantine 1981: 387)

This remains largely true. Though the literary image of the garden (at all scales) has been discussed in detail, alongside the meanings of garden design, and despite recent work on garden history and the history of gardening,[5] the stories of these smaller spaces have continued not only to be scarce but also, despite Constantine's lead, apolitical and ungendered (Hoyles 1995: 1–6). There are of course exceptions. Catherine Alexander's recent study of the leisured urban and suburban garden is observed through the lens of feminist critiques of public and private space (Alexander 2002: 857–871). There is a growing interest in the education and lived experiences of women horticulturalists and "lady" gardeners among women's historians.[6] And some of the latest rural and social histories deal with questions of access, recognise the history of the garden as a productive space and take account of the political and moral issues inherent in class.[7]

Nonetheless, while recognising the owned boundaries of the garden (Alexander 2002: 860), Alexander – in common with most of the women's historians, who mainly focus on middle-class women – passes by the working gardens of the poor and the material politics of the land.[8] Though (commendably) interested in the "transformative" power of the (well-to-do) garden she therefore assumes that the only domestic waste that went into the ground was from the kitchen; "[w]aste" she argues "that has been mediated by the human body is ejected from the entire domestic arena via drains" (Alexander 2002: 866). Yet, in rural labouring homes of the same period, gardens were often fertilised with human (and animal) ordure. The cottage garden, as a space that worked for the family – with all of the refuse of that family – therefore had different boundaries to that of the middle-class suburban or urban garden, even beyond the complexities of the legal title to its land. Alternatively, those historians, including Constantine, interested in the promotion of working-class recreation, the issues of property rights and radicalism have generally set aside the "ever-shifting social geometry of power and significance" of space (Massey 1994: 3).

This is problematic because, as Michael Reed has argued:

[the] fields, farms, churches and cottages, castles and gardens of the landscape are not discrete entities, but the constituent elements of a dense historical matrix, linked in the minds of men and women by assumptions, values, and preconceptions, sometimes formally codified, often nothing more than practices informally recognised through custom and long usage.

(Reed 1997: 265)

In this way, the cottage garden, just like the gardens of the elite and the middle class, belonged (and belongs) to a wider landscape, a palimpsest of power relations, the spatial morality of which constituted a set of historically specific identities. As such, gardens should not be treated as simple "staging grounds" within which identity, difference, politics or power are performed, but as part of the "real" space through which these things are enacted and created.[9] In exploring the supposedly essential and natural domesticity of the cottage garden, its enclosed introspection, its links to the community at large and the (formal and informal) practices of gardening, this chapter will investigate the construction of gender within this system.

The "historical matrix" of power relations, it should be noted, is both experiential and textual. This is to say that it is not enough to undertake a reading of the (or a) cottage garden as material space; it is also necessary to consider the conflicts around the space, the struggle for meaning.[10] The Victorian idea of a "cottage garden" itself was a construct created in part through art, literature and related discourses,[11] and in part through the efforts of landscape gardeners and designers such as J.C. Loudon,[12] William Robinson[13] and Gertrude Jekyll[14] who then built on these images. Loudon, for instance, wished "to see fruit-trees, ornamental shrubs, climbers, and flowers in every Cottager's garden, with bees, poultry, rabbits (if only for the children), pigeons and a cat" (Loudon 1840: 45). And, though keenly interested in the welfare of the labourer, in his view the "cultivation of a few Brompton and ten-week stocks, carnations, picotees, pinks, and other flowers ought never to be omitted: they are the means of pure and constant gratification which Providence has afforded alike to the rich and the poor" (Loudon 1840: 45). This compensatory mix became the paradigmatic cottage garden, promoted by reformers as well as garden writers, which in turn ensured, as Waters has observed, that the cottage garden maintained its place "within a signifying system of internally differentiated garden types" (Waters: 1988: 54).

This model has subsequently led both contemporary designers and garden historians to search for a wilder working-class authenticity and to set this against a cultivated bourgeois counterfeit. As one garden historian wrote in the 1960s:

> [the] cottage garden . . . can be seen to express a spirit of community. It is in short, a "folk" entity and harks back to a time when the community in which men have to live and are probably meant to live was not obscured by the cult of the individual mind and spirit.
>
> (Hyams 1966: 153)

Jekyll, it has often been noted, claimed that her inspiration was the labourer's cottage garden. "I have learnt much from little cottage gardens that help to make our English wayside the prettiest in the temperate world," she believed. "One can hardly go into the smallest cottage garden without learning or observing something new" (Jekyll, quoted in Hoyles 1991: 225). But she "simply did not see the yard-cum-garden of the average cottager, with its chicken-house, rabbit-hutches and outdoor earth-closet" (Ronald King, quoted in Hoyles 1991: 225), which is why Geoff Hamilton felt able to say of her that:

> [Her] [i]nfluence, and that of other great gardeners of her time and a little later, while masquerading as "cottage gardening", in my view missed the essence. ... Great gardeners, talented artists and original plantswomen they certainly were, but were they really cottage gardeners? For me the answer has to be no – the cottage garden is an *artisan's* creation, not an artist's.
>
> (Hamilton 1995: 31)[15]

This rather writes off the artistry of the artisan's work, and thereby re-inscribes the established hierarchy of gardens and gardeners.

Gillian Darley's assessment that the "cottage garden is an odd mixture of myth and reality" (Darley 1979: 151) is therefore currently the most commonplace. But, looking at the history of the cottage garden is not just a matter of setting "image" (or "myth") against "reality"; it is also a question of recognising the function of the image (Waters 1988: 53–55). As Lynda Nead argues, in the case of the rural idyll:

> [The] [w]ork of representation should not be seen in terms of a false construction of country life. Rather than a manipulation of reality, images of rural domesticity were the site where an ideological category designated the "rural labouring class" was defined and given visual form.
>
> (Nead 1988: 42)

It seems that today there is a certain fluidity to the cottage garden style, which allows the cottage garden itself to be re-created using up-to-date materials, in any location, without losing any of its "authenticity". Hence Hamilton in the populist *Geoff Hamilton's Cottage Gardens* asserts that the "cottage garden 'style' was not invented; it simply evolved. So you can just step into that process of evolution to make a traditional, romantic cottage garden – wherever you live" (Hamilton 1995: 5 (caption)). This is a view that highlights not only the fashion-driven nature of most gardening[16] but also the garden's intrinsic hybridity (Massey 1995: 186). Nonetheless, we should still note the particular stress on the "traditional" in Hamilton's statement, the organic "evolution" of the "cottage garden" "style" and the passing reference to the "romantic", which takes us back to the pre-industrial, the "folk" and the village community. The cottage garden "style", like the cottage, is as desirable today as it was in the nineteenth century because, ideologically, it is still supposed to provide a tranquil haven, a stable sanctuary from the world at large and safety from the vagaries of capital; and perhaps now post-colonialism. This, to follow Raymond Williams, is a matter, not of "historical error, but historical perspective" (Williams 1985: 10).

The artisan's "garden"

The "folk" or "artisan's" cottage garden historically consisted of a small plot of land, normally, though not always, attached to the labourer's house, mostly used for growing vegetables, fruit and herbs, and sometimes for keeping a pig, a few chickens or ducks, bees and, occasionally, some rabbits. Loudon suggested that a rood, i.e. a quarter of an acre, would be needed to supply the labourer, his wife and three small children, plus pigs and poultry, with vegetables and potatoes. He thought they should grow onions, leeks, carrots, beans, parsnips, cabbages and potatoes, currants, goose-berries, cherries, apples, pears and other soft fruit (Loudon 1840: 4, 41, 45). One such, according to Sir Thomas Bernard – a leading light in the Society for Bettering the Conditions and Increasing the Comforts of the Poor – tended by Britton Abbot (Burchardt 2002: 18, 33) was described in 1797:

> Two miles from Tadcaster, on the left-hand side of the road to York, stands a beautiful little cottage, with a garden, that has long caught the eye of the traveller. The slip of land is exactly a rood, inclosed [*sic*] by a cut quick hedge; and containing the cottage, fifteen apple trees, one green gage, and three winesour plum-trees, two apricot-trees, several gooseberry and currant bushes, abundance of common vegetables, and three hives of bees.
>
> (Anonymous 1806: 5)

Not that the productive labourer's garden necessarily lacked flowery interest (Hamilton 1995: 14). When tended by an enthusiast, a nineteenth-century English cottage garden might include dahlias, hollyhocks, delphiniums, sweet peas, roses, pinks, clematis, fritillaries, lilies, geraniums, mignonette and border carnations among its plant stock, as well as the requisite fruit, herbs and vegetables. Plants from the big house probably made their way to the small as fashions that became popular among the wealthy were adapted and adopted by the poor, who were therefore just as likely to plant up the rows of carpet bedding detested by the proponents of wild gardening, as they were to use "traditional" "old-fashioned" cottage garden plants (Hamilton 1995: 26–29; Hoyles 1991: 77, 226; Waters 1988: 53–54). But the evidence makes it clear that this small plot of land primarily had to provide the labourer's family with additional food, or a surplus, to help them spin out their wages. A garden, or an allotment, was in fact widely supposed to be essential to the survival and bodily comfort of the labourer by the first decades of the nineteenth century.[17] Hence Arthur Young, identifying an absence of land for gardening as problematic in 1804 when surveying Hertfordshire for the Board of Agriculture, observed:

SECT. II.– COTTAGES

I am sorry not to find any minutes in my notes upon this head, which is so truly important, except the remark so often recurring, that the cottagers have no where any land, more than the small amount of insufficient gardens. I twice went out of my way to make inquiries; where I was told that one or two labourers possessed enclosed land enough to support a cow; but the intelligence was unfounded: . . . The present system of supporting the labouring poor is certainly erroneous, both in practice and theory.

It appears to me as a matter of demonstration, from a multitude of facts, that the granting them land for cows, and an ample garden, is the only cheap mode of assisting them materially.

(Young 1971: 21–22)

Though it is difficult to assess what a cottage garden might have been worth to the labourer, this kind of early argument for land provision eventually became typical (Burchardt 2002: 11). It was suggested for instance that Britton Abbot, who reportedly earned 12 shillings to 18 shillings a week as an agricultural labourer, managed to get "from his [quarter-acre] garden, annually, about 40 bushels of potatoes, besides other vegetables"; while "his fruit, in a good year [was] worth from £3 to £4 a year" (Anonymous 1806: 8). This clearly supported Bernard's case in promoting the provision of gardens for the working class. Nonetheless, nearly a century later, Susan Silvester's mother, who, her daughter says, was a thrifty manager, helped keep her children well fed and well clothed in part because of the family's garden and allotment, pig and chickens. "Mother grew hyssop and horehound in the garden and made 'tea' with them when we had colds" (Silvester 1968). On moving with her (blacksmith) husband to a cottage in 1902 that had a large garden in which they could grow fruit and vegetables and had room for pigs, chickens, a pony and trap, Susan Silvester found herself well off (Silvester 1968: 3, 5, 27). A good-sized garden could therefore probably materially benefit those with access.

Having said this, it is hard to establish exactly how many rural homes had gardens attached (Constantine 1981: 392). There was greater availability of houses with gardens in some counties than others and when provided, the size of a garden plot varied considerably from region to region, due to hiring conditions, local custom and the availability of land. For "garden" we might more appropriately read "allotment" in many instances, especially as many "allotments" were themselves attached to the labourer's house. According to *The Penny Cyclopaedia of the Society for the Diffusion of Useful Knowledge* (1843):

> COTTAGE ALLOTMENTS may be considered as such portions
> of land hired by labourers, either attached to, or apart from, their
> dwellings, as they, assisted by their families, may be able to cultivate
> without ceasing to let out their services daily to others. . . . Various
> experiments have been tried, and the opinion of the persons best
> informed upon the subject appears now to be, that a quarter of an
> acre is about the quantity which, without prejudice to his other employ-
> ments, a labourer can in general thoroughly cultivate, and consequently
> derive the greatest profit from.
>
> (*The Penny Cyclopaedia* 1843: 88)

Despite enclosure, common land, or land used as such, did often survive
and this, plus small areas of sub-let land, edges and corners of fields,
were frequently available to the labourer for grazing or planting up with
potatoes, carrots and other staples in most regions even when a garden
was not available. In fact, many artisans and labourers rented small
acreages of land, worked on by the whole family, which were as important
to their subsistence as their cash wages, and which blur the occupational
divisions in the countryside, especially between small farmers, artisans and
farm workers (Reed 1984: 57–59). Meanwhile, *Chambers's Information for
the People* (1842) defined what it called "spade husbandry", otherwise known
as "cottage-farming" or "field-gardening" in terms of the labourer's require
ments (Chambers and Chambers 1842: 345).

The actual tools required to engage in "spade husbandry" were: "two
or three spades of different sizes, a pickaxe, three-pronged digging fork,
hoes, rake, light harrow which he can draw, scythe, reaping hooks, hay-
forks, flail, wheelbarrow, &c., according to means", which required quite
a large surplus income, for purchase and repair, unless inherited or
borrowed. It was useful if the labourer could sharpen his tools himself,
Chambers went on to say that the "labourer" was always assumed to be
male unless stated otherwise and in reading this, *he* was to become expert
and take control of the land – but he would never keep a horse:

> All the work is done by the manual labour of the farmer and his family.
> The only live stock is a cow or cows, pigs and poultry. The homestead
> consists of a cottage with several apartments – a cow-hose, pig-stye,
> and barn. The size of the farm is supposed to vary from four to six
> acres, and to be laid out in six or eight distinct fields, properly fenced.
>
> (Chambers and Chambers 1842: 345)

Chambers's then described the best methods, likely value of produce and
difficulties that might be experienced. It also referred to the Labourers'
Friendly Society – a leading organisation in the campaign for allotments,
founded in London in 1833 – as a "beneficial" body that was attempting
to secure "allotments" for the "labouring poor," and reproduced part of

its publication on keeping a cow and pig on an acre of land (Chambers and Chambers 1842: 345–348). In subsequent numbers – each cost 1½d – *Chambers's* outlined best practice in "The Kitchen Garden", "The Flower Garden", "The Fruit Garden", "Arboriculture", "Cattle and Dairy Husbandry", horse, sheep, pig, caged-bird and bee management; later it covered dogs and field sports, angling and "Out-Of-Doors Recreations" (headings in Chambers and Chambers 1842). In other words, it provided complete guidance on the best use of rural space, as well as advice on the management of the land, for the aspiring (upper working-class) reader.

Meanwhile, in the west of England, as for Britton Abbot, a cottage "garden" could be as large as a quarter of an acre and include a pigsty, while in Northumberland labourers and their families were often provided with a cottage, pigsty and garden rent free, including an additional 1,200 yards for planting potatoes and summer grazing rights (Howkins 1991: 20–21). Tom Mullins, a Staffordshire farm labourer, rented a smallholding of seven acres for £15 when he married in 1886 – totalling a rent of £20 a year for house and land (Burnett 1982: 67). Conditions varied, however, and by the mid-nineteenth century many labourers were left without either garden or allotment, especially in the south and east of England. This was in part because cottages were increasingly being built without any land attached to them – especially those thrown up by speculators – and because, according to campaigners, the land that had originally surrounded many older cottages was gradually being taken away (*The Penny Cyclopaedia* 1843: 89). As Britton Abbot said, as rhetorical mouthpiece for the campaigners:

> Sir, you have a pleasure in seeing my cottage and garden neat: and why should not other squires have the same pleasure, in seeing the cottages and gardens as nice about them? The poor would then be happy; and would love them, and the place where they lived: but now every little nook of land is to be let to the great farmers; and nothing left for the poor, but to go to the parish.
>
> (Anon 1806: 8, 9)

The cottage garden was a contested site.

It should also be borne in mind that during the late eighteenth to early nineteenth century, the enclosure of a piece of land as a garden was the first stage in a labourer's bid to become a squatter, build his/her own cottage and thereby gain possession, or title, to the land (Jefferies 1981: 56).[18] But social reformers (such as Bernard), commentators and clergymen continued to campaign throughout the nineteenth century for the extensive provision of gardens and allotments, in order to improve the overall condition of the labouring class. *Chambers's* itself engaged in this battle, finishing off its essay on "spade husbandry" by rebutting accusations in the *Encyclopaedia Britannica*'s entry on the "Cottage System" that cottage farming increased the number of paupers and distracted the labourer from his

proper labour (Chambers and Chambers 1842: 352). But, like the difference between "garden" and "allotment", the distinction between a smallholding and an allotment was itself a nice one. In keeping with the move
towards large-scale capitalist agriculture, even the most vociferous sup-
porters of allotment gardening were often highly critical of the idea of
small-scale subsistence farming publicised by *Chambers's* and *The Penny
Cyclopaedia*: "It has also been urged against the system of cottage allotments."
The Penny Cyclopaedia noted, for instance:

> That it tends to encourage early marriages, and to the production of
> a race of beggars. That cottage allotments may have this effect is true,
> but the objection is mainly applicable, and perhaps has been mainly
> applied, to cottage allotments which are of such magnitude as to render
> it impossible for the cottager to cultivate them without exchanging his
> character of a labourer for that of a farmer without adequate capital
> ... The multiplication of small farms is quite a different question from
> that of cottage allotments as here understood, and opinions can hardly
> be much divided as to the inexpediency of such increase.
>
> (*The Penny Cyclopaedia* 1843: 89)

As we might glean from the statement that an allotment was a piece of
ground that could be cultivated by the labourer "without ceasing to let
out their services daily to others", large farmers were generally found to
"have strong prejudices against the system" (*The Penny Cyclopaedia* 1843:
89). Whereas small farmers could get away with paying less than a subsistence wage if their labourers had an alternative source of income, at least
until the mid-nineteenth century (Reed 1984: 60–69), many larger farmers
were apparently reluctant to allow their hands land which might distract
them from their paid work on the farm, or make them less dependent on
a cash wage. There therefore seems to have been considerable confusion
between what might be seen as small-scale "peasant" farming and cottage
gardening – a confusion that has historically elided the difference between
work and leisure and within the discourse of Englishness currently provides
the cottage garden, which most histories trace back at least to the Norman
period, with "authenticity".[19] But, it is a confusion – and a debate – which
also centred on (sexual) morality, politics, "independence" and gender.

Setting the bounds

The combination of domesticity with a privileging of organic communal
relations provided a particularly powerful and persistent "Beau Ideal" in
the nineteenth century. As the lives of the new middle class became less
fixed, so they looked to a past for a present that would provide them with
both independence and the stability of long-standing tradition.[20] The
cottage garden, like all gardens, therefore appeared within most nineteenth-

century middle-class literature and art to be a place that was socially neutral, psychologically fixed and physically bounded by hedges, fences and walls (Davidoff and Hall 1987: 361). However, its links to, and its relationship with, the outside world, both concrete and cultural, overwrote it (Maclean *et al.* 1999: 1).

There is, in consequence, some irony in the etymology of the word "garden" derived as it is from the Old English "geard", i.e. yard or enclosed ground (Hoyles 1991: 1).[21] Even on the most literal level, many of the cottage garden's most popular plants were originally imports (Reed 1997: 253), thereby demonstrating the power of the English garden to take in, naturalise and domesticate what was once essentially "alien" to it.[22] The auricula, for instance, a common cottage plant by the mid-nineteenth century, "was brought to our sheltered lawns from the snowy moss of the Swiss Alps", while the humble *Ranunculus* was spread about by Mohammed IV (Willmott 1864: 165).[23] In this respect the history of gardening, even quintessentially English cottage gardening, is immutably bound up with the history of trade and the growth of empire, which clearly highlights the need to avoid what Doreen Massey has called "internalist and essentialist constructions" of place (Hoyles 1991: 78; Massey 1994: 183). As Martin Hoyles says in *The Story of Gardening*, "just as an examination of class relations, the ownership of land and the division of labour, is crucial to an understanding of gardening, so too is a look at international relations, particularly colonialism" (Hoyles 1991: 55). Abroad, the imposition of a country garden on a native landscape worked as visible demonstration of the civilising impact of the English,[24] while the cottage garden itself could be read as a tangible expression of the imperial project. Indeed, even today there is often an uneasy tension in gardening literature between the fantasy of the English garden as enclosed space, and the celebration of the English garden as harbour of exotic trophies.

As Agyeman and Spooner have noted, there are powerful associations made between "alien" and "native" peoples, and "alien" and "native" plants (Agyeman and Spooner 1997: 212.) Gardening and gardeners are far from being "innocent" in this respect and when in *The English Garden* (1966) Edward Hyams notes that William Robinson aimed to "'naturalise' into an English landscape plants more spectacular than any which the native temperate flora could furnish," (Hyams 1966: 129) one can see a continuing process not simply of writing botanical history but also of naturalising the colonial language of the "exotic" and "alien", the "temperate" and "native", into gardening discourse. Earlier on he describes Loudon as the gardener who defined the "English Garden", "which by that time had a definite meaning for foreigners; yet . . . was far from clearly defined . . . [Loudon helped establish it as] the English dream, the re-created paradise" (Hyams 1966: 120). When looking at the story of the cottage garden we therefore need to recognise that the politics of its formation and subsequent border crossings have largely been whitened and remained hidden within histories of the "natural" and leisured act of gardening (Hoyles 1991: 1, 5, 21)[25] – an act that, for the middle class, came to be seen as a singularly "rational recreation" in the nineteenth century, because it crucially allowed them to avoid the pitfall of idleness in their new-found leisure time (Constantine 1981: 389–390; Davidoff and Hall 1987: 373).

According to the journals of the day, such as *The Gardener's Magazine of Rural and Domestic Improvement* established in 1826 and edited by Loudon, or Joseph Paxton's *Horticultural Register* published from 1831, gardening required diligence and application, was physically and morally healthy, and stimulated the mind. In addition, given that the garden had gradually come to be thought of as a private space, as a part of the house and separated from the competitive and wilder world around it,[26] amateur gardening was widely supposed to be a domestic occupation, a home-centred activity that focused on the family (Davidoff and Hall 1987: 373–375). This pursuit, it was soon assumed, would have an equally beneficial effect on rich and poor alike. In particular it would ensure that the working class were employed in thrifty, healthful work, which would stop them from going to the pub and would strengthen their ties to home and family. The net result, according to those who promoted gardening among the poor, would be that the labourer would turn away from the dangers of community-based politics and mob rule (Constantine 1981: 389–391). As Robert Willmott noted, in his *A Journal of Summer Time in the Country* (1864), when suggesting that a history of gardening should be written, the "moral influence of a garden . . . is lively and lasting" (Willmott 1864: 155).

This rhetoric was focused on the male labourer, so that the condition of any cottage garden came to embody the thriftiness, industriousness, honesty, sobriety and continence – i.e. the material respectability – of the man who worked it. As another mid-century author in the *Cottage Gardener* put it, there "is moral beauty . . . in the cultivated cottage garden. Neatness and attendance bespeak activity, diligence, and care; neglect and untidiness tell of the *beer-house*" (cited in Hoyles 1991: 16). The material identity of the labourer, shaped by the work he had undertaken, was one with the garden that he worked. Not only did this kind of construction represent the cottage garden as a male domain, it also wrote off the political resonance/ radical potential of gardening for Geoff Hamilton's pet artisan:

> The labourer who possesses and delights in the garden appended to his Cottage is generally among the most decent of his class; he is seldom a frequenter of the ale-house; and there are few among them so senseless as not readily to engage in its cultivation when convinced of the comforts and gain derivable from it. When the lower order of a state are contented, the abettors of anarchy cabal for the destruction of its civil tranquillity in vain, for they have to efface the strongest of all earthly associations, home and its hallowed accompaniments, from the attachment of the labourer, before he will assist in tearing them from

others, in the struggle to effect which, he has nothing definite to gain, and all those flowers of life to lose.

(Johnson, cited in Hoyles 1991: 16)

Loudon agreed: "give the Cottager land that will reward his labours," he pleaded, "it will stimulate his industry, and ultimately tend to link each class of society in inseperable [*sic*] bonds for the preservation of national order and tranquillity" (Loudon 1840: 56). As Burchard outlines, "considerably more interest was shown in improving their [working class men's] moral character" than in philanthropy (Burchardt 2002: 17–19). The labourer's cottage garden, then, supposedly secured (male) working-class respectability and became a defence against agitation; it stood between chaos and civilisation, between licence and order. The country was secure as long as the labourer returned home to tend his flowers (metaphorical and literal) and dig over his quarter of an acre. The male labourer possessing and possessed by his garden was to be made moral through useful bodily toil. In the process a new working-class masculinity was produced by the garden, one which though "independent" was home-centred.

As a result, though demands by labourers such as Abbot that the poor have access to small plots of land in order to remain materially independent were partially permissible, because property-centred and individualist aspirations were always preferable to mass action (Constantine 1981: 391–392, 401), they were glossed over unless reference to them was thought expedient. For instance, in stating "now every little nook of land is to be let to the great farmers" Abbot seems to have been attacking the process of enclosure, but Bernard, who quotes him, is only interested in criticising the poor laws. His argument is premised on a considerable loss to national efficiency and encompasses some morally debilitating side effects, but by-passes Abbot's own concerns about enclosure and the rise of large-scale agriculture.

As Hoyles suggests, today it seems a little odd to talk of politics in reference to gardening. Gardening it seems – thanks to the Victorians – should be an escape from politics, a refuge; yet, if we mean the theories and practices of power relations between people, when we talk of "politics", then gardening has everything to do with it (Hoyles 1991: 1, 5–6). The politics of gardening have simply been hidden because the activity of gardening has come to be seen as "natural" (Hoyles 1991: 1, 5, 21), i.e. as universal and unchanging. The supposedly essential, innate domesticity of the cottage garden, its enclosed introspection – coupled with the moral benefits, the natural and leisured aspects of the act of gardening – has come to write off the political meanings of cultivation in dominant discourse and to obscure older forms of radical masculinity. As the garden came to be captured as a private space, as part of the house, so gardening was increasingly seen as a domestic occupation, home-centred and focused on the family; this despite the continued fight for land from working-class men themselves.[27]

Crossing the border

As Hoyles says, there "is a kaleidoscope of cultural meanings attached to gardening" (Hoyles 1991: 8), and gardening as an activity was gendered (Davidoff and Hall 1987: 310, 374–375). With reference to the working class, it was generally assumed that where the husband, or father of the family, would do the digging and care for the vegetables, his wife, or the mother and children, would look after the flowers and fowl. Yet, as can be gathered from Silvester's testimony, the bulk of cottage, even allotment, gardening often fell to the woman of the house. This was sometimes publicly recognised. Denison for one felt able to recommend the adoption of allotments when making a return for the 1843 *Reports of the Special Assistant Poor Law Commissioners on the Employment of Women and Children in Agriculture* (*PP* 1843 [510] XII) on the strength of work done by the family. In Beccles, Suffolk, the allotment system, which, he reported, largely employed women and children, had been found to be the best way to improve the moral, economic and social condition of the poor. Since the aim was to "dispauperize" them, each cottager was provided with 40–60 rods of land, and despite the loss of employment in the area, the condition of the labouring class was found to be good. The holders of allotments were industrious and orderly, cottage gardens were better looked after, because the young had begun to emulate their elders, and neighbours had begun to compete with each other for the quality and quantity of their produce (*Reports* 1843: 220, 259). The large amount of garden work that could be done by a labourer's wife is also outlined in an *Account of the Produce of a Cottager's Garden in Shropshire* (1806) published alongside the *Account of Britton Abbot's Cottage and Garden*. In this instance the labourer's wife largely managed the land – one and one-sixteenth acres – herself, planting most of it up with wheat and potatoes in rotation, fertilising it by keeping a pig and mostly cultivating it with her own labour, though her husband, a collier, generally helped with the digging. Her methods, the commentator noted were sound; her potatoes and wheat apparently being grown "in a way which has yielded good crops, and of late fully equal, or rather superior, to the produce of neighbouring farms, and with little or no expense" (Anonymous 1806: 6). Her work included manuring, hoeing, raking, planting and harvesting the crops. She also planted up peas, beans, cabbages, early potatoes and turnips on some of the land – used as a garden – and sold her early potatoes, peas and cabbages at market (Anonymous 1806: 5–12). This kind of very physical and dirty labour was given less and less public recognition, however, as the century wore on, because it increasingly came to be seen as an unfeminine and hence an inappropriate occupation (Davidoff and Hall 1987: 374; Sayer 1995).

It is therefore really only in autobiographical material that we find any evidence of the work that women did in their gardens, or the meanings they attached to that work. And, what this material suggests is that in

England women gradually learned to refuse to work on their family's allotments or vegetable gardens using the rhetoric of separate spheres. By the time Flora Thompson (English 1985)[28] was growing up, allotment work such as that undertaken by the collier's wife (like all outdoor work) was in the view of working-class women "men's work". What the women therefore hoped to have was a flower garden and "herb corner" in which to keep bees and grow herbs for cooking, laundry, medicine, mead and tea, and most homes "had at least a narrow border beside the pathway" (Thompson 1979: 114–116). Then, when tea had to be offered to:

> An important caller, or to friends from a distance, the women had their resources. . . . Thin bread and butter, . . . with a pot of home-made jam, which had been hidden away for such an occasion, and a dish of lettuce, fresh from the garden and garnished with little rosy radishes, made an attractive little meal, fit, as they said, to be put before anybody.
>
> (Thompson 1979: 118)

What we can see here is that the cottage garden could generate high status produce suitable for a visitor, as well as serviceable items for home use, and that the women's respectability, key to their gendered identity, rested on its products. But seeds and plants were expensive, therefore most women collected cuttings and roots from their neighbours. This meant that though their range was limited:

> They grew all the sweet old-fashioned cottage garden flowers, pinks and sweet williams and love-in-a-mist, wallflowers and forget-me-nots in spring and Michaelmas daisies in autumn. Then there was lavender and sweetbriar bushes, and southern-wood, sometimes called "lad's love", but known there as "old man".
> Almost every garden had its rose bush; but there were no coloured roses amongst them. Only Old Sally had those; the other people had to be content with that meek, old-fashioned white rose with a pink flush at the heart known as the "maiden's blush". Laura used to wonder who had imported that first bush, for evidently slips of it had been handed round from house to house.
>
> (Thompson 1979: 114–15)

In conjuring up for her urban reader an English idyll which conforms to the dominant image of the cottage garden in its sweet-smelling scents and "old-fashioned" plants (Waters 1988: 37–39, 48–58), Flora Thompson's account is striking in that the use of plants comes to stand in as a kind of material communication that takes place between gardens and ordinary (women) gardeners. Flora Thompson treats the garden as a fluid space, as part of the whole village, as a chronicle of resources and as communal, public space where the gardener swaps plants and gossips with her neighbour. The garden, as well as producing tangible benefits like jam – with the addition of hedgerow fruits – lettuce and radishes, could therefore be read as the visible expression of her community's history, as a spatial morality within which the women's gendered identities are formed. It is this understanding of the "artisan's" cottage garden that has continued to fascinate commentators and garden historians alike. And, it is here that we can see a direct relationship between the ascendancy of the ideal cottage and the development of the ideal cottage garden among gardeners. The images of rurality celebrated in the late nineteenth century by the artist Helen Allingham,[29] for instance – the domestic idyll, the vernacular, tradition, authenticity, romance, nostalgia – were put into gardening practice by designers like Gertrude Jekyll.

Within that art produced by the middle class and meant for upper working-class or bourgeois consumption, the cottage garden was constructed as a morally ideal, leisured and feminine space, which had little or nothing to do with the masculine world of politics or (paid) work (Huish 1903: 153). This is why so many of these images depict a woman at rest, children at play, or the hoped for labourer's return – the male labourer, despite Britton Abbott's example, is most often noticeable by his absence in the imaginary text. Meanwhile, the inclusion of "human incidents" meant that many rural pieces were deemed to be "pictures", i.e. snapshots of narratives, rather than mere "studies" (Huish 1903: 150). In Edwin Cockburn's *The Return from Market*,[30] for example, we see the return of a father to his family.

As we watch from a vantage point inside his cottage, he steps through the garden gate and is greeted by one child, while its mother holds up another and a toddler stands on tiptoe to see him out of the window. Here we are strongly reminded of those "strongest of all earthly associations, home and its hallowed accompaniments . . . all those flowers of life" which he might lose. As Nead has noted, "political struggle [was] defined as unhealthy and 'unpictoral'" in the Victorian period, and within this, "paintings of the rural idyll perpetuated hegemonic definitions of gender" (Nead 1988: 41). In this particular case, the window physically demarcates the woman's sphere inside while it determines that the man's is the world outside. "Woman," Nead argues, "is shown fulfilling her natural roles of wife and mother within the natural unit of the family home, situated in the natural domain of the countryside" (Nead 1988: 42). However, what we can also see in the eldest child's rush to greet its father is that the boundary of "home" to some degree extended out into the garden and only actually finished at the picket fence. The labourer therefore reaches "home" and safely escapes the grasp of caballing agitators before he even reaches the cottage door.

In most genre pieces the cottage garden is shown at its flourishing best, usually at the height of summer, though the fecundity we see as a "snap shot" promises to continue through the course of the seasons. It is usually full of tall, colourful (probably scented) flowers, healthy vegetables, trees weighted down with fruit, and a backdrop of climbers covering a wall or the cottage itself. Roses are, of course, commonplace in these scenes, signifying both an English idyll and femininity, there being a commonplace equivalence between flowers and women at the time (Waters 1988: 133–148). The resident cat and, commonly, caged bird stand in as signs of domesticity, while chicks and kittens equate with childhood and a nostalgic sense of the passing years. The remaining vegetables, occasional pig, bee hives and farm/garden implements, are suggestive of thrift, "independence" and industry, and assure the audience that the man of the family will be home soon. The female occupants of these lush gardens are normally engaged in some kind of household work such as sewing, or occupied in other suitably feminine activities such as flower-picking, convalescing, nursing, feeding ducks and fowl, kitten-taunting, courting or visiting.

Take for example, Sir Samuel Luke Fildes' *The Farmer's Daughter*, which appeared in the *Sunday Magazine* in 1868.[31] In this case, a young woman is seen standing just inside the wall of her cottage garden, watched by a cat on a chair in the house and a bird in a cage. Her garden blooms all around her, and soon, it seems, the climbers on the wall will cover and become part of the house, an effect that was actively sought by nineteenth-century architects who tried to bring Nature into the houses they designed through the use of large windows, pot plants, and conservatories (Tristram 1989: 239–245). The farmer's daughter – and as an idealised representative of women, she stands in for all farmers' and all cottagers' daughters – clearly only ever works within the most clear-cut definitions of household management, and within the strictest bounds of the domestic sphere. Because her garden is pleasant, prosperous, and suggests a self-contained, contented thriftiness, we "know" that the interior of her house is clean, tidy, neat, and decorated with shining pots and pans. She genteelly feeds her fowl – doves or pigeons, it is hard to tell, signifying peace and virginity – which mill around nearby. But, though her garden gate is open, she does not step beyond it. Rather, it seems to signify a constant return, or invitation, to her home. Her open gate always already leads back along the path through the open door of the house.

Within the codes of nineteenth-century genre painting, the wall, hedge[32] or fence provided an image of a physical and an ideological boundary between (sexual) innocence and experience – as such it was often used in courtship scenes, which had their own, frequently heady, implications of egress and congress[33] – between wider social relations and the individual family unit, and between the competitive, masculine, world of work and the comfortable, feminine, world of "spare time". This persisted at least until the Second World War. Because of this, according to Eleanor Rhode who was writing nostalgically in the 1930s, cottage gardens were associated:

In one's memories with cathedral and remote little country towns – gardens with *mellowed walls* where the broad, generous beds were full of fine *old hardy plants* and flowers with rich soft colours and *delicious scents* and *hoary with traditions of centuries*. Each garden had its characteristic atmosphere and the owner knew every tree and flower as a shepherd knows his sheep. It is to those gardeners and to cottagers that we owe the preservation of many treasures amongst the old herbaceous plants despised by the enthusiastic admirers of bedding-out plants.

(Hoyles 1991: 214, emphasis mine)

Here, every cottage garden is a unique space, while their historical authenticity is particularly emphasised in an evocative description of Englishness and vernacular domesticity. Their ability to remain distinct and to preserve treasures is dependent on each gardener's remote old-fashionedness and each garden's "mellowed walls". The walls around each cottage garden stand or mediate between Nature and Culture; the gardens themselves create a small, defensive clearing of Culture, or Nature tamed, as a haven within the wilderness of the wider world; a spatial morality (Ortner 1974: 67–87). The stress is on the peace and tranquillity of the country, its homeliness and privacy. But this is also a feminine Englishness, one less interested in the heroic exploits of official masculinity centred on "Great Britain" and "more inward-looking, more domestic and more private" (Light, quoted in Taylor 1994: 123).

The myth of the English cottage garden outlined in the epigraph was originally produced by and for a newly urbanised Edwardian middle class interested in what was becoming an alien territory: the countryside (Freeman 2003: 136, 149, 171–172). That myth adopted and adapted the ideological categories of the Victorian, so that gardening came to be seen as (largely) masculine, rational and apolitical, while the garden itself produced a "naturalised" femininity – whereby the nation's continued well-being was guaranteed. In this way, conflict was erased and the cottage garden became a haven to which both the honest labourer, and the weary Englishman abroad, could always return. It was the garden wall/gate, rather than the cottage, that secured the moral domesticity of the cottage's inhabitants, be they man or woman, adult or child. While the open gate invited casual visitors from the village and therefore served to link the cottage into the wider community made safe by gardening – or, rather, from the philanthropic middle class – it also called the husband back home to his wife and family while they waited patiently inside. The "mellowed wall", picket fence or privet hedge, therefore, physically demarcated the boundaries of that "home" and reassured a middle-class audience at least that within they would find bounty, security and a timeless backyard-England.

Notes

1 Paper originally delivered Gendered Landscapes, Penn State 31 May–1 June 1999; elements of this paper have appeared in Sayer 2000.
2 Ditchfield was a clergyman who claimed special knowledge of the labouring poor.
3 See Constantine 1981: 387–406; also Tristram 1989.
4 For "the famous five Rs: refuge, reflection, rescue, requiem and reconstruction" of pastoralism, cf. Sales 1983: 15–18.
5 For example: Waters 1988; Lang 2000; Hoyles 1991, 1995.
6 See, for example, Berger 1997: 107–116; Forrest and Ingram 1999: 206–218; Meredith 2001.
7 For example: Chase 1996; Howkins 2002, pp. 1–23; Burchardt 2002.
8 Alexander makes the point that "[w]here gardens have been studied, their purpose is usually for sustenance, and the analytical emphasis is often on the labor of production and the transformative element of such gardens". She then picks up this transformative element in her own analysis (Alexander 2002: 858).
9 For an elaboration of this re the city see Jacobs 1996: 3–5. Jacobs cites Appadurai re "staging grounds".
10 Jacobs makes a similar point re the post-colonial city in Jacobs 1996: introduction.
11 Waters makes a related point when he argues that artists and designers were "indebted to imaginative writers for inspiration, convenient frames of reference and examples" (Waters 1988: 53).
12 John Claudius Loudon (1783–1843) was a well-known and widely read British journalist and architect with a social bent, who produced a large number of gardening books and edited *The Gardener's Magazine*. Admired by Robinson, he had a considerable impact on all aspects of gardening practice through the whole of the period in question.
13 William Robinson (1838–1935) was originally an assistant gardener in Ireland and eventually herbaceous foreman in Regent's Park, London, England. He became a prolific garden writer and founded his own magazine, *The Garden: An Illustrated Weekly Journal of Horticulture In All Its Branches* in 1871. His best-known works are probably *The Wild Garden* (1870) and *The English Flower Garden* (1883). He disliked artificiality, including topiary and carpet bedding, and looked back to "pre-bedding" cottage gardening as an ideal.
14 Gertrude Jekyll (1843–1932), originally an artist, became a professional garden designer (the first woman to do so in Britain) when her eyesight began to fail. She wrote for and later became joint editor of *The Garden*, and often collaborated with architect Edwin Lutyens (1869–1944).
15 Hamilton 1995: 31 (emphasis in original). Geoff Hamilton, gardener and writer presented the BBC's "Gardener's World" – the longest running television gardening series in the UK. *Geoff Hamilton's Cottage Gardens* was published to accompany the highly popular six-part TV series of the same name.
16 "[F]ashion has been as fickle in gardening as in architecture – if anything more so" (Johnson 1979: 4).
17 See Burchardt 2002: 15–22.
18 Abbot, for one, only paid rent after negotiating for a piece of land that he could enclose and within which he could build his own cottage. See Howkins 2002 for an extensive discussion of this and its political aspects. Chase looks at one example of working-class attempts to regain access to the land in his article on the Chartist Land Plan.
19 See, for example, Hoyles 1991: 215. The practice of gardening as a form of recreation probably only began in earnest in the Elizabethan period. The sixteenth century, for example, saw the rise of topiary, an oft maligned craze

adopted by cottage gardeners which persisted into the twentieth century when Ditchfield was writing, examples of which can still be seen dotted around the country. See Ditchfield 1994: 92–93.
20 See Tristram 1989: 203–204; Davidoff *et al.* 1976: 139–175; Nead 1988: 40–44.
21 The *OED* defines a garden as "enclosed cultivated ground", or "an enclosed piece of ground devoted to the cultivation of flowers, fruit, or vegetables".
22 Zlotnick (1996: 51–68) makes a related point.
23 The *Ranunculus* family includes the meadow buttercup, the lesser celandine, and "bachelor's buttons", "fair maids of France" or "fair maids of Kent". *Ranunculus – ranunculus acris; ranunculus ficaria; ranunculus aconitifolius.* Bulbs cost between 4 to 6 pence per dozen, depending on the variety, in 1897. *Webbs' Spring Catalogue*, (1897), p. 99.
24 See McCracken 1997 for a discussion of the methods by which large-scale botanic, rather than cottage, gardens were established in British colonies.
25 Hence he stresses the importance of the labour, paid and unpaid, involved in the production of a garden.
26 See Lang 2000: 113.
27 See Chase 1996 and Howkins 2002.
28 Barbara English discusses the status of Flora Thompson's writing as fact or fiction in some detail and stresses that much of *Lark Rise* ought to be seen as having been written through the lens of nostalgia; the most painful memories are erased. *Lark Rise* therefore belongs to the genre of literary autobiography, written to adhere to the dominant forms of the day.
29 Helen Allingham, née Paterson, was born in 1848 in Derbyshire, England. Her aunt (Laura Herford) was a professional artist. Allingham attended the Royal Academy School in London, England, from 1867 and then worked as an illustrator for a range of novels, magazines and periodicals – she was the illustrator for Thomas Hardy's *Far From the Madding Crowd* serialised in the *Cornhill Magazine*. She also began painting watercolours in the 1870s and exhibited at the Royal Academy in 1874. She was elected a full member of the Royal Society of Painters in Water Colours in 1890. In 1879, she began painting the subjects that she became best known for, namely cottage gardens, country cottages and their inhabitants. For the latter, she often used professional models, including Mrs Stewart who also sat for du Maurier. Huish 1903.
30 Edwin Cockburn, *The Return from Market*, n.d., oil on panel, 24.9 × 30.5, private collection; reproduced in Nead 1988, Figure 9.
31 Reproduced in White 1897. Fildes produced many idyllic illustrations of the rural, such as *The Village Wedding* (1883) and *The Doctor* (1891), which were usually full of pathos, and represented middle-class life transposed to the village.
32 Fast-growing quick or privet hedges were commonly used to demarcate a piece of newly enclosed ground by labourers who wished to become cottagers with right of settlement.
33 See Brettel 1983.

Bibliography

Agyeman, J. and Spooner, R. (1997) "Ethnicity and the Rural Environment" in Cloke, P. and Little, J. (eds) *Contested Countryside Cultures: Otherness, Marginalisation and Rurality*, London: Routledge.
Alexander, C. (2002) "The Garden as Occasional Domestic Space", *Signs* 27(3) pp. 857–872.
Anonymous (1806) *Account of Britton Abbot's Cottage and Garden; and of a Cottager's Garden in Shropshire: to Which is Added Jonas Hobson's Advice to his Children: and the Contrast*

Between a Religious and Sinful Life (written in 1797 by Sir Thomas Bernard and reprinted by the Society for Bettering the Conditions and Increasing the Comforts of the Poor), London: Hatchard.

Berger, R. (1997) "Kitty Lloyd Jones: Lady Gardener and Nurserywoman", *Garden History* 25(1) pp. 107–116.

Brettel, R.R. and Brettel, C.B. (1983) *Painters and Peasants in the Nineteenth Century*, New York: Rizoli Skira.

Burchardt, J. (2002) *The Allotment Movement in England 1793–1873*, London: Royal Historical Society.

Burnett, J. (ed.) (1982) *Destiny Obscure: Autobiographies of Childhood, Education and Family from the 1820s to the 1920s*, Harmondsworth: Penguin.

Chambers, W. and Chambers, R. (eds) (1842) *Chambers's Information for the People*, Vol. II, Edinburgh: Chambers.

Chase, M. (1996) "'We Wish Only to Work for Ourselves': the Chartist Land Plan" in Chase, M. and Dyck, I. (eds) *Living and Learning: Essays in Honour of J. F. Harrison*, Aldershot: Scolar.

Constantine, S. (1981) "Amateur Gardening and Popular Recreation in the 19th and 20th Centuries", *Journal of Social History* 14(3) pp. 387–406.

Darley, G. (1979) "Cottage and Suburban Gardens" in Harris, J. (ed.) *The Garden: A Celebration of One Thousand Years of British Gardening*, London: New Perspectives Publishing.

Davidoff, L. and Hall, C. (1987) *Family Fortunes; men and women of the English middle class 1780–1850*, London: Hutchinson.

—— *et al.* (1976) "Landscape with Figures" in Mitchell, J. and Oakley, A. (eds) *The Rights and Wrongs of Women*, Harmondsworth: Penguin.

Ditchfield, P.H. (1994) *The Charm of the English Village* (first edition 1906), London: reprinted by Senate.

English, B. (1985) "Lark Rise and Juniper Hill: A Victorian Community in Literature and History", *Victorian Studies* 29(1) pp. 7–34.

Forrest, M. and Ingram, V. (1999) "Education for Lady Gardeners in Ireland", *Garden History* 27(2) pp. 206–218.

Freeman, M. (2003) *Social Investigation and Rural England 1870–1914*, Woodbridge: Royal Historical Society.

Hamilton, G. (1995) *Geoff Hamilton's Cottage Gardens*, London: BBC Books.

Howkins, A. (1991) *Reshaping Rural England: A Social History 1850–1925*, London: Harper Collins.

—— (2002) "From Diggers to Dongas: the land in English Radicalism, 1649–2000", *History Workshop Journal* 54 pp. 1–24.

Hoyles, M. (1991) *The Story of Gardening*, London: Journeyman Press.

—— (1995) *Bread and Roses: Gardening Books from 1560 to 1960*, Vol. 2, London: Pluto Press.

Huish, M.B. (ed.) (1903) *Happy England*, London: Adam & Charles Black.

Hyams, E. (1966) *The English Garden*, London: Thames & Hudson.

Jacobs, J.M. (1996) *Edge of Empire: Postcolonialism and the City*, London and New York: Routledge.

Jefferies, R. (1981) *The Toilers of the Field* (first edition 1892), London: Macdonald Futura.

Johnson, H. (1979) "Introduction" in Harris, J. (ed.) *The Garden: A Celebration of One Thousand Years of British Gardening*, London: New Perspectives Publishing.

Lang, K. (2000) "The Body in the Garden" in Birksted, J. (ed.) *Landscapes of Memory and Experience*, London and New York: Spon Press.

Loudon, J.C. (1840) *The Cottager's Manual of Husbandry, Architecture, Domestic Economy, and Gardening* (originally published in the *Gardener's Magazine)*, London: reprinted for The Society For The Diffusion of Useful Knowledge.

McCracken, D.P. (1997) *Gardens of Empire: Botanical Institutions of the Victorian British Empire*, London and Washington: Leicester University Press.

Maclean, G., Landry, D. and Ward, J.P. (eds) (1999) *The Country and the City Revisited: England and the Politics of Culture, 1550–1850*, Cambridge: Cambridge University Press.

Massey, D. (1994) *Space, Place and Gender*, Cambridge: Polity.

—— (1995) "Places and Their Pasts", *History Workshop Journal* 39 (Spring) pp. 182–192.

Meredith, A. (2001) *Middle-Class Women and Horticultural Education, 1890–1939*, unpublished thesis submitted for degree of Doctor of Philosophy, University of Sussex.

Nead, L. (1988) *Myths of Sexuality; Representations of Women in Victorian Britain*, Oxford: Basil Blackwell.

Ortner, S.B. (1974) "Is Female to Male as Nature is to Culture?" in Mitchelle, Z. *et al.* (eds) *Woman, Culture and Society*, Stanford CA: Stanford University Press.

The Penny Cyclopaedia of the Society for the Diffusion of Useful Knowledge (1843) Vol. VIII, London: Charles Knight and Co.

Reed, M. (1984) "The Peasantry of Nineteenth Century England: A Neglected Class?", *History Workshop Journal* (Spring) pp. 53–76.

—— (1997) *The Landscape of Britain from the Beginnings to 1914*, London: Routledge.

Reports of the Special Assistant Poor Law Commissioners on the Employment of Women and Children in Agriculture PP 1843: [510] XII, London.

Sales, R. (1983) *English Literature in History; 1780–1830, Pastoral and Politics*, London: Hutchinson.

Sayer, K. (1995) *Women of the Fields: Representations of Rural Women in the Nineteenth Century*, Manchester: Manchester University Press.

—— (2000) *Country Cottages: A Cultural History*, Manchester: Manchester University Press.

Silvester, S. (1968) *In a World that Has Gone*, Leicestershire: Brunel Library.

Taylor, J. (1995) *A Dream of England: Landscape, Photography and the Tourist's Imagination*, Manchester: Manchester University Press.

Thompson, F. (1979) *Lark Rise to Candleford: A Trilogy* (first editions, 1939, 1941, 1943), Harmondsworth: Penguin.

Tristram, P. (1989) *Living Space in Fact and Fiction*, London and New York: Routledge.

Waters, M. (1988) *The Garden in Victorian Literature*, Aldershot: Scolar Press.

White, G. (1897) *English Illustration: the Sixties: 1855–1870* (first edition), Westminster: Constable and Co.

Williams, R. (1985) *The Country and the City*, London: the Hogarth Press.

Willmott, R.A. (1864) *A Journal of Summer Time in the Country* (fourth edition), London: John Russell Smith.

Young, A. (1971) *General View of the Agriculture of the County of Hertfordshire, Drawn up for the Consideration of the Board of Agriculture and Internal Improvement* (first edition 1804), reprinted Newton Abbot: David and Charles.

Zlotnick, S. (1996) "Domesticating Imperialism: Curry and Cookbooks in Victorian England", *Frontiers* XVI(2/3) pp. 51–69.

3 Transplantation of the Picturesque

Emma Hamilton, English landscape, and redeeming the Picturesque

Shin-ichi Anzai

Le Pittoresque nous vient d'Angleterre.

(Stendhal *Mémoires d'un touriste*, 1838)

The Picturesque as a historical phenomenon emerged in eighteenth-century England.[1] It can be defined as a mode of vision which sees the world, more or less consciously, as a series of established pictures, such as that of Claude Lorrain (1600–82) and Salvator Rosa (1615–73). The most typical objects seen in this way were landscape gardens and natural scenery, but people also talked about Picturesque literature,[2] music, and human figures such as the poor and women. Then, in late eighteenth-century England, the Picturesque was formulated into aesthetic theories (Anzai 1992).

As has been pointed out (J. Burke 1976; Dobai 1975: 298), the Picturesque was such a highly multi-faceted cultural phenomenon that it is dangerous to generalize the concept. But, at the same time, one can safely say that the Picturesque mode of vision was broadly shared among the cultural elite of England and other nations in the eighteenth and nineteenth centuries, and that people were well aware that it was a sufficiently unified (and sometimes fashionable) one to be invoked under the single rubric of "picturesque." In addition, what is remarkable is its widespread influence, not only on English Romantic literature, but also, for example, on modern American landscape architecture (Conron 2000). In a recent, extensive study, Gena Crandell points out that as a result of the global diffusion of the Picturesque – via painting, animation, cinema, TV, video, computer technology, etc. – today "we take the world to be a picture":

> There is no doubt that the eighteenth-century English landscape garden has been the most influential force in the last two centuries of landscape design. ... Today when we think of nature we too often conjure up images borrowed from eighteenth-century England. ... *Undeniably, the landscape itself has become the repository of pictorial conventions and landscape architecture the perpetuator of the painterly vision.*
>
> (Crandell 1993: 11, 8, 165 [italic original])

Ever since its emergence in the eighteenth century, though, the Picturesque has been criticized and ridiculed (Watson 1970), because the Picturesque eye does not see the natural world directly – disregarding its moral aspects – and distorts or conceals it by the mediation of pictures. It assimilates the diversities and anomalies of the seen object to ready-made stereotypes. Especially when the Picturesque mode of vision is applied to the non-European world and the female body by a supposedly superior Western male subject, the implied violence and immorality of assimilation becomes obvious.[3]

However, I want not to condemn the Picturesque but to redeem it, by concentrating on the concept of "transplantation," a term which often appears in the literature. The Picturesque eye transplants pictures onto reality. As is always the case with transplantation, there certainly is a violence involved here, but transplantation can also provide healing. I will concentrate on this latter, neglected moment in the Picturesque, which has been transplanted into so many facets of today's world. In the following, therefore, my priority lies just in grasping the phenomenon of the Picturesque in general (hence the capital P) from eighteenth-century England on for the sake of redeeming it against the recent criticisms, not in going into the details of its historical ramifications, though in the last part I shall focus on a hidden aspect of the historical Picturesque around 1800.

Sir William Hamilton's "picture-madness": Emma and the English garden in Naples

First, let us look at the case of Emma Hamilton (1761?–1815), the second wife of Sir William Hamilton (1730–1803), the British Envoy Extraordinary to Naples, a geologist, and a dilettante.[4] Incidentally, Emma (as I shall refer to her throughout this chapter) later became well known as the mistress of Admiral Nelson who defeated Napoleon's navy.

Emma provides a typical example of the female body seen as Picturesque landscape, because she and her husband developed a performing art, named "attitude" (Figure 3.1; cf. Holmström 1967). In these "attitudes," Emma would pose in imitation of famous pictures, sometimes sitting in a box representing the frame of the picture. Her attitudes were seen by numerous grand tourists to Naples, some of whom were invited to draw her while posing. One such tourist-spectator was Johann Wolfgang von Goethe (1749–1832), who said about Emma: "Hamilton is a man of comprehensive taste, and after having searched all range of Nature [Schöpfung], he has arrived at a beautiful wife, the masterpiece of the great artist."[5]

This "attitude," which would lead to the popularity of tableaux vivants in the nineteenth century, is a most conspicuous way of acting out the Picturesque mode of vision in the sense that it consciously dramatizes the mechanism of seeing the reality as a picture. Sir William was such a lover of painting that Horace Walpole (1717–97), another famous dilettante of

Figure 3.1 Emma, Lady Hamilton, Tommaso Piroli, after Friedrich Rheberg, *Drawings faithfully copied from Nature at Naples and with permission dedicated to the Right Honorable Sir William Hamilton, 1794*

the time, had predicted before Sir William was dispatched to Naples: "[Hamilton] is picture-mad, and will ruin himself in virtuland [i.e. Italy]."[6] Sir William was also particularly familiar with the Picturesque aesthetic theories of his friends, Sir Uvedale Price (1747–1829) and Richard Payne Knight (1750–1824). Since Goethe and many paintings depicting Emma, in her time, referred to her as "nature" (Figure 3.1), she was obviously seen as a Picturesque natural object.[7] Sir William himself tended to identify her with Picturesque landscape, as is known from his letter reporting her arrival in Naples in 1786: "A beautiful plant called *Emma* has been transplanted here from England, and at least has not lost any of its beauty" (Jenkins and Sloan 1996: 20).[8]

This likening of Emma's arrival to horticultural transplantation was in many ways pertinent. First, the addressee of the letter is Joseph Banks

(1743–1820), the natural historian, who then as the head of Kew Gardens, London, and the president of the Royal Society, was hunting species of plants from all over the world (Desmond 1995: Ch. 6). Second, Sir William himself was at that time absorbed in making an English landscape garden in Naples. Third, the eagerly awaited English gardener chosen by Banks for constructing Sir William's garden, a John Andrew Graefer,[9] had arrived from England at Naples just a week before Emma. Fourth, he had lost his first wife four years before, mourning her death bitterly; and it was as a distraction from this very sorrow that he proposed to construct the English garden in Naples. Sir William writes to Banks:

> I promise myself great pleasure in this new occupation [of landscape gardening]. As one passion begins to fail, it is necessary to form another; for the whole art of going through life tollerably [*sic*] in my opinion is to keep oneself eager about anything. The moment one is indifferent *on s'ennuie*, and that is a misery to which I perceive even Kings are often subject.[10]

Clearly, Emma was, just like English gardening, a distraction from this "*ennui*" caused by his first wife's death. Hence, it must have been pertinent to the economy of his psychology to liken Emma's arrival to horticultural transplantation; in his mind Emma and his English garden commingle under the same category of the Picturesque.

It is totally legitimate here to accuse Sir William of sexism[11]; indeed, when he wrote "*Emma* has been transplanted," she was not even informed that she was brought to Naples only to become one of his distractions. Her lover at that time happened to be Sir William's nephew, who wanted to get rid of her to marry for money; thus, it was arranged that she be Sir William's mistress. Although Sir William would later marry Emma, she then got angry, naturally, when she came to realize his true intentions.

This transplantation, therefore, developed at least one symptom of rejection. However, one can say in general that there is always an element of rejection implied in the Picturesque, because the Picturesque inevitably involves transplantation, and every transplantation involves heterogeneity, as well as homogeneity, between what is transplanted and its recipient. A telling example of this situation is Sir William's English garden in Naples.

Originally, the Picturesque landscape garden had been produced by transplanting Italian landscape into English soil, via (mainly) seventeenth-century landscape paintings. This process is formulated by William Mason (1725–97) in a poem apostrophizing English elite grand tourists to Italy:

> . . . ye of Albion's sons
> Attend; Ye freeborn, ye ingenuous few,
> . . . .

Visit the Latian plain [i.e. the Roman Campagnia], fond to *transplant*
Those arts which Greece did, with her Liberty,
Resign to Rome. . . .

. . . .

. . . your eyes entranc'd
Shall catch those glowing scenes, that taught a CLAUDE [Lorrain]
To grace his canvass with Hesperian hues:
And scenes like these, on Memory's tablet drawn,
Bring back to Britain; there give local form
To each Idea; and, if Nature lend
Materials fit of torrent, rock, and shade,
Produce new TIVOLIS.

. . . .

In [great painters'] immortal works thou ne'er shalt find
Dull uniformity, contrivance quaint,
Or labour'd littleness; but contrasts broad,
And careless lines, whose undulating forms
Play thro' the varied canvass: these *transplant*
Again on Nature; take thy plastic spade,
It is thy pencil; take thy seeds, thy plants,
They are thy colours; . . .

(Mason 1772–81, 1783: 1: ll. 50f., 57ff.,
270–77 [the author's italic])

But what about an English Picturesque landscape garden transplanted
back into Italian soil, via an English ambassador? Does it not amount to
a tautology, as is shown by a painting of Sir William's garden, which looks
exactly like an Italian landscape itself depicted by a Claude Lorrain (Figure
3.2; cf. Greuter and Maier-Solgk 1997: 9). At least for Italian visitors, who
knew only the geometrical formal garden, a Picturesque English garden in
Naples was nothing but their everyday landscape, as Sir William deplores:

> To enjoy an English garden requires a previous education that is the
> case, no one can be sensible of the beauties of Homer coming to it
> directly from reading Tom Thumb & Jack the Giant Killer, so how
> many companies when they are in the Garden ask where is the English
> Garden? being told they are in it they say Lord! there is nothing but
> grass & trees that bear no fruit and often advise poor Graefer [the
> gardener of Sir William's English garden] to cut out some figures on
> his beautiful turf.[12]

(Jenkins and Sloan 1996: 288)

In other words, to appreciate the Picturesque, one has to dissimilate and
distance the familiar, making heterogeneity and Otherness emerge; and
this is more or less true of all the Picturesque transplantations.

Figure 3.2 View of the English Garden at Caserta, Jakob Philipp Hackert, 1793

To redeem the Picturesque, we should explore these elements of hetero-geneity and Otherness. In this light, I shall examine the Picturesque trans-plantation onto the non-European world, and then take a brief look at the aesthetic theories of the Picturesque in late eighteenth-century England, before re-examining Emma's case.

The Picturesque outside Europe: South Africa and Japan

It is easy – and legitimate – to criticize modern Western travelers and colonists for violently assimilating the non-Western world to their Picturesque stereotype. For example, the frontispiece of the widely read tour, François Le Vaillant's *Travels into the Interior Parts of Africa* (English trans. 1790; Figure 3.3), imposes the Claude Lorrainean style of composi-tion – where the picture plane is divided into foreground, middle ground, background, and sidescreen(s)[13] – onto the native land of Africa.

Even in this banal picture, however, one can point out some ambigu-ities not to be explained away only as violently colonialist. The author-cum-hunter is pointing to the interior of the book/Africa to be conquered. But this interior lies just *outside* the picture frame. The picture, therefore, suggests that imposing a ready-made formula on something else always

Figure 3.3 Encampment in the Great Namaqua Country, François Le Vaillant, *Travels into the Interior Parts of Africa*, frontispiece, 1790

already reveals the Other outside the formula and that the formula cannot cover the whole world. Hence, the superiority of the Western formula can be turned into its inferiority, as Le Vaillant himself asserts in the text:

> Ye English gardens twenty times changed with the wealth of the citizen! Why do your streams, your cascades, your pretty serpentine walks, your broken bridges, your ruins, your marbles, and all your fine inventions, disgust the taste and fatigue the eye, when we know the verdant and natural bower of the Pampoen-Kraal [in South Africa]?[14]

Here, the formula of the English garden finally turns out to be inferior to the reality of the African landscape. This inversion of superiority and inferiority can be observed in many other instances of the Picturesque transplantation around the world.

In the case of the Japanese landscape, it is another English ambassador, Sir Rutherford Alcock (1809–97), who first saw it as distinctively Picturesque (in the Western sense). Describing his 1861 travels in Japan, he adopts as a cultural elite the established formula of the English Picturesque tour:

While riding on alone, it would be easy to fancy the scene was in some picturesque English county. We came upon a very beautiful moss-rose growing by the side of a cottage, a variety of the English species apparently, . . . we passed through many scenes worthy of the artist's pencil; indeed, the number of tempting pictures was tantalising, since it was clearly impossible to take even the slightest sketch of all. . . . as we descended through a rocky pass into the valley below, and caught the first glimpse of the cultivated fields and terraced hills with another range of mountains towering beyond, picturesque Japanese figures filling up the foreground, it was difficult to pass and take no note.

(Alcock 1863: 2: 82f.)[15]

In the last scenery, the Japanese figures make up the foreground, the fields and hills are the middle ground, and the mountains, the background, while the rocky sides of the pass provide the sidescreens, thus constructing a typical Claude Lorrainean composition.

The superiority of the Western gaze, implied here, becomes manifest when Alcock describes Mt Fuji (Figure 3.4), the highest mountain in Japan, which Alcock climbed first as a foreigner, even though in pre-modern Japan foreigners were strictly prohibited from traveling:

[Fuji] may be seen from Yeddo [i.e. old Tokyo] at a distance of some eighty miles, on a bright summer evening, lifting its head high into the clouds, the western sun setting behind it and making a screen of gold on which its purple mass stands out in bold relief. Or, early in the morning, its glittering cone of snow, tipped with the rays of the rising orb; . . . and in either aspect it is certainly both singular and picturesque, . . . To the Japanese who are anything but cosmopolitan, it may be the "matchless" . . .

(Alcock 1863: 1: 406)[16]

The Japanese are too ignorant and narrow-minded to "match" Mt Fuji to anything else, whereas Alcock's cultivated eye can see it through the established medium of the picture.

Alcock's medium here, however, seems to be not Western oil painting, but the Japanese traditional folding-screen painting with gold leaves on the background,[17] which suggests that the Western formula is somehow inadequate or inferior for interpreting the Japanese landscape.[18] The inadequacy of Western media to depict Japanese scenes was already implied in the quotation above in which Alcock voiced his frustration about his incapacity to appropriate all the Japanese landscapes into sketches. Hence, the Western Picturesque formula itself can be utilized to emphasize the very "singularity" of the Japanese native landscape that cannot be subsumed by the ideology of the Picturesque.

VIEW OF FUSIYAMA FROM YOSIWARA

Figure 3.4 View of Fusiyama from Yosiwara, Alcock, 1863

This is exactly the tactic that the Japanese geographer Shigetaka Shiga (1863–1927) adopts in his encomium of Japan, *The Japanese Landscape* (1894), which was extremely popular and influential in the nationalist and militarist Japan just before and during the Second World War. Though a great reader of John Ruskin, Shiga belittles the English way of seeing nature, including the Picturesque:

> The English boast of their autumnal scenery so much, but they are almost ignorant of Japanese maples. . . . The beauty of Japanese nature poetry never occurred to even Wordsworth, the poet of the "Lake District," who observed minutely and loved dearly the English natural landscape. True, Sir Walter Scott, the genius with profound sagacity in depicting natural sceneries, had a tact of balancing the various colors of nature, nearing the masters of painting. But it is a pity that he rarely depicts maples. In short, English people not knowing the subtleties of maples, the English autumn is nothing compared to autumn in Japan.
>
> (Shiga 1894: 16f; author's translation)[19]

Thus, the Picturesque contains a (latent) dialectic or inversion between ideological superiority and inferiority. This dialectic, however, can be detected even in the English aesthetic theories of the Picturesque as early as in the eighteenth century.[20]

English aesthetic theories of the Picturesque: dialectic of superiority and inferiority

Certainly, those English Picturesque aesthetic theorists insisted on some superiority of the elite Picturesque eye over the common eye. For example, in an imaginary dialogue, Sir Uvedale Price, Sir William's friend, makes a layman to painting admit himself to be just a "vulgar observer" and inferior in appreciating nature: "Tell me, then, how you account for this strange difference between an eye accustomed to painting, and that of such a person as myself?" (Price 1810: 3: 273).

But at the same time this superiority of the Picturesque eye was based on some inferiority,[21] or at least on a humiliating recognition that the Picturesque mode of vision is limited, not able to comprehend all. William Gilpin (1724–1804), the founder of Picturesque aesthetics, had already realized this limitation peculiar to the Picturesque eye when he wrote:

> The case is, the immensity of nature is beyond human comprehension. She works on a *vast scale*; and, no doubt, harmoniously, if her schemes could be comprehended. The artist, in the mean time, is confined to a *span*. He lays his little rules therefore, which he calls the *principles of picturesque beauty*, merely to adapt such diminutive parts of nature's surfaces to his own eye, as come within its scope.
>
> (Gilpin 1782: 18 [italic original]; cf. Johnson 1753; Shenstone 1764: 142f.; Repton 1803: 221; Barrell 1983)

This very narrowness of the Picturesque vision and the agnostic inscrutability of the world resulting from it, however, is the source of the aesthetic pleasure that is peculiarly Picturesque. Another Picturesque theorist, Price, explains:

> the stimulus from whence the most constant and marked effects proceed, that which in a peculiar manner belongs to the picturesque, and distinguishes it from the beautiful, – arises principally from its two great characteristics, intricacy and variety, as produced by roughness and sudden deviation, and as opposed to the comparative monotony of smoothness and flowing lines.
>
> If we take any smooth [i.e., beautiful] object, whose lines are flowing, such as a down of the finest turf with gently swelling knolls and hillocks of every soft and undulating form, though the eye may repose on this with pleasure, yet the whole is seen at once, and no further curiosity is excited; but let those swelling knolls (without altering the scale) be changed into bold broken promontories, with rude overhanging rock; instead of the smooth turf, let there be furze, heath, or fern, with open patches between, and fragments of rocks and large stones lying in irregular masses, . . . the whole of the one may be comprehended immediately, and . . . if you traverse it in every

direction little new can occur; while in the other every step changes the whole of the composition. . . . All these deep coves, hollows, and fissures invite the eye to penetrate into their recesses, yet keep its curiosity alive and unsatisfied; . . .

(Price 1794: 105ff.)[22]

This tantalizing sensation – note the sexual overtones of the last sentence, which shall be explicated immediately below – caused by the impenetrability and inscrutability of the Picturesque landscape derives from some humiliating impotence of the Picturesque eye; since the Picturesque eye cannot penetrate all, there is always an excess or the Other, escaping its grasp, its comprehension, which, though frustrating, titillates it in a peculiar way.

Narcissism of Emma/nature: the Picturesque as healing

Now let us return to Emma's case, in order to complete our search for a redemptive moment in the Picturesque. Emma's husband, Sir William, a friend of several Picturesque aestheticians, was also aware that nature escapes human comprehension. Famous for his geological observations on volcanoes around Naples, Sir William criticizes the old type of natural history:

Nature acts slowly, it is difficult to catch her in the act. Those who have made this subject their study, have without scruple, undertaken at once, to write the Natural History of a whole province, or of an entire continent; not reflecting, that the longest life of man scarcely affords him time to give a perfect one of the smallest insect.[23]

(Jenkins and Sloan 1996: 68)

Does not this recognition of the inscrutability of nature apply to Emma herself, who was often identified with nature as was shown earlier? Is it not also true of her "attitude" that "it is difficult to catch her in the act"?

At the very least, the actual Emma tended to escape Sir William's control: her liaison with Nelson even forced Sir William to give up his diplomatic career. Surprisingly, though, Sir William had anticipated such scandals even before he decided to accept Emma, as he had said:

Tho' a great City, Naples has every defect of a Province and nothing you do is secret. It would be fine fun for the young English Travellers to endeavour to cuckold the old Gentleman their Ambassador, and whether they succeeded or not would surely give me uneasiness.

After his resignation, he recollects:

I am arrived at the age when some repose is really necessary, and I promised myself a quiet home, altho' I was sensible . . . that I shou'd be superannuated when my wife wou'd be in her full beauty and vigour of youth. That time is arrived, and we must make the best of it for the comfort of both parties.[24]

So, one can guess that he might have half-welcomed Emma's escape from his control. Indeed, according to his friend Price, the appeal of a Picturesque woman rests exactly on her characteristics of uncontrollable irregularity and inscrutability:

> it is also common to say of a woman – que sans être belle elle est *piquante* – a word, by the bye, that in many points answers very exactly to picturesque. The amusing history of Roxalana and the Sultan, is also the history of the piquant, which is fully exemplified in her person and her manners: Marmontel certainly did not intend to give the petit nez retroussé as beautiful feature; but to shew how much such a striking irregularity might accord and co-operate, with the same sort of *irregularity* in the character of the mind. The playful, unequal, coquetish Roxalana, full of sudden turns and caprices, is opposed to the beautiful, tender, and constant Elvira; and the effects of irritation, to those of softness and languor: the tendency of the qualities of beauty alone towards monotony, are no less happily insinuated.
>
> (Price 1810: 1: 73f. [italic original])[25]

A Picturesque woman as this is a "narcissistic" object in the Freudian sense; that is to say, it escapes external control and is autonomous to the degree of inscrutable.[26] Emma may have been such an object to Sir William, especially when acting "attitudes." In fact, the pictures depicting "attitudes," including caricatures (Figure 3.5), are all "absorptive," to use Michael Fried's dichotomy, compared to her other ordinary "theatrical" portraits[27]: i.e. in acting "attitudes," Emma is so completely captured by one or other emotion that she is not aware of the observers' gaze, showing narcissistic "self-contentment and inaccessibility" (Freud 1914; 1963: 14: 89).

This Picturesque narcissism of Emma could have provided some healing moment to Sir William. As was stated earlier, she was, just like the Picturesque English garden, a distraction from the "*ennui*" caused by his first wife's death. He devoted an "eager" "passion" to his first wife, or according to Freud's diction, the "complete object-love of the attachment type" (Freud 1914; 1963: 14: 88). But such strong attachment to the loved object can exhaust the subject's basic narcissism and living energy, hence the paradox that keeping the object consistently narcissistic can contribute to restoring the basic narcissism of the subject itself. Maybe, Emma's "attitude" was a contrivance to emphasize and enhance her narcissism, making her autonomous beyond the picture frame. This seems the more plausible because Sir William, even though aware of the possibility of scandals, dared to show her "attitudes" to many male grand tourists.[28]

Figure 3.5 Lady H****** Attitudes, Thomas Rowlandson (?), *c*.1800

To accept Emma's narcissism may not only be healing to Sir William but also prevent against his possible violence toward Emma. The Picturesque is characterized by some distancing of the object, which, by deferring the immediate satisfaction of the seeing subject's desire, can prevent the violent possession of the desired object.

This will become obvious by that comparison between the simply beautiful and the Picturesque which was one of the main topics among Picturesque aestheticians. On the one hand, according to Edmund Burke, the most influential precursor to Picturesque aesthetics, beauty is "that quality or those qualities in bodies by which they cause love, or some passion similar to it" (Burke 1757; 1759; 1987: 91). This "love" is closely connected with strong sexual desire and can lead to a violent, possibly destructive assimilation of the desired object by the desiring (male) subject, as Burke explicates:

> violent effects [are] produced by love, which has sometimes been even wrought up to madness, . . . the generation of mankind is a great purpose, and it is requisite that men should be animated to the pursuit of it by some great incentive. . . . [the object of love] can so quickly, so powerfully, or so surely produce its effect. . . . we like to have [beautiful objects] near us, and we enter willingly into a kind of relation with them, . . .
>
> (Burke 1757; 1759; 1987: 40–43)

On the other hand, the Picturesque results from abandoning such a satisfaction of desire. Thus Price contrasts the beautiful and the Picturesque:

> Soft, fresh, and beautiful colours . . . give us an inclination to try their effect on the touch; whereas . . . that inclination . . . in objects merely picturesque, and void of all beauty, is rarely excited.
>
> I have read, indeed, in some fairy tale, of a country, where [picturesque] age and wrinkles were loved and caressed, and [beautiful] youth and freshness neglected; but in real life, I fancy, the most picturesque old woman, however her admirer may ogle her on that account, is perfectly safe from his caresses.
>
> (Price 1810: 1: 71 and note)

To use an example favored by the Picturesque aestheticians, "the most beautiful objects will become [picturesque] from the effects of age, and decay."[29] The Picturesque is, therefore, produced by hampering the violent assimilation and possession of the object by the subject.

From this viewpoint of the prophylactic nature of the Picturesque, I will finally consider Thomas Rowlandson's caricature of the "attitudes" of Sir William and Emma (Figure 3.5). Here, Sir William is acting the role of a perverted old man (in the guise of a Pygmalion) who lets his wife be symbolically raped by the pen/is of a young tourist or artist. This certainly explicates the obscenity and sexist violence latent in the Picturesque gaze. But it is ironic that Emma's nude including the most obscene picture-in-a-picture was obviously drawn also by Rowlandson himself (as a portrait painter educated in The Academy, Rowlandson could have had a chance to paint a portait of Emma). So the laughter aimed at Sir William and the young artist implicates Rowlandson as well. In other words, the Picturesque gaze ridicules itself. This element of reflexivity and derogating oneself inherent in the Picturesque tradition can work against the violence of the Picturesque mode of vision, and provide a redemptive moment for healing.

These examples show that the tradition of the Picturesque is far from monolithic. It involves both assimilation and dissimilation, superiority and inferiority, violence and healing, and their inversions. This is because the Picturesque is a transplantation of a familiar, established formula on to the Other, making homogeneity and heterogeneity emerge in the same instance. Through this recognition, we may turn our own "picture-madness" transplanted in us into a redemption.[30]

Notes

1   The most relevant previous studies on the Picturesque are: Manwaring (1925); Hussey (1927, rpt. 1983); Hipple (1957); Watkin (1982); Andrews (1989); Robinson (1991); Copley and Garside (1994).
2   For Picturesque literature, see Hagstrum (1958); for Picturesque music, see Price (1810): 1: 43f.; for the Picturesque poor, see, for example, Price (1810): 1: 63; I shall discuss Picturesque women below.
3   For recent criticisms of the political ideologies of the Picturesque (including its colonialist tendency), see Barrell (1972; 1980); Bermingham (1986); Liu (1989): Ch. 3; Robinson (1991): 47–89; Copley and Garside (1994). For the complicity between colonialism and the Picturesque English garden and aesthetics, see also McLeod (1999): esp. 145f., 165ff., 221–29; Bohls (1999).
4   For information about Emma and William Hamilton, I am indebted to Jenkins and Sloan (1996).
5   *Italienische Reise* (Goethe 1988, 11: 217 (the author's translation); cf. Goethe (1988), 11: 209, 330f. Goethe uses such tableaux vivants at a pivotal point in his novel, *Die Wahlverwandtschaften* (1809), II, 5, 6 (Goethe 1988: 6: 391–94, 402–05). For Goethe and the Picturesque, see also his *Dichtund und Wahrheit*, II (1812), 8 (Goethe 1988: 9: 320f.).
6   Horace Walpole, to Mann, 8 June 1764 (Walpole 1937–83: 22: 243).
7   Cf. Jenkins and Sloan (1996): Figures 160, 168, 193. Emma is also seen as a "statuesque" object, as Walpole comments: "Sir William Hamilton has actually married his gallery of statues" (To Mary Berry, 11 September 1791; Walpole 1937–83: 11: 349). Goethe says he witnessed Emma's attitudes lit by fire, which was in accordance with the contemporary custom for watching statues (Bätschmann 1985). In Figure 3.5 of this chapter, Sir William is clearly identified as Pygmalion the sculptor.
8   Letter to Joseph Banks, BL. Add. MS 34,048, f. 30 (cit. Jenkins and Sloan 1996: 20).
9   Graefer was a disciple of William Kent (1685–1748), the pioneer of English landscape gardening (Jenkins and Sloan 1996: 18).
10   Letter to Banks, 3 May 1785, BL. Add. MS 34,048, ff. 24–25 (cit. Jenkins and Sloan 1996: 18 [italic original]).
11   For feminist readings of the Picturesque, see Fabricant (1977, 1985); and especially Bermingham (1994), Bohls (1995).
12   Letter to Banks, BL Add. MS 34,048, f. 89; cit. Jenkins and Sloan (1996): 288; cf. Venturi (1981).
13   For this formula, see especially William Gilpin's first published Picturesque tour (Gilpin 1782: 8). Jane Austen, who sometimes uses Gilpin's ideas, refers to this formula in her *Northanger Abbey* (Austen 1818: Ch. 14: 125).
14   François Le Vaillant, *Travels into the Interior Parts of Africa*, trans., anon. (London, 1790): 1: 165; cit., in Bunn (1994): 133.

15 Alcock regards Japan as the most "garden-like" in the world besides England and dreams of transplanting the Picturesque Japanese scenery into English gardens: "[Japanese plants] give studies for the landscape painter of unrivalled beauty. There is an infinite variety of form, character, and colouring, in the masses of foliage that everywhere meet the eye, grouped in the midst of well-kept fields and verdant slopes which any English gentleman might envy for his park" (1863: 1: 201).

16 In Japanese, "Fuji" could originally mean "matchless." His own sketch of Mt Fuji (Figure 3.4 of this chapter) is exactly based on the conventional Claude Lorrainean composition.

17 In the paragraph following this quotation, Alcock himself says that he witnessed abundant Japanese illustrations of Fuji in "the ornament of tea-cups or cabinets" and woodcut prints (Alcock 1863: 1: 407). Generally he estimates the traditional Japanese fine arts very highly, stating that the Japanese "have an eye for form and picturesque grouping; and understand effects of light and shade," though he doesn't appreciate the Japanese landscape painting, because "Their knowledge of perspective is too limited" (2: 281). Alcock's collection of Japanese fine arts would be shown at the second Great Exhibition in London (1862), leading to the upsurge of Japanism in Britain.

18 Alcock himself partly admits this: "Like Don Quixote, . . . [Western] writers on Japan have hitherto seen everything through highly coloured glasses, and generally of a Claude Lorraine hue" (Alcock 1863: 1: 46). Here Alcock is referring to the so-called "Claude Lorrain glass" used by English Picturesque tourists to tone down the real landscape into a picture (Andrews 1989: 67–73).

19 One of the most important modern Japanese novelists, Sôseki Natsume (1867–1916), having studied English literature in London, wrote a consciously Picturesque novel, *Kusamakura* (Natsume 1906), which proved to be a best seller at the time. In the early part of the novel, the hero, a Picturesque tourist-painter representing the author proclaims: "I . . . from now on will regard everyone I meet . . . as no more than a component feature of the overall canvas of Nature. . . . Three feet away from the canvas you can look at it calmly . . . you are not robbed of your faculties by considerations of self interest, and are therefore able to devote all your energies to observing the movements of the figures from an artistic point of view" (Natsume 1906 [1965]: 23f.). But he considers this disinterestedness inherent in the Picturesque as not so much Western but traditionally *oriental*: "[Western poets] are content to deal merely in such commodities as sympathy, love, justice and freedom, all of which may be found in that transient bazaar which we call life. . . . Happily, oriental poets have on occasion gained sufficient insight to enable them to enter the realm of pure poetry" (19ff.). Indeed, there had been an important technique of "mitate" (literally, "seeing-as" or "surrogation") in the traditional Japanese arts similar to that of the Picturesque, where some real thing is identified as, or likened to, another (famous) thing usually depicted in painting or literature. Cf. Tsuji (1986): 63–66. As Tsuji emphasizes, "mitate" sometimes produces the effect of playfulness or parody, which element I shall detect in the Western Picturesque itself based on Figure 3.5 of this chapter.

20 Generally speaking, such dialectic can be said to be peculiar to the entirety of modernity as "the age of the world picture." Cf. Heidegger (1938; 1977). And again, it is ultimately based on the fundamental ambiguity of the "representation" (Vorstellung) itself. When reality is recognized through representation, the representation at once reveals and conceals reality; then, on the one hand, if the former moment of revealing is emphasized, the revealing eye will be privileged into a superior position of the discoverer of truth; on the other hand, if the latter moment of concealing is emphasized, the representation will be denigrated as an inferior stiff stereotype, as in the Romantic reaction against the Picturesque. Cf. Mitchell (1990).

21 This includes the cultural inferiority of England to Italy, which was the precondition for the vogue of the grand tour among the English in the eighteenth century. It was probably as a reaction against this humiliating fact that the patriotic novelist Henry Fielding (1707–54) opposed the fashionable grand tour in *Joseph Andrews*, etc. See also Turnbull (1740): xv–xviii, 99.

22 For clarity's sake, I quote from the first edition, instead of the 1810 edition cited in the first paragraph of this section (Price 1810: 1: 122f.). Here, Price is criticizing the "smooth" landscape gardening of "Capability" Brown (1716–83). In mid-eighteenth-century England, however, such clear "prospect" and Rousseauistic transparency were favored, probably as a reaction against the ongoing obscurity and inscrutability of the society as a whole brought about by the division of labor and societal compartmentalization resulting from the commercialization and industrialization of Britain. Cf. Barrell (1990, 1983); Anzai (2000). See also Starobinski (1988).

23 Sir William Hamilton, *Campi Phlegraei, Observations on the Volcanos of the Two Sicilies* (Naples, 1776): 1: 54; cit. Jenkins and Sloan (1996): 68. Of course, this is merely a cliché in the agnostic, experimentalist milieu of eighteenth-century England. In the context of landscape aesthetics, see, for example, Johnson (1753): 4: 95; Shenstone (1764): 2: 129, 142; Knight (1805; 1808): 474ff. Hamilton's *Campi Phlegraei* is unique among the genre because of its exceptionally numerous and beautiful illustrations, i.e. its Picturesqueness.

24 Letter written in 1785; BL, Add. MS 42,071, f. 4; cit. Jenkins and Sloan (1996): 20; A. Morrison, *Catalogue of the Collection of Autograph Letters and Historical Documents formed between 1865 and 1882 by A. Morrison. The Hamilton and Nelson Papers* (London: Thibideau, 1893–94), No. 684; cit., Jenkins and Sloan (1996): 22.

25 Price also says the Picturesque is "the coquetry of nature" (Price 1794: 86). He is well aware that his sexual (and tactile) characterization of the Picturesque will cause ridicule. See Price (1794): 77f. and compare it with the revised version, Price (1810): 1: 114, where the sexual allusion has been eliminated.

26 Freud (1914; 1963), 14: 73–107. For the concept of narcissism applied to the Picturesque, see Ferguson (1992), Modiano (1994).

27 Cf. Fried (1980). For example, Jenkins and Sloan (1996): figures 155–61, 166–69, 192–93.

28 At least Horace Walpole was well aware of the possibility of scandals, as he said: "I shall not be so generous as Sir William, and exhibit my wives in pantomime to the public" (To Mary Berry, 11 September 1791; Walpole 1937–83: 11: 350).

29 Price (1810): 1: 82f; cf. 1: 170; 1: 203f; 3: 291–94. Hence, the Picturesque cult of ruins. Price's theory of the Picturesque is an explicit reaction against the French Revolution. This reactionary character seems more conspicuous in the first edition of his treatise, published just after the Revolution, than in its later revisions. His reaction is also closely connected with that against the bourgeois "levelling" of the social hierarchy within Britain, represented by the "levelled" smooth garden of "Capability" Brown (Price 1794: 28f. and note: here Price may be referring to the Puritan radicals, "Levellers," who were supposed to claim demolition of unequal distribution of landed property in the mid-seventeenth century). Thus, his aesthetics reflects the loss of self-confidence and the impotence on the side of the cultural elite, caused by the traumatic upsurge of bourgeois hegemony (cf. Andrews 1989: esp. 40f.) and can be interpreted as the elite's attempt to regain their living energy (and possibly, superiority) through distancing the bourgeois and granting them a narcissistic self-contentment (or confinement).

30 For another redeeming reading of the Picturesque, see Everett (1994): esp. Ch. 3, which asserts the Picturesque coexisted with benevolence, philanthropy, and morality.

Bibliography

Alcock, R. (1863) *The Capital of the Tycoon: A Narrative of a Three Years' Residence in Japan*, 2 vols, London: Longman.

Andrews, M. (1989) *The Search for the Picturesque*, Stanford CA: Stanford University Press.

Anzai, S. (1992) "Gilpin, Price, and Knight: A Critical Survey of the Aesthetics of the Picturesque," The Japanese Society for Aesthetics (ed.) *Aesthetics*, 5: 65–76.

—— (2000) "Illusionismus und Enttäuschung in englischen Gartentheorien um 1800" in G.-H. Vogel and B. Baumüller (eds) *Carl Blechen (1798–1840): Grenzerfahrungen-Grenzüberschreitungen*, Greifswald: Steinbecker Verlag.

Austen, J. (1818; 1972) *Northanger Abbey*, Harmondsworth: Penguin Books.

Barrell, J. (1972) *The Idea of Landscape and the Sense of Place 1730–1840*, Cambridge: Cambridge University Press.

—— (1980) *The Dark Side of the Landscape: The Rural Poor in English Painting 1730–1840*, Cambridge: Cambridge University Press.

—— (1983) *English Literature in History 1730–80: An Equal, Wide Survey*, London: Hutchinson.

—— (1990) "The Public Prospect and the Private View" in S. Pugh (ed.) *Reading Landscape: Country-City-Capital*, Manchester: Manchester University Press.

Bätschmann, O. (1985) "Pygmalion als Betrachter: Die Rezeption von Plastik und Malerei in der zweiten Hälfte des 18. Jahrhunderts" in W. Kemp (ed.) *Der Betrachter ist im Bild*, Köln: DuMont.

Bermingham, A. (1986) *Landscape and Ideology: The English Rustic Tradition, 1740–1860*, Berkeley CA: University of California Press.

—— (1994) "The Picturesque and Ready-to-wear Femininity" in S. Copley and P. Garside (eds) *The Politics of the Picturesque: Literature, Landscape and Aesthetics since 1770*, Cambridge: Cambridge University Press: 81–119.

Bohls, E.A. (1995) *Women Travel Writers and the Language of Aesthetics 1716–1818*, London: Cambridge University Press

—— (1999) "The Gentleman Planter and the Metropole: Long's History of Jamaica (1774)" in G. MacLean, D. Landry, and J. P. Ward (eds) *The Country and the City Revisited: England and the Politics of Culture, 1550–1850*, Cambridge: Cambridge University Press.

Bunn, D. (1994) "'Our Wattled Cot': Mercantile and Domestic Space in Thomas Pringle's African Landscapes" in W.J.T. Mitchell (ed.) *Landscape and Power*, Chicago: The University of Chicago Press.

Burke, E. (1757; 1759; 1987) *A Philosophical Enquiry into the Origin of our Ideas of the Sublime and Beautiful*, J.T. Boulton (ed.), Oxford: Basil Blackwell.

Burke, J. (1976) *English Art: 1714–1800*, London: Oxford University Press.

Conron, J. (2000) *American Picturesque*, University Park PA: The Pennsylvania State University Press.

Copley, S. and Garside, P. (eds) (1994) *The Politics of the Picturesque: Literature, Landscape and Aesthetics since 1770*, Cambridge: Cambridge University Press.

Crandell, G. (1993) *Nature Pictorialized: "The View" in Landscape History*, Baltimore MD: The Johns Hopkins University Press.

Desmond, R. (1995) *Kew: The History of the Royal Botanic Gardens*, London: Harvill.

Dobai, J. (1975) *Die Kunstliteratur des Klassizismus und der Romantik in England*, vol. 2, Bern: Benteli.

Everett, N. (1994) *The Tory View of Landscape*, New Haven CT: Yale University Press.

Fabricant, C. (1977) "Binding and Dressing Nature's Loose Tresses: The Ideology of Augustan Landscape Design," *Studies in Eighteenth-Century Culture*, 8: 109–33.

—— (1985) "The Aesthetics and Politics of Landscape in the Eighteenth Century" in R. Cohen (ed.) *Studies in Eighteenth-Century British Art and Aesthetics*, Berkeley CA: University of California Press.

Ferguson, F. (1992) "In Search of the Natural Sublime: The Face on the Forest Floor" in *Solitude and the Sublime: Romanticism and the Aesthetics of Individuation*, New York: Routledge.

Freud, S. (1914; 1963) "On Narcissism: An Introduction" in *The Standard Edition of the Complete Psychological Works of Sigmund Freud*, London: Hogarth.

Fried, M. (1980) *Absorption and Theatricality: Painting and Beholder in the Age of Diderot*, Chicago: The University of Chicago Press.

Gilpin, W. (1782; rpt. 1973) *Observations on the River Wye, and Several Parts of South Wales, &c. Relative chiefly to Picturesque Beauty*, Richmond: The Richmond Publishing.

Goethe, J.W. (1988), *Werke: Hamburger Ausgabe*, München: Beck.

Greuter, A. and Maier-Solgk, F. (1997) *Landschaftsgärten in Deutschland*, Darmstadt: Wissenschaftliche Buchgesellschaft.

Hagstrum, J.H. (1958) *The Sister Arts: The Tradition of Literary Pictorialism and English Poetry from Dryden to Gray*, Chicago: The University of Chicago Press.

Heidegger, M. (1938; 1977) "The Age of the World Picture" in W. Lovitt (trans.) *The Question concerning Technology and Other Essays*, New York: Harper & Row.

Hipple, W.J. (1957) *The Beautiful, the Sublime, and the Picturesque in Eighteenth-Century British Aesthetic Theory*, Carbondale IL: The Southern Illinois University Press.

Holmström, K.G. (1967) *Monodrama, Attitudes, Tableaux Vivants: Studies on Some Trends of Theatrical Fashion 1770–1815*, Stockholm: Almqvist & Wiksell.

Hussey, C. (1927; rpt. 1983) *The Picturesque: Studies in a Point of View*, London: Frank Cass.

Jenkins, I. and Sloan, K. (eds) (1996) *Vases and Volcanoes: Sir William Hamilton and his Collection*, London: British Museum Press.

Johnson, S. (1753; 1825; rpt. 1970) *The Adventurer*, No. 107 (1753), in *Dr Johnson's Works*, New York: AMS.

Knight, R.P. (1805; 4th edn 1808; rpt. 1972) *Analytical Inquiry into the Principles of Taste*, Farnborough, Hants: Gregg International.

Liu, A. (1989) *Wordsworth: The Sense of History*, Stanford CA: Stanford University Press.

McLeod, B. (1999) *The Geography of Empire in English Literature, 1580–1745*, Cambridge: Cambridge University Press.

Manwaring, E.W. (1925) *Italian Landscape in Eighteenth Century England: A Study chiefly of the Influence of Claude Lorrain and Salvator Rosa on English Taste 1700–1800*, New York: Oxford University Press.

Mason, W. (1772–81; 1783; rpt. 1982) *The English Garden: A Poem*, New York: Garland.

Mitchell, W.J.T. (1990) "Representation" in F. Lentricchia and T. McLaughlin (eds) *Critical Terms for Literary Study*, Chicago: The University of Chicago Press.

Modiano, R. (1994) "The Legacy of the Picturesque: Landscape, Property and the Ruin" in S. Copley and P. Garside (eds) *The Politics of the Picturesque: Literature, Landscape and Aesthetics since 1770*, Cambridge: Cambridge University Press: 196–219.

Natsume, S. (1906) *Kusamakura*; A. Turney (trans.) (1965), *The Three-Cornered World*, London: Peter Owen.

Price, U. (1794) *An Essay on the Picturesque, as compared with the Sublime and the Beautiful; and, on the Use of Studying Pictures, for the Purpose of Improving Real Landscape*, 1st edition, London: J. Robson.

—— (1810; rpt. 1971) *Essays on the Picturesque, as compared with the Sublime and the Beautiful; and, on the Use of Studying Pictures, for the Purpose of Improving Real Landscape*, 3 vols, Farnborough, Hants: Gregg International.

Repton, H. (1803; rpt. 1980) *Observations on the Theory and Practice of Landscape Gardening*, Oxford: Phaidon.

Robinson, S.K. (1991) *Inquiry into the Picturesque*, Chicago: The University of Chicago Press.

Shenstone, W. (1764; rpt. 1982) *Unconnected Thoughts on Gardening*, in *The Works in Verse and Prose*, New York: Garland.

Shiga, S. (1894; rpt. 1995) *Nihon Fûkei Ron* (*A Treatise on the Japanese Landscape*; in Japanese), Tokyo: Iwanami.

Starobinski, J. (1988) *Jean-Jacques Rousseau, Transparency and Obstruction*, trans. A. Goldhammer, Chicago: Chicago University Press.

Tsuji, N. (1986) *Playfulness in Japanese Art*, Lawrence KS: The Spencer Museum of Art, University of Kansas.

Turnbull, G. (1740) *A Treatise on Ancient Painting*, London: A. Millen.

Venturi, G. (1981) "The Landscape Garden in Lombardy: Utopia, Politics and Art at the Beginning of the 19th Century," *Lotus International*, 30: 38–45.

Walpole, H. (1937–83) *The Yale Edition of Horace Walpole's Correspondence*, W.S. Lewis (ed.), New Haven CT: Yale University Press.

Watkin, D. (1982) *The English Vision: The Picturesque in Architecture, Landscape and Garden Design*, London: John Murray.

Watson, J.R. (1970) *Picturesque Landscape and English Romantic Poetry*, London: Hutchinson Educational.

Part II
Mobile homes

4   At home aboard

The American railroad and the changing ideal of public domesticity

Amy G. Richter

In April 1877, Mrs. Frank Leslie set out on "a pleasure trip from Gotham to Golden Gate." As her train readied for departure from New York City, she took stock of "the charming little residence" in which she found herself – a Wagner Palace Car. She decided it "shall be called a home," and noted that:

> [it] very soon assumed the pleasant aspect of the word, as the bouquets, shawls, rugs, sofa-cushions, and various personalities of the three ladies of the party were developed and arranged upon or around a table in the center division of the car, which was to represent the general salon . . .
>
> (Leslie 1877: 19)

Mrs. Leslie was not alone in her portrayal of trains as domestic spaces. The language of home pervaded passengers' descriptions of late nineteenth-century rail travel. Once settled on board, passengers frequently engaged in activities usually identified with domestic leisure. According to the journalist Benjamin Franklin Taylor, the "flying drawing-room" enabled travelers to "go about in your revolving chair . . . read quietly, write comfortably, converse easily. . . . It is home adrift" (Taylor 1874: 158). The association between train and home was so strong that at least one group of appreciative travelers joined together to sing a chorus of "Home, Sweet Home" while aboard (*The Transcontinental* 28 May 1870: 2). By the 1880s, railroad companies and other travel promoters routinely boasted that one could be "at home" on the rails. For example, in 1882 *The Pacific Tourist and Guide of Travel Across the Continent* touted the domesticity of the Pacific Railroad's Palace Cars, assuring readers that "one lives at home in the Palace Car with as much true enjoyment as in the home drawing room. The little section and berth allotted to you, so neat and clean, so nicely kept becomes your home" (Shearer 1882: 5–6). Likewise, the Chicago, Milwaukee & St. Paul Railway espoused its "first aim" as seeking to "provide [the passenger with] every luxury to which one is accustomed in his home" (Chicago, Milwaukee & St. Paul Railway 1900: 2–4). In 1910, the

Pennsylvania Railroad referred to its "Mexico Special" simply as "this comfortably appointed train-home . . ."(Pennsylvania Railroad 1910: 5).

Many historians have identified this nineteenth-century tendency to recast public spaces in domestic terms. Although these works do not constitute a recognized historiographic field, they speak to and past one another in significant ways. To varying degrees, all underscore women's contribution to public life and argue that women were active participants in, if not creators of, public settings designed for their protection. All, moreover, draw connections between the domestication of public life and women's – especially white women's – continued identification with and emancipation from the private sphere.[1] Taken together, they chronicle numerous efforts between the 1830s and the 1920s to make women feel "at home" (and therefore respectable) by domesticating diverse public settings. The existing literature suggests that as middle-class women gained access to more numerous and varied spaces beyond the home (hotels, stores, colleges, various workplaces, and even entire cities), domesticated public facilities proliferated to serve them. Embedded within this argument is the belief that because such women were associated with the home, they could move into public comfortably only if these settings were rendered "homelike." Even accounts emphasizing the subversive nature of women's entry into public suggest that this movement was cloaked in the protective language of the domestic realm.[2]

Individual works convincingly make a case for the transfer of the domestic ideal to public life. But in different studies, the connotations of "domestic" range from the proper placement of a comfortable sofa to guarantees of white women's physical and moral safety; from luxurious opulence to a true belief in the uplifting influence of Victorian womanhood.[3] Using the same vocabulary to describe different transformations, these studies create a historical narrative in which women are forever entering and domesticating public life, but are never quite considered as public actors beside men. Public life is always new to women and the impact of their presence is "domestication." In order to contextualize these studies and understand more broadly the significance of domesticated public spaces, historians must scrutinize what Victorian Americans meant when they imagined public spaces as "homes." A variety of scholars have shown that the domestication of public spaces occurred again and again throughout the nineteenth century.[4] This process occurred repeatedly not simply because women laid claim to new spaces, but because the meaning of "home" and "public" were themselves mutable. This essay, therefore, focuses on the multiple and changing meanings of "home" as applied to the public spaces of the nineteenth-century railroad. Only by understanding what Victorian Americans meant by "home" can we come to understand the significance of the domesticated public spaces they created.

During the second half of the nineteenth century, the meanings of home abounded, blending into one another with subtle shifts in emphasis. The

American home maintained its significance as the primary agent of moral instruction while increasingly acting as a site for consumer display and the expression of individual taste. At mid-century, Victorian homes stood for ethical excellence and, according to historian Clifford Clark, were "described in moral terms as good or bad, honest or dishonest" (Clark 1986: 19). Karen Halttunen makes a similar point when she argues that Victorian domestic culture considered "the parlor . . . the most important arena for the demonstration of character" (Halttunen 1989: 158). The mid-century moral home was defined by the popular belief that the proper surroundings fostered self-discipline and gentility in its inhabitants – what Louise Stevenson has identified as the Victorian hope that "the material world . . . direct people into desired behavior" (Stevenson 1991: 3). Reformers and advice books urged that in the moral home the right architectural forms and domestic goods imparted Christian values of self-restraint and order to household inhabitants. More importantly, these values once instilled at home passed beyond individual thresholds and into the larger community. For many of its exponents then the moral home exerted a civilizing influence, fostered stability and social harmony, and represented the best antidote to the disruptive effects of westward expansion, the spread of the factory system, urban growth, and immigration.

But by the century's end, these same historians argue, domestic goods also reflected individual taste, creativity, and the ability to appreciate and afford the best.[5] Domestic objects proved unstable moralizing agents that too easily yielded to concerns for self-expression, physical comfort, and even luxury. If the moral home sought to bring order and stability to disorderly urban settings, this self-expressive home telegraphed the individual taste of its inhabitants to a homogeneous collection of neighbors. According to Clark, this new ideal put "more emphasis on comfort and consumption" than on lessons of self-restraint (Clark 1986: 104). Domestic treasures previously intended to offer moral instruction stood instead as evidence of a family's superior taste. Halttunen characterizes this shift slightly differently when she describes a movement "from parlor to living room" (Halttunen 1989: 157). As the middle class moved out of heterogeneous cities to more homogenized suburbs, they forsook the rigidity and high morals of parlor life and embraced a less ordered home characterized by an open design and fewer specialized spaces. Living among people of similar social and economic class, the middle class of the late Victorian period relaxed and created homes that emphasized self-expression. The moral home yielded to the comfortable home in which the presence or absence of domestic amenities came to connote personal taste and class difference.

Shifts in the meaning of the Victorian home throughout the nineteenth century suggest that the domestication of public life was more complex than earlier studies have argued. As travelers imagined railway cars as homes, they carried the cultural baggage of changing domestic meanings with them. The differences and tensions between the moral and the self-expressive home were played out publicly aboard the trains and influenced the nature of the domesticated public spaces that emerged on the rails. Passengers and railway companies hoped that the domesticated train could reinscribe the values of the moral home onto public life, even as they embraced the train's luxurious interiors as sites of expressive consumption. The train, therefore, underwent at least two types of domestication. The "train as home" metaphor implied both a public space infused with the values of self-restraint usually associated with the home and also a comfortable setting that gave expression to consumer longings and the tasteful manifestation of individual purchasing power. Neither "domestication" truly turned railroad cars into homes but instead reflected the selective ways Americans – women and men, passengers and railroad representatives – moved qualities of private life onto the train and thereby amended the meaning of both home and public. They created instead a "public domesticity" – a social ideal that was neither private like home nor socially unruly like a public street – that attempted to bring the moral associations and behaviors of home life to bear upon social interactions among strangers.

An examination of the "train as home" metaphor, moreover, makes clear that the domestication of public life was not the static end-product of women's entry into public but an on-going process involving new attitudes toward consumption, gender roles, and the articulation of class and racial difference. Instead of tentatively entering public life under the guise of domesticating it, here women emerge as actors alongside men within a larger history of "public domesticity." Aboard the trains, male and female passengers struggled to resolve the tensions between women's role as moral agents and their role as consumers, between the home's significance as an instrument of public morality and as a site of personal fulfillment and expression. At times, railroad companies and passengers presumed women would extend their roles as moral keepers of the home into the cars and bring their influence to bear upon men's public conduct; at others, they expected women to act as skillful consumers evaluating the domestic amenities of train life and displaying their appreciation of comfort and good taste. In short, shifting gender roles undergirded the competing domestic visions of nineteenth-century Americans and shaped the varied models of public domesticity enacted on the railroad.

Before the train could draw upon the values of the Victorian home, it needed to evolve as a physical and technological space. The cars of the 1830s and 1840s little resembled the restrained moral enclaves associated with the Victorian ideal of the private sphere. If one merely defined a home by its ability to provide both shelter from the elements and some degree of physical comfort, the earliest railway cars were anything but homelike. Travel upon these early trains was filthy, inconvenient, and dangerous. The cars were too warm in summer, too cold in winter, and poorly ventilated year-round. Fumes from axle lubricants mingled with smoke, dust, soot, and sparks from the locomotive to create a stifling and nauseating atmosphere. Passengers inhaled this mix as they sat upon hard, low-backed benches. Jostled and jarred by unreliable track work, sharp curves, and bumpy grades, railway travelers exposed themselves to new injuries like "railway spine" or to the dangers of derailment. Indeed, derailments occurred so frequently that, according to one historian of the passenger car, some railroad tickets stipulated that "the passenger was subject to call if needed to help replace the engine or cars on the rail" (Menken 1957: 7). The best that could be said for train travel during the 1830s was that the journeys were short and, by nineteenth-century standards, quick.

Physically uncomfortable and very public, these early cars nonetheless provided a starting point for considering efforts to domesticate the trains' physical appearance as they were increasingly called upon to provide the conveniences of home. Although few people yet spoke of trains in domestic terms, technological innovations in car design and the rail network increasingly permitted efforts to translate domestic architectural forms and furnishings to the cars. During the 1830s, railroad cars reached a design plateau from which improvements could proceed. The introduction of the eight-wheel, double-truck car, for example, permitted an increase in car size, ensured a smoother motion, and resulted in fewer broken axles. Eight-wheeled cars quickly replaced the earlier generations of cars that had resembled stagecoaches and omnibuses mounted on four wheels. Now that the cars were larger and more stable, car designers could pay increased attention to the comfort and decoration of their interior spaces. Improvements in the rail network further encouraged these efforts. As journeys of up to 500 miles became possible, railroad companies, car designers, and passengers increasingly emphasized the importance of comfort during travel. Prolonged periods of time aboard – in effect living on the trains – created needs and expectations that placed the cars in dialogue with the comforts of home. Certainly, sitting up all night in a railway car afforded passengers a painful and memorable lesson in the existing problems of railway interiors and fared badly when compared with a night's sleep in one's own bed or a nap in a favorite chair.

With many of the basic technological achievements in place by the late 1830s, railroad companies and car designers paid increased attention to car interiors. By 1847, a Frenchman traveling in the United States could write that American railway cars "are actually houses where nothing, absolutely nothing is lacking for the necessity of life" (Sweet 1923: n.p.). He then described a sleeping car on the Baltimore & Ohio Railroad that approximated the specialization of space in a Victorian home with several "rooms, some for men and some for women alone" (Sweet 1923: n.p.). The rooms each contained six beds or couches. From such rudimentary beginnings, the replication of home interiors within railway cars brought increased domestic comfort to the rails, as the interior designs grew more elaborate and luxurious throughout the century. In addition to sleeping cars equipped with convertible berths, dressing rooms and lavatories, car designers and railroad companies added comfortable cars of reclining chairs, parlor cars, dining cars, library cars, and even barbershops. When in 1887 the invention of the vestibule reduced the violent jerking motion of car platforms and made it easier to pass safely between cars, the train was ready to act as one large domestic space. A passenger willing and able to pay for his comfort could spend part of his journey reading the latest periodicals in the library car or gazing out the window of a parlor car, then eat with his family at a comfortable table in the well-stocked dining car, retire to the smoker for an evening of male camaraderie, and finally pass a comfortable night's sleep in his own berth. Or in the words of the Pullman Company in 1893: "the traveler may pass from his dining-room to his sitting-room, or to his sleeping-room, as in his own home" (Pullman Company 1893: 12).

In the 1870s and after, guidebooks and railroad brochures closely documented the luxurious amenities that continued to transform the previously uncomfortable railway car into a temporary home. The publications offered descriptions of the fine woods, elaborate hangings, rich upholsteries, and silver-plated metal work employed in the best cars. One 1883 guidebook published by the Passenger Department of the Savannah, Florida and Western Railway Company boasted that the buffet in one of the line's sleeping cars harmonized with its surroundings to an extraordinary degree. According to the author, because "the buffet's finish, both in woodwork and marquetry, corresponds with the interior of the car, the effect is highly pleasing, suggesting in convenience and luxury an elegant sideboard in a richly appointed mansion" (Graves 1883: 11). Another brochure for the Pennsylvania Railroad praised the "harmonious colors . . . snowy linen, cut glass and silver" of one of the railroad's dining cars (Pennsylvania Railroad 1895: n.p.). Guidebooks and railway brochures also made note of the variety of decorating styles. Some cars, for example, were decorated in an "Oriental" style and others in "Persian" or "Renaissance," and often ladies' rest rooms were decorated in a style different from that employed in the rest of the car.

As the technological improvements in car construction made the addition of domestic touches imaginable, the domestic amenities of the cars made the technological achievements of the railroad legible to many of the train's passengers. Although the new technological and engineering achievements engendered pride, they also fostered anxiety and fear. The domestic comforts of the American railway cars presented travelers with a less ambiguous symbol of progress. To Victorians, improvements in comfort reflected improved technology and suggested an increase in railway safety. An 1875 guidebook, *Popular Resorts and How to Reach Them*, praised the achievements of the new hotel cars: "The safety, pleasure and comforts of railroad travel have been wonderfully improved during the last few years, but this, one of the latest, will unquestionably be pronounced the best of comfort-seeking inventions yet produced" (Bachelder 1875: 339). In 1885, *The New York Times* reported Americans' pride in the speed of their trains. The article also reported that the weight of the luxurious Pullman cars limited the speed of American railroads. This handicap, however, was a source of national pride: "The Pullman car is, of course, much more comfortable than the English carriages ... No one would propose to supersede it here by cars of the English pattern" (*The New York Times* 23 June 1885: 4). The domestic luxury of the parlor cars, as much as the speed of the trains, represented progress and convinced Americans that their trains were technologically sound. Moreover, the familiar domestic accents and furnishings very likely reinforced this sense of safety by mimicking the secure environment of the home and suggesting the permanence of a heavily constructed dwelling. Together, the domestic train and the technological train created a single and comprehensible narrative of American progress. Side by side with the technological innovations of the railroad, the domestic interiors of the cars asserted that American progress was orderly and safe.[6]

Just as a well-appointed home ideally reflected the morality and character of its inhabitants, the elegant trains represented popular hopes for the character of the American people. The materials of the comfortable home transplanted onto railway cars – the sideboards, upholstery, and paneling – imparted stabilizing cultural meanings to American technological progress and reinscribed the values of the moral home onto public life. They offered travelers social cues for their own behavior and reassurances concerning the conduct of those around them. Civility would not be sacrificed in the face of rapid social and technological change, but instead would regulate the pace and shape of that change. In addition to rendering train travel physically comfortable, the domestic interiors of the cars might minimize the social disorganization wrought by new technological innovations that increased mobility and loosened social ties. If, as home manuals advised, "home is the central pivot upon which depends the weal or the woe of families and communities," perhaps the home could also serve as model for national public life: If private homes built communities, public ones might be able to do the same (Holloway 1883: iv). By behaving in public as they did in private, travelers could create a public life suffused with the values of order and morality.[7] Together in the well-decorated cars, strangers might come together and, bound by an appreciation of home's moralizing influence, assert the best values of American civilization. *Godey's Lady's Book* gave voice to this hope when it observed that "our public conveyances are not only schools of public instruction in ethics of etiquette, but they also testify to the high state of civilization our free institutions have reached" (*Godey's Lady's Book* May 1867: 467–468).

More than any other railroad man George Mortimer Pullman was invested in and popularly associated with the vision of a public home on rails as an agent of civilization. The story of Pullman's entry into and later domination of the specialty railroad car business has been often and well told, and will not be recounted here. Both contemporary and later railroad enthusiasts have praised his vision and anointed him a great civilizer – a man who brought comfort to the rails and taught Americans the meaning of luxury (Husband 1917). The expensive and opulent decoration of his sleeping cars, hotel cars, dining cars, and parlor cars was unprecedented (Beebe 1961). Other critics, by contrast, have vilified Pullman as a man who valued too strongly his system of car management and production and foolishly ignored his workers' well-justified demands. His company town is frequently invoked to underscore the limitations of industrial paternalism (Buder 1967). An appreciation of the domestic nature of Pullman's vision, however, is lost in these contradictory interpretations of his significance. In both his cars and his company town Pullman acted upon his belief that a beautiful and domestic environment could bring moral order to public and commercial life.[8]

Pullman, for example, brought the values of the moral home into his business dealings when he set out to build a more comfortable passenger car in the 1860s. Throughout his professional life Pullman acted on his belief that a well-designed home environment would foster moral conduct and public harmony. Explaining the reasoning behind his spending so much money designing elaborate car interiors, Pullman justified the decision to put carpets on the floors, provide clean bed linen, and decorate with costly ornamentation and upholstery: "I have always held that people are very greatly influenced by their physical surroundings, and bring [a man] into a room elegantly carpeted and furnished and the effect upon his bearing is immediate" (Doty 1893: 61). If men behaved badly in public but were refined at home, then bring the influence of home into public; to Pullman, "the more artistic and refined the mere external surroundings, in other words, the better and more refined the man" (Doty 1893: 61). Like the authors of domestic advice manuals Pullman asserted that civility could be learned, and he designed and decorated his cars in order to educate passengers for a more harmonious public life. In the perfect world of Pullman's vision, not simply the interiors of the cars but more importantly the resulting conduct of the passengers lent credence to the belief that civilization was literally rolling across the North American continent.

Women's presence and their ability to travel in the cars seemed to confirm the realization of Pullman's vision of the railway car as a domesticated, and therefore moral, public setting. Lady travelers, many Victorians presumed, inspired self-restraint in others and their comfort aboard confirmed the success of their influence.[9] The best way to understand such women's significance aboard is as an extension of their role in the home; ideally, they exerted their moral influence by acting as ornaments refining and beautifying the space. More simply, in the words of an 1875 article on wives as traveling companions: "Home is the pleasantest place in the world, and that is not home where the wife is not" (*The New York Times*

28 November 1875: 4). Traveling women carried the physical and moral comforts of home with them on their journeys. Women, after all, tended to the lunch basket, gathered flowers to decorate the cars, or, like Mrs. Leslie and her companions, arranged pillows and shawls to make the train pleasing and homey. Promotional images drew upon and reinforced this relationship and in so doing identified the cars as a public space suffused by women's moral and comforting influence.[10] Railroad company brochures depicted women at ease upon comfortable sofas as they engaged in domestic and familial pursuits, and popular fiction echoed these themes, underscoring the broad acceptance of the train as a setting in which men and women lived out elements of domestic life.[11]

But in many respects this depiction of the train as a feminized and moral enclave was truly a work of fiction. Domestic furnishings and decorations were unstable conveyers of moral meaning. As in a private home, the beauty and positive influence of these domestic items did not always reflect the careful moral choices of the car's inhabitants but instead announced the passengers' ability to pay for them. By transplanting domestic comforts and artifacts – cute dressers and cozy seats, writing tables and upholstered chairs – into the cars, rail and car companies hoped to bring the controlled conduct of the moral home to bear upon public interactions. At the same time, by relocating these domestic goods they removed them from their context within the everyday lives of individual families and further drained them of their didactic significance. The domestic touches that Pullman employed to encourage self-restraint and uplift in "the roughest man" could also serve as an expression of a discriminating passenger's individual refinement. Similarly, the amenities that permitted a sense of ease and fostered family feeling among passengers frequently denoted economic privilege or class-standing rather than impeccable moral conduct. The sofas, carpets, and wood engravings intended to promote self-restraint by reminding travelers to act as they would at home also acted as badges of luxury and taste for the passengers that could afford to travel in the best parlors or sleeping cars. In the end, then, Pullman passengers may have refrained from spitting on the carpets because they had paid for the privilege of riding in a beautiful car, not because they had been morally uplifted by their surroundings.

Just as the focus of private domestic life shifted from moral instruction to comfort and consumption over the course of the nineteenth century, so did the meaning of public domesticity shift aboard the train. Even as railroad companies, car designers, and passengers embraced and propagated the ideal of a moral home on rails, that particular domestic vision was challenged by the more consumption-driven home of comfort and self-expression. Increasingly the ideal of being "at home aboard" implied the presence of the best and most comfortable amenities – even amenities many passengers could never dream of possessing in their own homes. William Dean Howells' 1876 farce *The Parlor Car* captured the changing nature of the train's public invocation of domestic life. Howells depicted a young woman traveling in the cars with her fiancé. Soon after mistaking a routine recoupling of cars for a threatening collision, she settles down and enjoys her journey. She is most taken by the domestic elements of the train, and wonders aloud whether Mr. Pullman would sell her the car. When asked why she would like to own it, she responds that "it's perfectly lovely, and I should like to live in it always" (Howells 1876: 46). She then imagines the train car "fitted up for a sort of summer-house" in the garden where her future husband could go to smoke (Howells 1876: 46). Howells translated the woman's enjoyment of the parlor car into a dream of domestic consumption.

If the ideal of the moral home on rails was intended to bind diverse groups of passengers together into a harmonious collection of well-behaved individuals, the consumer home proved divisive. Over time the domestic amenities of travel separated travelers into a stratified consuming public with different types of public "homes." Although the seemingly democratic nature of American rail accommodations had long been a source of national pride and was frequently invoked as evidence of American exceptionalism, by the 1890s social critics and rail promoters hesitantly acknowledged the growing separation of passengers along class lines.[12] An 1897 article on the art of travel noted that increasingly Americans made allowances for classes albeit in an indirect way:

> Abroad, carriages of the first, second, and third, and even fourth, class (where you stand up) are provided, and plainly marked. Here we have the corresponding divisions without such harsh names: The Pullman is first, the day-coach second, the smoker third-class; and perhaps, a seat in the caboose of a freight train may be called fourth-class.
>
> (Iddings 1897: 353)

Belying earlier praise of America's classless society, in 1896 *The Railway Age* and *Northwestern Railroader* openly called attention to the class stratification taking place "under other names." The periodical explained, "the construction of sleeping, parlor and emigrant cars has practically established class distinctions among travelers no less clearly defined and even more numerous than were ever in use in England" (*The Railway Age* 8 February 1896: 67). The same domestic amenities and images that sustained the possibility of a public moral home on rails ironically provided a uniquely American vocabulary for describing increased social stratification aboard. Before embracing the language of class difference or Jim Crow aboard the trains Americans used the language of domestic goods to sort passengers – to denote not only respectability but also class and, increasingly, racial difference.[13]

Such social stratification transformed the moral home on rails. Passengers and companies embraced the sale of luxury and the resulting class stratification, instead of a unified social order transcending difference. The cars no longer sought to uplift but to provide physical and social comfort. A public home informed by the forms of etiquette, not the values of morality, became the best accommodations the train could offer. But only some passengers could afford the most comfortable and domesticated physical settings and with these settings, many believed, they bought the best conduct. An article in *The New York Times* described without apology the genteel home on rails. According to the *Times*, "the idea of the parlor car [and, by extension, other specialty cars] is a place where people who wish to be treated as they know they ought to treat others while traveling can sit together by the payment of an extra fare" (*The New York Times* 10 July 1892: 4). In the parlor car, the golden rule was for sale and only those who could afford the fee gained access. Moreover, unlike the moral home, this particular vision of public domesticity was intended for those who already knew the rules rather than those who needed to be taught. Although still speaking the language of domesticity, the railroad cars now carried a different message.

The separation of domestic amenities from the goals of moral uplift and social harmony was especially apparent in the conception and transformation of the emigrant cars that transported newly arrived Europeans across the continent. From their beginnings in the 1850s, no tasteful decoration or domestic touches sought to encode self-restraint and refined conduct into the physical design of the emigrant cars. Instead these cars were unadorned boxcars with windows. No upholstery fostered a sense of ease; no thick carpets discouraged spitting on the floors. Emigrants were given bare wooden berths upon which they could place their own bedding and cooking stoves so that they could prepare their meals. Unlike first-class passengers, emigrants had neither porters nor waiters to meet their needs; instead the burdens of domestic labor traveled with them in the cars.[14]

Even when competition among rail lines and car companies encouraged the production of more comfortable emigrant cars during the 1870s, little talk of moral uplift accompanied these improvements. In fact, the old-fashioned moralists at *Godey's Lady's Book* fretted that the addition of meager amenities removed all hardship from the long transcontinental journey. Instead of praising the uplifting powers of the improved domestic comforts, *Godey's* warned that the covered wagon had "filled the great West with its self-reliant and hardy population; and perhaps it developed some fine qualities, which the [emigrant] family car, with its ease and convenience, may fail to bring out" (*Godey's Lady's Book* June 1874: 563). Despite such concerns, during the 1880s the emigrant car evolved into the tourist sleeper – a scaled-down version of the more luxurious and expensive sleeping cars.[15] These cars provided bed-linen and the attentions of a porter to, what one brochure described as, "those who are contented with good accommodations at a lower price" (Atchison, Topeka and Santa Fe Railroad 1892: 4). With rattan, rather than upholstered seats, but at two-fifths the cost of a regular sleeper, the tourist sleepers provided an improved level of domestic comfort free of moral lessons for those unable or unwilling to pay for the best.

By the opening decade of the twentieth century, those passengers who could afford to paid a premium for comfort and refinement. Others simply enjoyed a less exclusive, albeit homey, physical environment in which they continued to rub up against strangers. A 1903 article on "The Comforts of Railroad Travel" captured this stratified yet seemingly domestic public life. In the private car, the most exclusive, luxurious, and expensive domestic setting, the travel writer described the arrival of the genteel home aboard. Here, amid great luxury, the rules of genteel propriety held as "the magnate's wife is settling herself beside the broad rear plate-glass window, and the busy man is dictating letters to a stenographer beside the center table" (Cunniff 1903: 3577). The author likewise beckoned the reader to "look into the second-class tourist cars of a transcontinental flyer at a prairie station" where one could view a more boisterous and diversified scene. Women and children mingled with single men and "whole families bound for Rocky Mountain resorts loll about on the wicker or leather seats; one or two people are heating coffee on the range at the end of the car; three or four straw-hatted men are clustered smoking on the vestibuled platforms; heads project from open windows; everybody is happy" (Cunniff 1903: 3577). Thus by the end of the century, the line of cars moving across the continent transported a variety of groups each in its own domestic setting instead of a single group unified by a shared appreciation of moral domestic life. Initially intended as an organizing and integrating force for a commercial society on the move, public domesticity was itself turned into a commercial good.

Significantly, the proliferation of specialty cars and emigrant sleepers coincided with the emergence of Jim Crow cars devoid of domestic amenities. Indeed, racial segregation was justified in the name of growing demands for public comfort, as the mixing of the races was argued to discomfort white passengers and endanger the safety of black travelers. This framing of Jim Crow within the context of public domesticity not only emphasizes the racial homogeneity of the home on rails but also reveals the play between gendered and racial transformations of Victorian public culture. Even as the domestic amenities of the train were for sale in a variety of forms, African Americans were increasingly excluded from this marketplace. African-American travelers including W.E.B. Du Bois, Ida B. Wells, and Mary Church Terrell recalled their exclusion from the domesticated spaces of rail travel. Even when they had paid for a first-class ticket, blacks were routinely relegated to part of the baggage car or a divided-off section in a sleeper. Not even the possession of a Pullman-Car ticket could consistently insulate an African-American traveler from the insults of white racism.[16] In the world of commodified public domesticity, the most common role for blacks was to serve as porters and waiters (Hale 1998). Blacks partook of railroad domesticity by acting as public domestic servants

– yet another amenity for sale to white passengers as a marker of their respectability and comfort.

The domestication of public life that took place on the American railroad during the second half of the nineteenth century thus reveals itself as a complex process that mobilized various meanings of home to order the spatial and social landscape of travel. At times, domestication was about binding diverse groups together into a cohesive and moral public; at other times, this same process served to separate passengers along lines of class and racial difference. Although white women often served as the markers, means, and justification for this process, they were not its sole agents or targets. Public domesticity was more than a way of making middle-class women comfortable in public. Its emergence as a social ideal marks a major transformation in American cultural life. For a culture committed to the moral and structural values of separate spheres, the belief that people should be "at home" in public stands as a striking innovation.

Notes

1 While race and class certainly shaped the separate spheres ideal, this gendered dichotomy potentially masked class and racial divisions by crossing the high and low boundaries of the American class structure and transcending racial lines. A working-class immigrant could learn from a ten-cent pamphlet the same rules of etiquette that an African-American woman read as part of her education at normal school or that a white middle-class woman read in a beautifully bound etiquette manual. Following these rules, all might consider themselves "respectable" within Victorian social conventions. Although a diverse group of women may have shared the values of Victorian respectability, white middle-class women were more likely than working-class women or women of color to have their status respected in the heterogeneous public life of the cars.

2 See, for example, Stansell (1987) and Ryan (1990). On department stores as feminine enclaves see: Abelson (1989); Benson (1986); Leach (1984: 319–342). For a study of tea-rooms see Brandimarte (1995: 1–19).

3 For example, the domestication of an office building or a work culture in the 1870s is not easily compared with the "domestication of politics" in the 1890s, even though each act seemingly drew upon a shared body of cultural associations. On the domestication of the corporate office see Kwolek-Folland (1994) and Aron (1987). Compare this approach with Baker (1984: 620–647).

4 Much in the way that other historians have described successive breakdowns in some ideal notion of community, these scholars depict respectable women domesticating public life again and again throughout the nineteenth century. For a similar analysis of the community decline model see Bender (1982). Just as Bender praises individual works employing what he calls "the community breakdown model" but questions the usefulness of their analytical model, I admire the quality of the studies considered here, but challenge them to clarify the meaning of a homelike public.

5 As with any cultural change, the transfer from ideal home to the next was uneven. In the words of architectural historian Gwendolyn Wright: "More than one model for the home and family have usually co-existed although seldom in harmony" (Wright 1983).

6 Over time, new visions of progress would make themselves felt in the design of car interiors. In the final decade of the nineteenth century and the opening

decades of the twentieth, the style of car interiors began to change as older styles with their heavy draperies and thick carpets became associated with germs and other health threats. The late nineteenth- and early twentieth-century drive for hygiene pushed out the Pullman car plush and ushered in more stark, albeit luxurious interiors.

7   Etiquette books repeatedly advised readers to conduct themselves in public as they would at home. See *How to Make Home Happy: A Housekeeper's Hand Book* (Anonymous 1884: 218). According to the author, "Discretion should be used in forming acquaintances while traveling. ... Any attempt at familiarity should be checked at once ... The flirting and freedom often indulged in by young people in public conveyances is unworthy of them – if, indeed, it does not indicate low breeding, and often leads to evil consequences. Whether at home or abroad, the same rules of good behavior should prevail."

8   It is not my intention to downplay Pullman's love of profit, but only to suggest that the ways he sought to make his money reflect his beliefs about the importance of home as a social and cultural force.

9   Women's ability to travel freely by train enabled the cars to be homelike and underscored the domesticity of American women and the superior character of American men who created a safe public setting for their women. See Bederman (1995). According to Bederman, for late nineteenth-century Americans "gender ... was an essential component of civilization. Indeed, one could identify advanced civilizations by the degree of their sexual differentiation. Savage (that is, nonwhite) men and women were believed to be almost identical, but men and women of the civilized races had evolved pronounced sexual differences. Civilized women were womanly – delicate, spiritual, dedicated to the home. And civilized men were the most manly ever evolved – firm of character; self-controlled; protectors of women and children."

10   An article on the proliferation of railroad advertisements in magazines and newspapers notes the self-conscious use of such domestic and feminized images in attempts to woo travelers. Describing the first railroad line to place such an ad, the author imagines a scenario in which "some enterprising traffic manager tried the venture of advertising his line – say, for California travel – in one of the great magazines, with large type and an attractive picture of a girl or a bridal pair, seated blissfully in a luxurious car and viewing the flying summerland scenery." See *The Railway Age* (17 March 1905: 358).

11   This observation is based on a survey of *Godey's Lady's Book* from 1860 through the end of the magazine's run in the 1890s and *The Railway Age Gazette* from 1880 to 1920. I have also relied upon articles and stories found in Poole's Index of nineteenth-century periodicals.

12   This issue of class on the rails is difficult to sort out because the enthusiasm for the train as an agent of democracy distorts contemporary accounts as well as the more celebratory secondary studies of the American railroad. The belief that the trains were democratic social spaces has been reinscribed without critique in many secondary accounts of nineteenth-century rail travel. See for example Alvarez (1974: 126). According to Alvarez, "The early railroad car was a conspicuous example of the American concept of an egalitarian society. Unlike his European counterpart, the American traveler was almost totally lacking in the type of class-consciousness that would have resulted in different cars for each social class. With the exception of Negroes, some immigrants, and those cars reserved for ladies, all passengers mingled together and shared equally the uncertainties of the road." Alvarez notes the exclusion of African Americans, immigrants, and women from this democratic space with no apparent irony.

13 As early as 1870, some guidebooks were advising travelers of the different accommodations available in first-, second-, and even third-class cars. See Hart (1870: 11).
14 There is not a great deal of information on the history and physical design of emigrant cars. See White (1978: Ch. 6). For primary observations of the condition of the passengers traveling in the emigrant cars see Leslie (1877: 284) and *Frank Leslie's Illustrated Newspaper*, vol. XLV, no. 1167, 9 February 1878, 389–390. For a more positive depiction of the emigrant cars see Daly (1886: 38).
15 The assumption that domestic labor should follow immigrants into public was also reflected in the design of station waiting rooms for their use. See Droege (1916: 29): "A facility which has become a necessity in large stations is a waiting room for immigrants. It should be in an out of the way place, and may even be in a separate building. It should be furnished with complete sanitary equipment, and if possible, tubs for washing clothes and dryers."
16 In Terrell's words: "There are few ordeals more nerve-racking than the one which confronts a colored woman when she tries to secure a Pullman reservation in the South and even in some parts of the North" (Terrell 1940: 295).

Bibliography

Abelson, E. (1989) *When Ladies Go A-Thieving: Middle-Class Shoplifters in the Victorian Department Store*, New York: Oxford University Press.

Alvarez, E. (1974) *Travel on Southern Antebellum Railroads, 1828–1860*, Tuscaloosa AL: University of Alabama Press: 126.

Anonymous (1884) *How to Make Home Happy: A Housekeeper's Hand Book*, New York: John W. Lovell: 218.

Aron, C. (1987) *Ladies and Gentlemen of the Civil Service: Middle-Class Workers in Victorian America*, New York: Oxford University Press.

Atchison, "Topeka and Santa Fe Railroad" (1892) *In a Tourist Sleeper to California via Santa Fe Route*.

Bachelder, J.B. (1875) *Popular Resorts and How to Reach Them*, Boston MA: John B. Bachelder.

Baker, P. (1984) "The Domestication of Politics: Women and the American Political Society, 1780–1920," *American Historical Review*, 89(3): 620–647.

Bederman, G. (1995) *Manliness and Civilization: A Cultural History of Gender and Race in the United States, 1880–1917*, Chicago: University of Chicago Press: 25.

Beebe, L. (1961) *Mr. Pullman's Elegant Palace Car*, Garden City, New York: Doubleday.

Bender, T. (1982) *Community and Social Change in America*, Baltimore MD: Johns Hopkins University Press.

Benson, S.P. (1986) *Counter Cultures: Saleswomen, Managers, and Customers in American Department Stores, 1890–1940*, Urbana IL: University of Illinois Press.

Brandimarte, C. (1995) "'To Make the World More Homelike': Gender, Space, and America's Tea Room Movement," *Winterthur Portfolio*, 30(1): 1–19.

Buder, S. (1967) *Pullman An Experiment in Industrial Order and Community Planning, 1880–1930*, New York: Oxford University Press.

Chicago, Milwaukee & St. Paul Railway (1900) *Between Chicago & Omaha: The Chicago and Omaha Short Line*, Chicago: C.M. & St P. Ry.

Clark, C.E. (1986) *The American Family Home*, Chapel Hill CA: University of North Carolina.

Cunniff, M.G. (1903) "The Comforts of Railroad Travel," *World's Work*, 6: 3576–3580.

Daly, J. (1886) *"For Love & Bears"; a Description of a Recent Hunting Trip with a Romantic Finale. A True Story, by "James Daly". Profusely illustrated by pencil sketches, also a cabinet photograph of Grace Horton*, Chicago: F.S. Gray: 38.

Doty, Mrs D. (1893) *The Town of Pullman*, Pullman IL: T.P. Struhsacker.

Droege, J.A. (1916) *Passenger Terminals and Trains*, New York: McGraw Hill: 29.

Godey's Lady's Book (1867) "Women and Children in America," May: 467–468.

—— (1874) "The Family Car," June: 563.

Graves, J.T. (1883) *The Winter Resorts of Florida, South Georgia, Louisiana, Texas, California, Mexico and Cuba . . . and How to Reach Them*, Passenger Department, Savannah, Florida, and Western Railway Co.

Hale, G.E. (1998) *Making Whiteness: The Culture of Segregation in the South, 1890–1940*, New York: Pantheon.

Halttunen, K. (1989) "From Parlor to Living Room: Domestic Space, Interior Decoration, and the Culture of Personality," in S. Bronner (ed.) *Consuming Visions: Accumulation and Display of Goods in America: 1880–1920*, Winterthur DE: The Henry Francis du Pont Winterthur Museum.

Hart, A.A. (1870) *The Traveler's Own Book, A Souvenir of Overland Travel, via The Great and Attractive Route, Chicago, Burlington & Quincy to Burlington. Burlington & Missouri River R.R. to Omaha. Union Pacific Railroad to Ogden. Utah Central Railroad to Salt Lake City. Central Pacific Railroad to Sacramento. Western Pacific Railroad to San Francisco*, p. 11.

Holloway, L.C. (1883) *The Hearthstone; or, Life at Home. A Household Manual*, Philadelphia PA: Bradley Garretson.

Howells, W.D. (1876) "The Parlor-Car," in *The Sleeping-Car and Other Farces*, New York: Houghton Mifflin.

Husband, J. (1917) *The Story of the Pullman Car*, Chicago: A.C. McClurg & Co.

Iddings, L.M. (1897) "The Art of Travel," *Scribners*, 21: 356–367.

Kwolek-Folland, A. (1994) *Engendering Business: Men and Women in the Corporate Office, 1870–1930*, Baltimore MD: Johns Hopkins University Press.

Leach, W. (1984) "Transformations in a Culture of Consumption: Women and Department Stores, 1890–1925," *Journal of American History*, 71: 319–342.

Leslie, M.F. (1877) *California; A Pleasure Trip from Gotham to the Golden Gate*, New York: G.W. Carleton, Publishers.

Menken, A. (1957) *The Railroad Passenger Car; An Illustrated History of the First Hundred Years with Accounts by Contemporary Passengers*, Baltimore MD: Johns Hopkins Press.

New York Times (1875) "The Wife As A Traveler," 28 November: 4.

—— (1885) "Handicapped," 23 June: 4.

—— (1892) "Parlor-Car Manners," 10 July: 4.

Pennsylvania Railroad (1895) *Pennsylvania R.R. Around the World Via Washington and Transcontinental America*.

—— (1910) *Tour to the Mardi Gras, Mexico and Grand Canyon*, Philadelphia PA: Allen, Land & Scott.

Pullman Company (1893) *The Story of Pullman*, Chicago.

Railway Age (1896) [no title] 21(6) (8 February): 67.

—— (1905) "Railway Advertising in Newspapers," 39 (17 March): 358.

Ryan, M. (1990) *Women in Public: Between Banners and Ballots, 1825–1880*, Baltimore MD: Johns Hopkins University Press.

Shearer, F.E. (1882) *The Pacific Tourist and Guide of Travel Across the Continent*, New York: J.R. Bowman.

Stansell, C. (1987) *City of Women: Sex and Class in New York, 1789–1860*, Urbana IL: University of Chicago Press.

Stevenson, L.L. (1991)*The Victorian Homefront: American Thought and Culture, 1860–1880*, New York: Twayne Publishers.

Sweet, C.S. (1923) *History of the Sleeping Car*, unpublished manuscript.

Taylor, B.F. (1874) *The World on Wheels and Other Sketches*, Chicago: S.C. Griggs.

Terrell, M.C. (1940) *A Colored Woman in a White World*, Washington DC: Randsdell.

Transcontinental (1870) [no title] 1(3) (28 May): 2.

White, J.H. (1978) *The American Railroad Passenger Car*, Baltimore MD: Johns Hopkins University Press.

Wright, G. (1983) *Building the Dream: A Social History of Housing in America*, Cambridge MA: MIT Press: xvi.

5 "The salt water washes away all impropriety"

Mass culture and the middle-class body on the beach in turn-of-the-century Atlantic City[1]

Debbie Ann Doyle

In the 1880s, a business in Atlantic City, New Jersey, distributed a trade card depicting a beach scene. A folded flap on the right side of the card featured a drawing of a wooden door marked "Ladies' Bath House – Don't Open." Opening the "door" revealed the interior of the bathhouse, allowing the viewer to spy on the young women inside. However, the women in bathing suits splashing in the water at the left of the image wore as little clothing as the two women in stockings and chemises made visible by this imaginary invasion of privacy. The women in bathing suits frolicked in full view of the strolling spectators on the beach.[2] The card depicted the beach as a place where private acts such as changing became a public spectacle, and the normally concealed female form became visible in a playful yet titillating manner. The advertiser who distributed the card hoped to associate his or her business with a popular landscape of leisure identified with the sight of the middle-class body at play. At the beach, the trade card implied, looking and being looked at were all part of the fun.

Like many consumer spaces that emerged during the late nineteenth century, including department stores, amusement parks, and world's fairs, the seaside resort blurred the strict dichotomies between public and private that characterized early nineteenth-century middle-class culture (See Miller 1981; Lewis 1983; Nava and O'Shea 1996; Rabinovitz 1998). The tourism industry, which began catering to the middle class in the mid-nineteenth century, produced numerous hybrid public and private places, such as downtown hotels or railroad sleeping cars (Cocks 2001; Aron 1999; Richter 2003). Most historians suggest that these new commercial spaces generated anxiety about the safety and morality of middle-class women, anxieties tamed as women were transformed into passive objects of male visual desire (Rabinovitz 1998; Nord 1995; Latham 1995). Scholars of tourism argue that it, too, relied on voyeurism to appropriate and control unfamiliar landscapes and cultures through vision and photography (Urry 1990; Rojek and Urry 1997; Kinnaird and Hall 1994). However, at the beach the difference between viewer and viewed, between the pleasures of looking and the pleasures of being looked at, were as indistinct as the difference between public and private.

Tourists of both genders traveled to beach resorts to put themselves on display, not as passive objects but as active participants in a new and thrilling form of public behavior. The late nineteenth-century seaside became a zone of play and freedom, where the pleasures and appearance of the body took center stage. Few critics complained about public immorality at resorts, a complicated moral landscape where people flirted with the line between playful sensuality and improper behavior while the presence of spectators ensured the ultimate morality of the scene. Atlantic City became a key site in the evolution of a new beach culture. This in turn redefined the beach as a landscape of collective bodily display. At the beach, vacationers clad in revealing bathing costumes gleefully displayed their bodies to their fellow bathers. They also abandoned the respectability that defined the early nineteenth-century bourgeoisie and reveled in playful, undisciplined motion that contrasted sharply with the stiff and formal posture the middle class maintained in most other contexts. The beach, with its own temporary morality, became a place apart from everyday life.[3]

Turn-of-the-century Atlantic City also reveals the dynamic connection between tourism, landscape, and photography that emerged in the late nineteenth century. The Atlantic City business community aggressively promoted the resort in the emerging national media. Local entrepreneurs advertised the town in newspapers in cities from Philadelphia to Chicago, sent advertising agents armed with photographs and illustrated guidebooks on tours of the Northeast, and set up permanent information bureaus in New York and London to attract tourists to the town. Encouraged by local boosters, newspapers in Philadelphia and New York, as well as popular national periodicals, devoted multiple column inches to praising the charms of Atlantic City.[4] As the resort's fame grew, both local and national authors produced guidebooks and souvenir photo albums describing its attractions, while national postcard distributors such as the Curt Teich Company and the Detroit Publishing Company marketed numerous views of Atlantic City. Displaying one's body on the beach became a way to actively participate in a new culture characterized by the circulation of images (Charney and Schwartz 1995: 1; Schwartz 1998). Like the newly invented cinema, the beach presented "physicality and sexuality" as "both entertaining for witnesses and fun for participants" (Ullman 1997: 27). The playful moral landscape of the beach, with its temporary permission to make a spectacle of oneself, introduced the middle class to an emerging modern culture in which the boundaries between private acts and public spectacle became increasingly permeable (Ullman 1997: 136; Charney and Schwartz 1995: 1).

As sea bathing became a popular form of recreation in the second half of the nineteenth century, resorts sprang up along the coasts of the United

States, Great Britain, and France (Corbin 1994; Haug 1982; Walton 1983). While many resorts remained semi-rural cottage communities, several, including Atlantic City, Coney Island in New York, and Blackpool in England, evolved into urbanized landscapes adjacent to the sea. Atlantic City was among the largest and most popular seaside resorts in the turn-of-the-century United States. The resort's four-mile Boardwalk, forty feet wide and ten feet high, divided the beach and the town. On one side of the esplanade stretched the panorama of sea and sand, occasionally broken by one of the city's five amusement piers, while huge luxury hotels as many as ten stories high, theaters, photography studios, restaurants, confectioners, and other shops lined the land side of the promenade. Electric lights lined the Boardwalk at night, while huge electric signs made the resort second only to Times Square as a showcase for illuminated advertising (Jakle 2001; Taylor 1991). The paradise of commercial entertainment highlighted the connection between the spectacle on the beach and the numerous other commercial spectacles that drew patrons to contemporary urban entertainment districts.

Most visitors to Atlantic City were business owners, professionals, or clerks and managers employed in business, manufacturing, or government offices, the core of the growing middle class (Funnell 1975: 24–8). In the early nineteenth century, middle-class Americans clung to the manners and morals that, they believed, distinguished them from the working class. Bourgeois identity hinged on hard work and adherence to strict rules of deportment, demanding mastery over the body and its sexual and sensual desires. Men and women socialized separately and elaborate codes of manners, posture, and social behavior structured the behavior of both sexes (Halttunen 1982; Kasson 1990; Yosifon and Stearns 1998). Scholars frequently argue that this culture endured until relaxed posture, informality, and mixed-sex leisure became central components of middle-class identity in the 1920s (Erenberg 1981; Peiss 1986; Snyder 1989). However, at the seaside, where the ability to afford an extended stay away from home replaced work and comportment as a sign of middle-class status, tourists experimented with new kinds of public behavior and new rules about bodily display (Aron 1999: 35).

Men and women played together with relative freedom on American beaches, where "mixed bathing" had been common since the mid-nineteenth century. Middle-class Americans considered the seashore a kind of outdoor version of the private ballrooms where respectable unmarried men and women mingled without compromising their reputations. This permissive attitude toward physical contact between the sexes shocked some European observers. Swedish author Frederica Bremer, visiting Cape May in 1850, observed with surprise that:

> "Miss —— , may I have the pleasure of taking a bath with you, or of bathing you?" is an invitation which one often hears at this place from

a gentleman to a lady, just as at a ball the invitation is to a quadrille
or a waltz, . . .[5]

Likewise, an 1877 guide to seaside resorts assured readers that "a gentleman
may escort a lady into the surf . . . with as much propriety and grace as
he can display in leading her to a place in the ball-room in the evening"
(Norton 1877: 6). Linking mixed bathing to sanctioned contact between
the elite in the public areas of their private homes rendered it respectable
rather than risqué. Nevertheless, outdoor mingling of the sexes remained
sufficiently unusual to fascinate observers like Bremer, who noted with
wonder that "the salt water washes away all impropriety."[6]

Surf bathing encouraged playful, uncontrolled movements that contrasted sharply with the stiff and formal posture usually adopted by the
middle class. Nineteenth-century Americans, few of whom received any
formalized swimming instruction, "bathed" rather than "swam" in the sea,
splashing and tumbling in the waves (Lencek and Bosker 1998: 179).
Writing about surf bathing in *Scribners* magazine, writer Duffield Osborne
explained that the tumultuous surf endangered the most controlled
demeanor:

> A big, foamy crest curls over [the bathers] and falls with a roar; and,
> as it rolls in, you think you see a foot reaching up pathetically out of
> its depth, and now a hand some yards away, until at last, from out of
> the shallows of the spent wave two dazed and bedraggled shapes stagger
> to their feet and look, first for themselves, and then for each other.
> . . . [Meanwhile] a roar of laughter floats shoreward as a demoralized
> form is seen to gather itself up, almost upon the beach . . . he cannot
> tell you just how many somersaults he has turned since the ocean
> proceeded to take him in hand . . .
>
> (Osborne 1890: 103)

While Osborne implied that onlookers enjoyed such accidents more than
the victims, he acknowledged that even the experienced surf bather considered "an overthrow of both his person and his pride" as part of the thrill
of the sport (Osborne 1890: 102). Atlantic City resident and guidebook
author Alfred Heston described ocean bathing as an ebullient celebration
of chaotic movement and the occasional loss of dignity, encouraging his
readers to:

> bounce into the surf with a hop, skip, and a jump, . . . Now dance,
> leap, tumble, swim, float, kick, or make any other motions that seem
> good to you. Keep in motion. . . . If your teeth are of the kind which
> did not grow in your mouth, beware lest a wave knock them out.
>
> (Heston 1888: 120)

Both Osborne and Heston encouraged beachgoers to relinquish control over their bodies to the ocean, for the enjoyment of both themselves and others.

The relaxed behavior permitted in the surf extended, to some extent, to the posture of those on the sand. Photographs of the beach show that visitors in bathing costume and street clothes sat cross-legged or reclined casually on the sand. For example, in a postcard entitled "Having a Rest at the Beach, Atlantic City, NJ," a group of young men and women in bathing suits sprawled in a semi-circle, two of the men resting their heads on the shoulder of a seated female companion while another young woman reclined in her beau's arms. A fourth young man lay prone, his head in the lap of a young woman in walking clothes. The loose and intimate posture of this group of young people advertised the relaxed atmosphere of the beach while adding elements of collective relaxation and implied sexual intimacy. While not quite as ostentatiously relaxed as the couples in the foreground, the other tourists in the photograph sat casually on the ground or waded in the water. A line of beach chairs appeared to the left of the image. These reclining canvas swings suspended beneath striped awnings encouraged their occupants to recline rather then to maintain the upright posture demanded in the parlor or on a park bench.[7] Seated together on the sand, vacationers collectively replaced the upright carriage of their daily lives with a casual posture that physically signaled the distinction between the seaside and everyday life.

By the turn of the century, the combination of close contact between unmarried men and women and the loosening of the physical restraints of everyday life earned the seaside a reputation as a place of playful flirtation and sensuality. Songs and postcards cheerfully advertised the potential for flirtation at the seashore. One vaudeville song described a group of "pretty maids" who called out:

> I'd love to go bathing with someone,
> Won't some boy please teach me to swim?
> I'll let you show me
> Just how to hold me
> If you will only take me in.
>   (Huckins and Palmer 1907)

Postcards like "Relaxing on the Beach" or a picture of a man and woman embracing entitled "A Tight Squeeze at Atlantic City, NJ," depicted the seaside as a fertile ground for casual contact between the sexes. These images encouraged vacationers to think of the seashore as a place where young people enjoyed a respite from the rules that normally governed middle-class courtship (Bailey 1988).

The beach permitted a specialized form of public undress – the bathing costume. Beginning in the 1880s, men wore knitted wool trunks and tunics that fit trunk and torso as closely as a tailored suit while exposing the arms and lower legs. Women adopted the short-sleeved, fitted "princess" bathing dress, worn over short bloomers and opaque stockings. Many women wore this costume over corsets, so that their bodies retained their familiar, clothed shape, even as they revealed their forearms and the shape of their legs, normally concealed under clothing.[8] While today nineteenth-century bathing costumes seem both puritanical and highly impractical, contemporaries perceived bathing dress as a form of quasi-nudity. As the trade card discussed above revealed, bathing costumes approached the intimacy of undergarments. Indeed, a correspondent for the *Ladies' Home Journal* claimed to have heard a small girl "ask her mother if she might take off her dress and play in her underclothes like the ladies did on the beach."[9] Like underwear, which created a new erotic focus on "the body neither public nor absolutely private, . . . neither dressed nor exactly undressed," bathing costumes, particularly when wet and clinging, suggested an erotic appeal that transformed the seaside into a highly-charged spectacle of public undress (Finch 1991: 337–63).

Mass-produced images helped spread the idea that bathing costume could be erotic, often by exaggerating the way a typical bathing costume revealed the body. For example, a series of tobacco cards intended for an audience of male smokers represented the beaches of the world with pictures of young women in improbably clinging bathing costumes closer to the costumes of burlesque performers than actual swimwear.[10] Atlantic City became known as an ideal place to experiment with audacious beach wear; the author of an 1895 *Cosmopolitan* article noted that "I never saw at any European sea-shore resort costumes that were as suggestive or indecent as the thin, white suits which a dozen women bathers wear every summer at Atlantic City"(Adams 1895a: 324). Though he seemed to disapprove of some American women's taste in bathing costumes, the author noted elsewhere that by swimming from bathing machines that concealed the swimmer from other beachgoers, the English "miss the essence of [bathing's] great attraction"(Adams 1895b: 397). Most observers considered beach costume an interesting, somewhat risqué novelty rather than a serious moral threat. The visibility of the body on the beach generated a tension between dress and undress, concealment and revelation, public and private that lent an aura of daring and risk to the seaside.

The difficulty of maintaining strict control over one's outward appearance in a wet wool bathing costume contributed to both the titillation and the fundamental morality of ocean bathing (Aron 1999: 77; Halttunen 1982: 65–90). Bathers already displaying their bodies to an unusual degree sometimes found themselves displaying more skin than they intended, a sight both shocking and, because it was accidental, innocent. Women wringing out the wet skirts of their bathing dresses, adjusting sagging, drenched stockings, or helping a companion whose belt or corset had come undone as she played in the surf revealed their bodies even as they attempted to restore their composed appearance.[11] The impact of water on women's hairstyles provides the best example of the simultaneous innocence and audacity of such gestures. Etiquette demanded that hair be kept neatly pinned up in all public contexts, yet magazine articles teaching young men how to escort a lady into the surf gravely warned "never promise her, either expressly or by implication, that you will not let her hair get wet," because the waves could destroy an elaborate hairstyle in an instant.[12] Women combing out and rearranging their wet hair frequently appeared on postcards and other views of Atlantic City.[13] A practical response to the impact of the ocean, hairdressing on the beach was also a public performance of a grooming ritual usually confined to the bedroom, suggesting sensuality and abandon (Clarke 1985: 136).[14] Yet these images purported to depict respectable middle-class women who never intended to make such a spectacle of themselves, although they often posed with a smile that suggested they were amused rather than horrified by the situation. Respectable people behaved differently at the beach, where they quite literally made spectacles of themselves for an eager audience.

The messages tourists wrote when they mailed postcards of women grooming themselves on the beach suggest that vacationers regarded accidental exposure and the destruction of a composed public appearance as part of the appeal of a day at the beach. One female visitor mailed a postcard of a woman adjusting her stockings to her niece, commenting without evident embarrassment, "lost mine in the water yesterday."[15] Other tourists found humor in the contrast between the idealized commercial images and their own appearance, as did the woman who ironically noted on the back of a postcard of a model coyly wringing out her skirts that "I have a brown bathing suit but don't look anything like this, especially after coming out of the water."[16] These women used commercial images of the female body to commemorate the experience of displaying themselves on the beach.

The pleasure of displaying oneself in a daring fashion was combined with the popular pastime of watching other vacationers as they splashed in the surf and sprawled on the sand. Atlantic City became famous more for the spectacle of its beach than the quality of its bathing. Postcards, stereo views, souvenir photograph collections, and illustrations for magazines, newspapers, and guidebooks depicted the popularity of watching the bathers at Atlantic City, where the Boardwalk and the amusement piers provided an ideal vantage point for taking in the scene on the beach[17] (Figure 5.1). A journalist noted that during the midday "bathing hour," "thousands of spectators . . . crowd the piers for hours, that they may secure better views of a most fascinating picture."[18] "Beach strollers" in street clothes mingled with the bathers on the sand, where people spent as much time watching other beachgoers as swimming.[19] Guidebooks assured readers that it was worth the price of an Atlantic City vacation to enjoy the sight of the "ten to twenty thousand men, women, and children [who] may be seen any day during the bathing season, disporting in the foaming

Figure 5.1 "Watching the Bathers from the Boardwalk," *c.*1900–1908
Postcard in the author's collection

breakers, creating a living picture which the most gifted artists have not
equaled on canvas, which talented pens have failed to fully describe and
which no other watering place on the planet can approach."[20] The opportunity to obtain an unusual vantage point from which to observe
middle-class bodies at play became an important attraction of the seaside
resort. The beach and Boardwalk together became the two most famous
Atlantic City "sights," and the appeal of each depended in part on proximity to the other. The presence of the urban spectacles of the Boardwalk
and amusement piers placed the scene on the beach in the context of a
new world, a culture based on spectacle, sensation, and change.

While the unusual dress and behavior of beachgoers explains part of the
attraction of watching the bathers, the sheer number of people who packed
the Atlantic City beach on summer weekends also contributed to the visual
appeal of the beach. Postcards and photographs of the beach almost always
depicted large crowds of vacationers. These images advertised Atlantic City
as a landscape crowded with middle-class Americans united by the
communal pursuit of pleasure. Individually, the behavior of the bathers
may have seemed shocking, but the sight of a large crowd of presumably
respectable citizens flaunting the rules of propriety identified the seaside as
the site of a new and exciting type of collective pleasure. National postcard companies, including the Detroit Publishing Company and the Curt
Teich Company, produced numerous crowd views, often simply titled

"A Holiday Crowd on the Beach," "The Crowd of Bathers at Richards Baths," or "The Crowd on the Beach at South Carolina Avenue."[21] These images intensified the experience of visiting the resort, since vacationers became both the audience and the chief sight and tourist attraction at the beach. Vacationers came to Atlantic City to watch themselves acting in ways they only acted at the beach.

Postcards of crowds on the beach also connected Atlantic City to the emerging national circulation of photographic images. Postcards imply a link between the viewer or sender and the subject, so shots of the crowd contained an implicit promise that the viewer might one day become part of the most interesting scene in Atlantic City (Schor 1992: 237; Stewart 1993: 138). Traveling to Atlantic City enabled tourists to literally put themselves into the picture. Vacationers used postcards of the crowd to identify themselves as participants in the depicted scene. Some vacationers rhetorically located themselves within the crowd with comments such as "Can you see us in the crowd [?]" or "if you find me here tell me."[22] Others drew arrows and "X"s on the images to indicate their position, a playful gesture that located themselves both in the crowd and in the fictional space depicted on the postcard.[23] Postcard manufacturers accommodated tourists' desire to announce their presence among the multitude, selling views of the crowd with captions such as "How They Met Me on My Arrival," and "$5.00 Reward if you find me in this crowd of bathers"[24] (Figure 5.2). On another postcard publishers promised "If you want to smile and be happy like this crowd, then you had better come to the World's Play Ground, Atlantic City, NJ."[25] By mailing these postcards home from Atlantic City, tourists declared a temporary new identity as members of the horde of pleasure-seekers gathered at the seaside. A tourist who sent one of these postcards to a friend, relative, or coworker invited the recipient to imagine joining the throng on the beach as well as the more abstract audience of people looking at mass-produced pictures of Atlantic City and, by extension, the audience for all mass-marketed images. The Atlantic City crowd thus became a symbol of the tourist's connection to modern mass culture.

Atlantic City beach culture was part of the same cultural shift that produced the cinema (Charney and Schwartz 1995). In a significant cultural transformation, images of the body, both clothed and partially unclothed, became common and once private behavior became a public spectacle (Ullman 1997; Rabinovitz 1998). A 1903 film entitled *Seashore Frolics* played with the tension between looking and being seen on the beach and looking at and being seen by the nascent film industry.[26] The director cut together four separate scenes shot with a single stationary camera. The film opened with a shot of six young couples posing on the sand. In the next scene, the men formed a line and bent over. The women ran in from off camera and jumped onto their backs, after which the men jumped up and down as the women rode them like horses, falling one by one to the ground. The following scene depicted eight couples posing in front of the ocean.

Figure 5.2 "$5.00 Reward if you find me in this crowd of bathers, Atlantic City, NJ," *c.*1916

Postcard in the author's collection

The men hoisted the women on to their shoulders and carried them out into the water, then ran back out and dropped the women onto the sand. These scenes recreated the common postcard theme of young people horsing around on the beach, setting the action in motion. The final scene metaphorically linked the action on the beach to the film industry as the young women interrupted a game of leapfrog to run off screen and pull the cameraman into the frame before throwing him into the surf. The appearance of the cameraman both reminded viewers that they were watching a technological representation of the Atlantic City beach and highlighted the relationship between mass-produced images and the pleasures of the resort. The film depicted the beach as a place of casual and playful behavior, bodily display, and heightened intimacy between the sexes where the distinction between the "real life" behavior of middle-class vacationers and the images marketed by the early film industry disappeared.

Tourists traveled to the beach resort to see middle-class individuals engaged in a collective public performance of leisure and to take part in that performance; to watch others display themselves and to put themselves on display. The relaxed attitude that prevailed at the seashore created a hybrid landscape that blurred not just the distinction between public and private but the difference between clothed and unclothed, respectable and immoral, and viewer and spectacle. These elisions set the seaside apart from everyday life and identified the beach as a daring, thrilling environ-ment well worth the price of a visit. At well-advertised, urban Atlantic City, the beach also blurred the boundary between the audience and the show, between consumers of mass culture and mass-produced photographic images. This connection intensified the excitement of experimenting with the new moral landscape of the seashore, allowing vacationers to tempor-arily take part in the nationwide moral landscape produced by the cinema. Significantly, vacationers did not passively observe the spectacle, but were active participants in its creation. Historians often assume that the cinema developed among the working class and was later tamed for middle-class consumption (Sklar 1975; Peiss 1986; Friedman 1995; Cripps 1997). Atlantic City beach culture suggests that well before the invention of downtown movie palaces in the 1920s, the middle class took part in the spectacular modern culture that produced the cinema.

Notes

1 Bremer, quoted in Studley (1964).
2 Trade card, Warshaw Collection of Business Americana, Archives Center, National Museum of American History, Smithsonian Institution, Washington DC. The same trade card appears in Martin (n.d.).
3 Tony Bennett and Cindy Aron have previously pointed out the unusual morality of the seaside. Bennett describes Blackpool, the English beach resort, as a place that "inscrib[es] the body in relations different from . . . everyday life" (Bennett 1995: 238). Cindy Aron notes that middle-class women at the beach "revealed their bodies in ways that were not only unusual, but in other circumstances improper" in an environment where "some of the rules of middle-class propriety were suspended, or at least relaxed" (Aron 1999: 77, 79).
4 Smith (1923: 17); "Scrapbook of Menus," (n.d.); Bureau of Information and Publicity (1906).
5 Bremer in Studley (1964: 133); and "Letter From Atlantic City, July 4, 1856" in Martin (n.d.).
6 Quoted in "Letter from Atlantic City, July 4, 1856," in Martin (n.d.).
7 "Having a Rest at the Beach, Atlantic City, NJ", postcard *c*.1910, Heston Local History Collection, Atlantic City Free Public Library, Atlantic City, NJ. John Kasson observed similarly relaxed postures among the working-class patrons of Coney Island in Kasson (1978: 45). While the three clothed women hold their backs stiffly upright, this may reflect the restrictions of corsets designed for more formal situations rather than a reluctance to relax as completely as their compatriots.
8 See Kidwell (1965); *New York Times* (7 July 1897), 6 comments on the practice of swimming in corsets.
9 Holt (1890: 6). Aron, who also quotes Holt, discusses the unusual spectacle of bathing costumes in Aron (1999: 77). Aron finds that many mid-nineteenth-century vacationers associated the seashore with impropriety and immorality. This stigma waned in the 1890s.
10 Tobacco cards, *c*.1880, Warshaw Collection of Business Americana, Archives Center, National Museum of American History, Smithsonian Institution, Wash-ington DC; on the "princess" bathing dress, see Kidwell (1968); on burlesque, see Allen (1991).

11 "Wringing Wet, Atlantic City," stereo view, 1897, Division of Photographic History, National Museum of American History, Smithsonian Institution, Washington DC; "Atlantic City, A Bell(e) (W)Ringing Wet," stereo view, n.d., Division of Photographic History, National Museum of American History, Smithsonian Institution, Washington DC; "I miss the Rubbers down here," postcard, 1906, Heston Local History Collection, Atlantic City Free Public Library, Atlantic City NJ; "Have you ever been ripped up the back?" postcard, 1906, Heston Local History Collection, Atlantic City Free Public Library, Atlantic City NJ. Souvenir books, including *Views of Atlantic City* (1894); *Atlantic City, The Gem of the Coast* (1902); and *Fifty Views of Atlantic City, Americas Greatest Health Resort* (n.d.) often included variations on this theme.

12 Osborne (1890: 106); also see Osborne (1902: 523) and Wells (1900: 549). On women's hairstyles, see Halttunen (1982: 105); Peiss (1998: 155 and 186).

13 "An Emergency Hair Dresser," *c.*1900, postcard in the author's collection; "Bathing Girls from Watercolor by Frank A. Nankivell," *Philadelphia Inquirer* (13 August 1899), cover of section 4.

14 T.J. Clarke argues that unbound hair symbolized female sensuality, luxuriousness, and wantonness (Clarke 1985: 136).

15 Postcard, 1906, Postcard Collection, Heston Local History Collection, Atlantic City Free Public Library, Atlantic City NJ.

16 "The Girl in Brown," in the authors' collection, mailed 1910.

17 "Watching the Bathers," postcard in the author's collection; image also included in *Views of Atlantic City* (1907, 1909).

18 Patten (n.d.).

19 *Philadelphia Inquirer* (9 August 1896: 18); "again we have the beach stroller, who does not even don bathing apparel, but whose chief object seems to be to observe and to be observed." Photograph, 1900, Atlantic County Historical Society, Somers Point, NJ.

20 *Harpers Weekly* (30 August 1890: 676); Hall and Bloodgood (1899: 71).

21 All *c.*1911, Curt Teich Postcard Archives, Lake County Museum, Wauconda IL. While postcard companies reproduced photographs of crowds in other cities, crowd scenes were unusually common in postcards of Atlantic City. Twenty-five percent of the "crowd" postcards issued by the Detroit Publishing Company, for example, depicted the New Jersey resort. Detroit Publishing Company Collection, Division of Prints and Photographs, Library of Congress, Washington DC.

22 Postcard, "Boardwalk and Beach, Atlantic City, NJ," *c.*1905, in the collection of the author; postcard, *c.*1915, Kutschera Postcard Collection, Atlantic City Free Public Library, Atlantic City NJ.

23 Examples include a 1912 postcard in the Kutschera Collection, Atlantic City Free Public Library, Atlantic City NJ and a 1911 postcard in the collection of the Atlantic County Historical Society, Somers Point NJ.

24 Postcard *c.*1910, Heston Local History Collection, Atlantic City Free Public Library, Atlantic City NJ; Postcard *c.*1910, Kutschera Collection, Atlantic City Free Public Library, Atlantic City NJ; postcard *c.*1916, in the author's collection.

25 Postcard *c.*1916, in the author's collection.

26 *Seashore Frolics* (Atlantic City: Edison, 1903) Paper Print Collection, Library of Congress, Washington DC. The film was directed and filmed by cinematic innovator Edwin S. Porter, who worked for the Edison film company. Porter, known for his use of editing to propel the narrative of a film, also made *The Life of an American Fireman*, *Uncle Tom's Cabin*, and *The Great Train Robbery*, three of the most popular and influential contemporary films in 1903. However, in the days before movie magazines, audiences probably did not connect the films, except to note that they were all released by Edison. See Cripps (1997: 18–22) and Sklar (1975: 24–7).

Bibliography

Adams, J.H. (1895a) "Bathing at the American Sea-Shore Resorts," *Cosmopolitan* 19 (July).

—— (1895b) "Bathing at the English Sea-Shore Resorts," *Cosmopolitan* 19 (August).

Allen, R.C. (1991) *Horrible Prettiness: Burlesque and American Culture*, Chapel Hill NC: University of North Carolina Press.

Aron, C.S. (1999) *Working At Play: A History of Vacations in the United States*, New York: Oxford University Press.

Atlantic City, The Gem of the Coast (1902) Philadelphia PA: J. Howard Avil.

Bailey, B.L. (1988) *From Front Porch To Back Seat: Courtship in Twentieth-Century America*, Baltimore MD: Johns Hopkins University Press.

Bennett, T. (1995) *The Birth of the Museum: History, Theory, Politics*, New York: Routledge.

Bremer, F. quoted in M.V. Studley (1964) *Historic New Jersey Through Visitors' Eyes*, New Jersey Historical Series, Princeton NJ: D. Van Nostrand Company: 133.

Bureau of Information and Publicity (1906) "Summary of First Annual Report," Atlantic City NJ: Bureau of Information and Publicity.

Charney, L. and Schwartz, V.R. (eds) (1995) *Cinema and the Invention of Modern Life*, Berkeley CA: University of California Press.

Clarke, T.J. (1985) *The Painting of Modern Life: Paris in the Art of Manet and His Followers*, New York: Alfred A. Knopf.

Cocks, C. (2001) *Doing the Town: The Rise of Urban Tourism in the United States, 1850–1915*, Berkeley CA: University of California Press.

Corbin, A. (1994) *The Lure of the Sea: The Discovery of the Seaside in the Western World, 1750–1840*, Berkeley CA: University of California Press.

Cripps, Thomas (1997) *Hollywood's High Noon: Moviemaking & Society before Television*, Baltimore MD: Johns Hopkins University Press.

Erenberg, L.A. (1981) *Steppin' Out: New York Nightlife and the Transformation of American Culture, 1890–1930*, Westport CT: Greenwood Press.

Fifty Views of Atlantic City, Americas Greatest Health Resort (n.d.) Atlantic City NJ: John A. Clement.

Finch, C. (1991) "'Hooked and Buttoned Together': Victorian Underwear and Representations of the Female Body," *Victorian Studies* 34 (Spring): 337–63.

Friedman, A.S. (1995) "Prurient Interests: Anti-Obscenity Campaigns in New York City, 1909–1945," PhD dissertation, University of Wisconsin at Madison WI.

Funnell, C.E. (1975) *By The Beautiful Sea: The Rise and High Times of That Great American Resort, Atlantic City*, New Brunswick NJ: Rutgers University Press.

Hall, J.F. and Bloodgood, G.W. (1899) *The Daily Union History of Atlantic City, New Jersey*, Atlantic City: Daily Union

Halttunen, K. (1982) *Confidence Men and Painted Women: A Study of Middle Class Culture in America, 1830–1870*, New Haven CT: Yale University Press.

Harpers Weekly (1890) "The Bathing Hour – Atlantic City," 30 August: 676.

Haug, J. (1982) *Leisure and Urbanism in Nineteenth-Century Nice*, Lawrence KS: Regents Press of Kansas.

Heston, A.M. (1888) *Hand-book of Atlantic City, New Jersey*, Atlantic City NJ: A.M. Heston.

Holt, F. (1890) "Promiscuous Bathing," *Ladies' Home Journal*, August: 6.

Huckins, P.L. and Palmer, I.C. (1907) *Bathing Waltz Song*, Chicago: Music Post Card.

Jackle, J.A. (2001) *City Lights: Illuminating the American Night*, Baltimore MD: Johns Hopkins University Press.

Kasson, J. (1978) *Amusing the Million: Coney Island at the Turn of the Century*, New York: Hill & Wang.

—— (1990) *Rudeness & Civility: Manners in Nineteenth-Century Urban America*, New York: Hill & Wang.

Kidwell, C.B. (1968) "Women's Bathing and Swimming Costume in the United States," *United States National Museum Bulletin 25*: Paper 65.

Kinnaird, V. and Hall, D. (1994) *Tourism: A Gender Analysis*, New York: John Wiley.

Latham, A.J. (1995) "Packaging Women: The Concurrent Rise of Beauty Pageants, Public Bathing, and Other Performances of Female Nudity," *Journal of Popular Culture* 29 (Winter): 149–67.

Lencek, L. and Bosker, G. (1989) *Making Waves: Swimsuits and the Undressing of America*, San Francisco: Chronicle Books.

Lewis, R. (1983) "Everything Under One Roof: World's Fairs and Department Stores in Paris and Chicago," *Chicago History* 12 (Fall): 28–47.

Löfgren, O. (1999) *On Holiday: A History of Vacationing*, Berkeley CA: University of California Press.

Mackaman, D.P. (1998) *Leisure Settings: Bourgeois Culture, Medicine, and the Spa in Modern France*, Chicago: University of Chicago Press.

Martin, J.H. *Sketches of Atlantic City, New Jersey, 1856–1885*, Philadelphia PA: The Historical Society of Pennsylvania.

Miller, M. (1981) *The Bon Marché: Bourgeois Culture and the Department Store, 1869–1920*, Princeton NJ: Princeton University Press.

Nava, M. and O'Shea, A. (eds) (1996) *Modern Times: Reflections on a Century of English Modernity*, New York: Routledge.

New York Times (1897) "Types at Atlantic City," 7 July.

Nord, D.E. (1995) *Walking the Victorian Streets: Women, Representation, and the City*, Ithaca NY: Cornell University Press.

Norton, Charles Ledyard (1877) *American Seaside Resorts: A Hand-Book for Health and Pleasure Seekers, Describing the Atlantic Coast, from the St. Lawrence River to the Gulf of Mexico*, New York: Taintor Brothers, Merrill.

Osborne, D. (1890) "Surf and Surf Bathing," *Scribners* 8 (July).

—— (1902) "Surf Bathing," *Outing* 40 (August): 523.

Patten, D.A. (n.d.) "Atlantic City: America's Greatest Resort," *International Gazette*, clipping, vertical files, Atlantic County Historical Society, Somers Point NJ.

Peiss, Kathy (1986) *Cheap Amusements: Working Women and Leisure in Turn-of-the-Century New York*, Philadelphia PA: Temple University Press.

—— (1998) *Hope In A Jar: The Making of America's Beauty Culture*, New York: Metropolitan Books.

Philadelphia Inquirer (1896) "Atlantic City," 9 August: 18.

—— (1899) "Bathing Girls from Watercolor by Frank A. Nankivell," 13 August, cover of section 4.

Rabinovitz, L. (1998) *For the Love of Pleasure: Women, Movies, and Culture in Turn-of-the-Century Chicago*, New Brunswick NJ: Rutgers University Press.

Rojek, C. and Urry, J. (1997) *Touring Cultures: Transformations of Travel and Theory*, New York: Routledge.

Schor, N. (1992) "*Cartés Postales*: Representing Paris 1900," *Critical Inquiry* (Winter): 188–241.

Schwartz, V.R. (1998) *Spectacular Realities: Early Mass Culture in Fin-de-Siècle Paris*, Berkeley CA: University of California Press.

"Scrapbook of Menus and Newspaper Clippings, Hotel Dennis, 1901–1917," (n.d.) Somers Point NJ: Atlantic County Historical Society.

Seashore Frolics (1903) [film], Atlantic City NJ: Edison, Paper Print Collection, Library of Congress, Washington DC.

Sklar, R. (1975) *Movie-Made America: A Cultural History of American Movies*, New York: Random House.

Smith, H.E. (1923) "Advertising Atlantic City During the Year 1890: A Pioneer Publicity Agent Tells Methods in Old Days," *Atlantic City* (June): 17.

Snyder, R. (1989) *The Voice of the City: Vaudeville and Popular Culture in New York*, New York: Oxford University Press.

Stewart, S. (1993) *On Longing: Narratives of the Miniature, the Gigantic, the Souvenir, and the Collector*, Durham NC: Duke University Press.

Taylor, W.R. (ed.) (1991) *Inventing Times Square: Commerce and Culture at the Crossroads of the World*, Baltimore MD: Johns Hopkins University Press.

Ullman, S. (1997) *Sex Seen: The Emergence of Modern Sexuality in America*, Berkeley CA: University of California Press.

Urry, J. (1990) *The Tourist Gaze: Leisure and Travel in Contemporary Societies*, London: Sage Publications.

Views of Atlantic City (1894) Philadelphia PA: William H. Rau.

—— (1907, 1909) Portland MA: L.H. Nelson.

Walton, J.K. (1983) *The English Seaside Resort: A Social History, 1750–1914*, New York: St Martins Press.

Wells, F.J. (1900) "Surf Bathing," *Outing* 36 (August): 549.

Yosifon, D. and Stearns, P.N. (1998) "The Rise and Fall of American Posture," *American Historical Review* 103 (October): 1057–95.

6 How to travel with a male

Thomas M. Heaney

From the production of travel destinations such as National Park "wilder-nesses" or sacred national sites like Mount Vernon to the development of the road with its motels and fast-food outlets, to sibling conflicts over control of space in the backseat of the station wagon, the American family vacation presents an intriguing web of connections between different forms of space and their gendered representations as aspects of the moral land-scape. But vacations are not experiences removed from or out of the cultural context of a society. In the decade following the end of the Second World War, popular magazines promoted gendered definitions of travel spaces and landscapes that reflected and reinforced popular ideas about gender, family, and work in ways that bound leisure closely to a culture of consumerism and work discipline.

Popular discourses about leisure, vacations, and travel during the second half of the 1940s and the first half of the 1950s were embedded within the larger cultural context with its specific moral constructions of gender and family. For example, Elaine Tyler May, in her *Homeward Bound: American Families in the Cold War*, argues that an ideology of "domestic containment" emerged in the years after World War Two (May 1988: 14). That ideology, according to May, sought to enclose tightly the socially disruptive power of sexuality within the family home while providing the individuals in the family with a sense of security from the chaos of the external world. May argues that this hegemonic ideology of domestic containment pro-claimed the family home the place of individual fulfillment, individual security, and contained sexuality. Popular culture was a primary trans-mitter for this ideology of the family, and the discourses about leisure travel echoed and reinforced this ideology through gendered definitions of space and landscape.

As promoted and celebrated in the postwar decades, the family vaca-tion was essentially a private excursion into public space – an excursion that effectively privatized public space. Although many families might visit a travel destination at the same time, they did not do so communally; rather, they did so as discrete, hermetically sealed units. How, the ques-tion seemed to be, was a family to privately negotiate the public spaces of transportation, and how should it experience that movement through space? This tension between the private goals of leisure travel and the public spaces of the modes of travel and destinations played out in promotional literature through the prism of gender where it was mediated by women's labor.

Articles and advertisements in popular magazines made it clear, for example, that certain forms of travel – airplanes, trains, and ships – and certain destinations – resorts in general and beaches specifically – were particularly feminine. Other forms of transportation – especially automobile travel – and destinations – the mountains and camping – were masculine and discursively proscribed for women because they implied an active engagement in directing the travel. These definitions appeared sometimes as implicit, subtle, or ambiguous and at other times as explicit, obvious, or unequivocal. But, however ambiguous, this popular discourse about travel was charged with moral significance and relied on highly gendered representations of space and movement.

Advertising and advice columns effectively categorized all destinations into two forms: the beach and the mountains. The "beach," associated with relaxed inactivity, sunbathing, and revealing bathing suits, was specifically recognized as being both feminine in nature and a site of female activity. In contrast, popular discourse associated the "mountains" with ruggedness, camping, and fishing, and defined them as masculine and a male domain. A 1947 Canadian Railway advertisement in *Sunset*, a Western travel and garden magazine, explicitly demonstrated this division with pictures of a man and a woman each describing what they desired in a vacation: The man announces that he wants "a mountain vacation" with "fighting trout" and "bracing North Woods air" whereas the woman declares that she wants "a holiday by the sea" with "picturesque fishing villages, white, sun swept sands," and "cool ocean breezes" (*Sunset* 1947: 12). During the 1940s and 1950s this dichotomy appeared in the image of the "divided couple," a husband and wife arguing over which of the two locations to spend their vacation. An excellent example of this image is the July 1948 cover of *Westways*, the magazine of the Automobile Club of Southern California, which showed a young woman in beach wear with a pamphlet marked "BEACH" arguing with her husband who is dressed to go fishing and armed with a brochure about fishing in the Sierra (Figure 6.1) (*Westways* 1948: cover). Trans World Airlines used almost the same image in a *Holiday* ad with a slightly older man and woman, and considering the age difference between the two couples and the two ads, one might assume they were the same imaginary couple still arguing "beach vs mountains" four years later (*Holiday* July 1952: 93). In the same issue of *Holiday*, American Airlines featured a two page ad with the divided couple. Further, the advertisement claimed to offer air travel as "a modern answer to that age-old vacation conflict between beach and mountain vacations," as if couples had been arguing over vacation destinations for

Figure 6.1 The "divided couple"
Westways July 1948: cover

centuries (*Holiday* July 1952: 18–19). The gendered representations of travel
destinations were so common, so embedded, that they were understood to
be almost interchangeable with the genders that represented them.

Articles and advertisements maintained that women unaccompanied by
men could travel on planes, ships, and trains to a variety of destinations,
although "the beach" remained a principally feminine vacation destination. Articles in such popular magazines as *Holiday*, *Sunset*, and *Westways*
about women traveling alone or with other women were rare during
the late 1940s and the 1950s, although one from a 1946 issue of *Holiday*
provides an example. Ewell Sale followed forty female Smith College
students as they traveled to Bermuda for ten days, and reported on their

chartered vacation. The tour organizer had arranged all air transportation, accommodations, and meals so the students had nothing to do but enjoy themselves. Sale noted that the tour organizer had planned "a new twist in the tour formula" by "giving [the] girls a free rein." "Formerly," Sale said, "groups went traveling to see things, were shuttled into sight-seeing buses and driven to each spot of interest." But the tour guide explained that "the island is small and transportation easy, so if" the young women were "interested, they'll find all the places they want to see." Bermuda – "the beach" – was small enough, safe enough, and organized enough to allow women "free rein" (Sale 1946: 51).

Modes of transport were also described in gendered ways. During the postwar decade, the popular discourse about leisure travel made it clear that certain forms of travel such as airplanes, trains, and ships were particularly feminine. Rather than actually traveling, women were understood to *be transported*; they needed forms of transportation that mediated their travel and the environment. In contrast, men used transportation that was under their direct control and provided greater freedom of movement. Feminine transportation emphasized safety and comfort, professionalized technical assistance, and a lack of individual control, while masculine travel emphasized personal autonomy and control, ruggedness, and adventure.

Railroads especially touted their ability to transport women over long distances in complete comfort and safety while providing incomparable views of the landscape. In contrast to railroads' advertisements in the 1930s that had stressed "minimum daylight hours en route" (*Travel* 1930: 5) postwar railroad advertisements highlighted the passenger's new ability to see "spectacular daylight views" in the new Vista-Dome cars *(Sunset* 1950: 6). But they also emphasized comfort and privacy. One Southern Pacific advertisement that featured a photo of two women traveling together, emphasized the scenery by announcing that "you make virtually the entire trip in broad daylight," but nevertheless also noted the privacy of the bedrooms, each attached to its own bathroom (*Sunset* 1950: 6). The railroad car builders themselves published ads highlighting that "every passenger has privacy" in a room "with its own broad window, its private toilet and lavatory, its individual heat and light control, with air conditioning" (*Holiday* 1946a: 97). Some railroads even provided special cars for women and children only. One particularly telling ad showed a little girl and her mother emerging from the *Olympian Hiawatha* as the girl declared to her father, "Gee, Daddy, they had a private car just for us ladies," suggesting that women, like children, could not travel without assistance and protection (*Holiday* 1947: 127). The railroad car in effect provided "spectacular" views of the open landscape from within a moving, morally appropriate feminine space.[1]

Airline travel during the 1940s and 1950s, with its luxury and relatively short travel times, is a particularly fascinating nexus between gender and morality, and the negotiation of space. In an airplane, one could travel in a matter of hours what otherwise took days, and do so in comfort and style. But airline ad writers seemed confused about how to promote this means of movement through space. For example, both Pan American and United Airlines immediately after the Second World War promoted their new flights to Hawaii as an extraordinary and thrilling combination of luxury and adventure. United's ads at first highlighted "sumptuous" lunches, socializing, and an evening buffet as examples of the luxury which the ad said made the flight "a memorable part of your Hawaii vacation" (*Sunset* 1948: 27). But after a brief time both airlines began emphasizing not the memory of the flight but an amnesia about it through their overnight sleeper service. Both airlines produced, and reproduced, a very common image of what could be called the "pampered woman" in which a stewardess administers to a woman lying in either a berth or a reclining chair while she is whisked across the globe, conveying the mediation of space and time through pampered luxury, professional assistance, and sleep-induced amnesia (Figure 6.2) (*Sunset* 1949: 3; 1952: 13). Rather than highlight the exotic and exciting nature of air travel, such ads emphasized how time and space are removed or alleviated; the travel itself, and its possible adventures, become irrelevant. Women were thus able to fly to Hawaii "lying down," only to wake up in Honolulu asking "You mean I've slept all the way to Hawaii?" (*Sunset* 1951: inside cover). Hibernation becomes the answer to the problem of mediating the private movement through public space.[2]

In dramatic contrast to the numerous images of women in the Pan Am ads is the almost complete absence of men. Out of the twenty-six ads, only four contain any images of men on board Pan Am airplanes, and three of those ads are a repetition of the same image of men and women enjoying the cocktail lounge available on Pan Am's Stratocruisers. Although one might have expected Pan Am, in trying to promote vacations in Hawaii, to show men reclining and relaxing in the luxury of a Stratocruiser chair, such images never appeared in any ad during this time. Instead, the image of the pampered woman communicated to the reader that air travel was so easy, even women could do it in style. But it also communicates that although advertisers certainly sought male customers, they did not wish to present upper-class men reclining in airline chairs being assisted by stewardesses since such an image threatened to feminize or emasculate their target audience.

While air and rail travel were understood to be feminine in nature, the discourse about vacation travel promoted the automobile as both a masculine means of travel and as the superior form of navigating the public spaces of the nation because it provided greater control, immediate access to the landscape, and the ability to move the family as a single unit. Unlike women who rode in comfort, even luxury, through the landscape toward their goal – relaxing in the sun on the beach – men, behind the wheel and in control of their car, penetrated the landscape on the road to adventure

Figure 6.2 The "pampered woman"
Sunset March 1949: 3

in the great outdoors. In contrast to the discourse regarding other forms
of vacation travel, automobile travel became a specific discursive formation of masculine out-of-doors leisure, feminine domestic labor, and family
togetherness. Although articles and advertisements declared that it was the
family that went on the road, children seemed nearly incidental to the practice, even barriers to a successful family vacation while women appeared
primarily to minister to the needs of men and children.

These magazines did not describe the act of driving as being a strictly
male activity (images of female drivers were relatively common); instead it
was the automobile trip for leisure that became constructed as male through
its association with a variety of ideas and practices. Articles and advertisements contrasted road trips with travel by train, plane, and ship. They

joined masculine camping with automobile vacations; described the car as a means to mostly masculine activities such as fishing and hunting; and differentiated the feminine home from the masculine outdoors. In contrast to the other methods of travel, the automobile traveler could go anywhere he pleased – he was the one in control, rather than simply a passenger, and there was no mediator between the automobile's driver and the road. Safety was also a means of gendering. Articles, and certainly advertisements, never mentioned accidents or fatalities when describing forms of mass transportation – planes never crashed or suffered mechanical failure, trains never jumped the track or collided, and ships never sank or tossed about in storms. Indeed, it was the apparent safety of mediated forms of travel that made them appealing and gendered as feminine. This mediated, protected atmosphere conveyed women to their destination in moral safety as well as physical safety. In comparison, authors often referred to the inherent dangers of automobile driving, highlighting death tolls and emphasizing the need for a driver to remain in total control at all times.

The automobile, according to travel writers, combined personal control with the ability to see everything closely with nothing between the viewer and the landscape. Al Hine, writing in *Holiday*, declared that "you really don't get the full feeling and thrill of your own America till you cross it yourself, ideally by car and not in a hurry, so that its immensity and variety unfold before you, with you" (Hine 1952: 54). Writing for the same magazine, Hal Borland, a poet and writer about the outdoors and rural America, affirmed not only the superiority of the car as a means of exploring the country but poetically emphasized the importance of the side road, saying that "when we would see and hear and feel America, we turn to the lesser road." Once you have turned off the highway, he said:

> you are ambling along a winding road that hugs the hills and snuggles in the valleys, a road as much a part of the land as the fields and the fences and the oak groves on the hillsides. It is as though you had left a mode of detached, impersonal space travel and were now seeing the land for the first time, meeting it and listening to its own leisurely language.
>
> (Borland 1946: 95)

Thus, writers claimed the automobile not only provided the driver with a sense of true freedom, but also asserted that it was the only way to see America as one traveled.

According to these self-identified road-travel experts, the freedom of driving and seeing the country from behind the wheel could not compare with other forms of transportation which placed barriers – both physical and human – between the passenger and the world outside. In his own 1947 *Holiday* article, Lloyd Shearer, later publisher of the immensely successful Sunday newspaper supplement *Parade*, pronounced that the automobile was "the only way ... to see God's country" (Shearer 1947: 52). Shearer declared that he could see the landscape without seeing the "cities en route as blurs from track or stratosphere. I can detour along any byways that strike my fancy without worrying over stopover rights and lost reservations. I'm my own boss when I cross the US by car" (Shearer 1947: 52). Just as popular culture described leisure as the site of individual freedom, so Shearer described the automobile as the superior form of leisure travel because of the freedom it granted. The car provides a man with the real freedom to decide what to do and where to go:

> I can come to a halt without pulling an emergency cord and I can set off on my travels without rereading my insurance policy to check whether it has an air clause. There are no uniformed hostesses aboard my car, but I can always take my wife.
>
> (Shearer 1947: 53)

Shearer implies that the automobile does not provide women with freedom; rather, it allows them to administer to the needs of the driver whom the writer assumes is male.

Popular magazines also differentiated automobile travel from other forms of travel for leisure by the terrible dangers associated with cars and the road – another method of gendering the road. *Westways*, in part because it was the house organ of the Automobile Club of Southern California, constantly cautioned readers about the vast number of deaths that occurred yearly in automobile accidents.[3] While celebrating the ending of postwar gas rationing and the opening of a new era of travel in its "Call of the Open Road" issue of April 1946, *Westways* warned of the impending death toll on the highways by invoking the recent memories of battles in the Pacific:

> Much has been written on the statistics of the mounting American traffic death toll. Motorists have been warned that cold figures prove that our highways are more deadly than were the beachheads at Tarawa or Iwo Jima. Yet the American motorist speeds on to his destruction – and his going takes others with him.
>
> (*Westways* April 1946: 7)

Shearer advised drivers to keep their speed below sixty miles per hour to avoid getting into accidents with other drivers and to watch out for worn roads that "were neglected during the war" (Shearer 1947: 53). But one's own car always threatened to career out of control: he recommended that drivers keep "a firm grip on the wheel at all times; there is ever-present danger of a blowout." The road was a place where a driver took his own life into his hands. Writers advised rotating drivers every hour or two when traveling long distances, and several writers drove with their wives and took turns with them at the wheel. Yet, even here there were explicit gender roles; Hine's wife drove during "stretches of country driving," with Hine himself "taking over for the confusion of cities" (Hine 1952: 55).[4]

There was also the ever-present danger of hitchhikers. Both *Westways* and *Holiday* advised that if a hitchhiker was injured in a driver's automobile, the hitchhiker could sue for damages.[5] Shearer went even further; he warned readers about "a good many holdups" committed by "befriended hitchhikers" and hitchhiking couples who always had a "sob story." Most importantly, he cautioned male drivers not to pick up:

> a female thumber, unless of course, you feel violently that the Mann Act is unconstitutional. It is safe to assume that any woman thumbing a ride on a highway is in trouble, and as my father used to tell me – drying a woman's tears is one of the most dangerous occupations known to man.
>
> (Shearer 1947: 53)

Shearer not only advised travelers of the dangerous nature of hitch-hikers, but in his image of the unattached woman on the road he asserted that women had no place being there. The female hitchhiker was not just in trouble, she was trouble. The road, strewn with potholes, populated with poor drivers, and inhabited by menacing hitchhikers of both sexes, was far too dangerous for women to take the wheel alone. Further, the proscription on picking up hitchhikers can also be seen as reemphasizing the need for privacy and individuality on the road. The family was to travel as a secure unit, and picking up hitchhikers might be akin to letting strangers into the family home. Conventions of the road, such as these, provide a sketch of the moral landscape of the time.

Television as well as magazines reflected this discourse about automobile travel. In a 1956 episode of *I Love Lucy*, Lucy and Ethel, after losing their train tickets, hitch a ride at a tourist bureau with a woman driving her car from New York to Florida. Although the female driver is experienced and knowledgeable – she brings along her own "watercress sandwiches" to avoid stopping for meals and camps out on the side of the road to save money on lodgings – Lucy and Ethel have no idea what they are doing. When asked to replace a flat, they manage to both pierce the car's fender with the jack and take off the flat tire only to put it back on by mistake. But after they hear a news bulletin about a prison escape, they become convinced that their driver is a ruthless murderer. Meanwhile the driver hears a similar announcement, and thinks Lucy is the murderer traveling with a member of her criminal gang, Ethel. Paranoia between the driver and her strange, red-haired passenger runs rampant as Lucy and Ethel express regrets about taking to the road without their husbands. At a truck stop, the frightened owner of the car abandons Lucy and Ethel who eventually reunite with Ricky and Fred. Even though the episode presented a competent female driver experienced with the road, it also portrayed her as eccentric, even bizarre. The mostly implied, but at times explicit, meaning of the program was that women should take mediated forms of travel, and stay off the road (*I Love Lucy* 1956).

While women were helpless on the road, the discourse of automobile travel inextricably connected together masculinity, the road, camping, and the rugged outdoors. The cover of *Westways'* April 1946 "The Call of the Open Road" issue demonstrates the gendered nature of the outdoors (Figure 6.3). A family has driven to the mountains, and pitched camp. Father, pipe in mouth, proudly displays his fishing prize as his son observes from a distance and his wife records the event on film (*Westways* 1946: cover). The main function of the road vacation was clear: it may have been the "family car," but it was Dad's vacation, and the great out-doors, in effect, was where man belonged. Dr Morris Fishbein, the editor of the *Journal of the American Medical Association*, expressed this connection very straightforwardly in a *Holiday* article by asserting that Calvin Coolidge "knew the value of the out-of-doors and of physical activity as an antidote to the kind of life he lived in the White House" (Fishbein 1946: 51). Although President Coolidge, Fishbein admitted, "did not climb moun-tains, he gained both from the relaxation and the stimulation that come from a mountain environment. He enjoyed motoring along mountain roads deep into the forests". The road thus did not simply lead to the mountains; mountain driving itself was a vigorous outdoor activity and an antidote for a sedentary occupation. L. Callahan, in his tongue-in-cheek *Westways* article "Don't Take to the Woods," used humor to make clear the connection between masculinity and the rugged outdoors:

> You dream of getting out and communing with Nature. You picture yourself throwing up your tent, throwing down your mattress, build-ing your campfire, frying bacon and eggs – pausing between-times, of course, to grab deep breaths of pine-scented air.
>
> But wait a minute! It doesn't work that way, not if you've a wife and kids to complicate things.
>
> (Callahan 1948: 10)

Rather than specifically being a family vacation in the forest or desert, it was father's vacation in the outdoors with the rest of the family tagging along. Callahan's essay was accompanied by several cartoons, including one in which Father imagines himself getting away from it all into the solitude of nature only to find himself at the mercy of his wife and chil-dren, tied to the car like another piece of equipment. Just as vacation destinations were gendered into beach and mountains, so the masculine mountain became attached to the masculine road, often in ways that made the division between the two indistinct.

Figure 6.3 "The call of the open road"
Westways April 1946: cover

Indeed, the masculine sites of "the mountains" and "the road" merged in the activity of camping which, ironically, became a reconstruction of the family home in the woods with the automobile making this possible. A 1957 guide book entitled *Sunset Ideas for Family Camping* further demonstrates the centrality of the automobile in the reconstruction of the family home on the road. The book's editors pronounced that "the car may be parked in an improved campground or off in the wilds, but it is an essential member of the party" (*Sunset Ideas for Family Camping* 1957: 5). Here the car was not part of the camping equipment; it was part of the family. The car also permitted others besides men to penetrate the wilderness as suggested by the introduction's history of "family camping":

> Not too many years ago, camping was looked upon as an adventure for only the hardy, or as a suspect way of vacationing in discomfort.

> The typical camping party was considered to be a group of men who enjoyed living in gentlemanly squalor.
>   ... The dominant party today in Western forest camps is the camping family – husband, wife, and assorted children. Playpens are now seen in remote campgrounds that at one time rarely felt the tread of a feminine boot.
>
>   (*Sunset Ideas for Family Camping* 1957: 7)

The outdoors, the mountains, and the road were gendered as masculine, but the auto created the discursive possibility of the family in the woods.[6]

The automobile, according to popular discourse about car travel, permitted individuals, especially women and children, to enter the wilderness, but it also allowed a family to carry along the vast amount of commodities needed to make the camping experience comfortable, essentially by remaking the home in the woods. *Sunset Ideas for Family Camping* provided a picture of an ideal campsite with all the possible goods that allowed a family to "rough it" as well as how it had to be packed into the family car. Covering two pages, *Sunset*'s panorama of camping perfection divided the campsite into spaces for washing, cooking, living, and relaxing and includes a medicine cabinet, dining table, larder, icebox, cooler, and wardrobe (*Sunset Ideas for Family Camping* 1957: 14–15). Father relaxes with his fishing pole, while junior's gaze is fixed on the center of the image where Mother is cooking a meal over an open campfire. Implied was the idea that these goods permitted wives to travel with their husbands, and, as the book's photographs point out, administer to the needs of their male spouses. (Out of the ten photos in *Sunset Ideas for Family Camping* that show someone cooking, women are at the fire or stove in eight of them.) In effect, through women's work, automobiles permitted a compartmentalizing of the space on the highway and in the wilderness where campgrounds became suburban subdivisions.

Sunset's ideas about camping point out a vital aspect of the automobile for the family vacation: since it allowed the family to travel the public roads as a private unit, it also permitted the family to reproduce the moral structures of the nuclear family home while on the road. Recreating the home on the road with a strictly gendered division of labor was strongly promoted by travel writers in the postwar decade. A good example is a 1949 *Good Housekeeping* article entitled "How to Travel with a Male," in which writer Augusta Wilkins declared that "the pleasure" of travel "depends largely on [the] female of the species" (Wilkins 1949: 118). She warned the woman who likes "to gallivant in company with her spouse" that she "must make allowances for the fact that when he opens a strange bureau drawer and finds that the blue shirt is not in its accustomed corner, he's going to be upset." "The remedy for this pitfall," Wilkins advised, was "to reproduce, as nearly as possible ... the setup he's used to at home." "Housewifely duties persist to a certain extent in travel," she insisted, and it was up to the wife to pack and unpack, clean, and press "clothing, linen, and shoes" (Wilkins 1949: 120). It may have been a "family vacation," but mothers and wives were expected to reproduce their household labor while on the road.

Lucille Popenoe, in a 1948 *Westways* essay, made clear this contradiction of work and leisure in the family vacation. Assuming a female reader, she presumed that when her husband suggests taking the family camping, the reader will respond, "'Must we?' you groan, knowing it won't be any vacation for you. Doing the daily routine of cooking and child tending under primitive conditions is not your idea of fun. It's just housework the hard way" (Popenoe 1948: 13–14). She offers hints for reducing the amount of work required, but more importantly, she advised taking at least one day off by making menus for each day and then occasionally requiring the husband and kids to follow the menu and prepare the day's meals. "After all," she declared, "it's my vacation too." Similarly, a *Sunset* article noted that at least one family occasionally "gave the mother a break from camping out by treating her to fine resort hotels" while traveling between campsites. Popular discourse about automobile travel assumed that women's labor was so intrinsic to the family vacation that it actively sought to mitigate these demands to some extent (*Sunset* July 1950: 22).

The postwar discourse about travel and vacations produced by such magazines as *Holiday*, *Westways*, and *Sunset* transmitted and reinforced the postwar ideology of domestic containment in many ways. It did so in part by constructing specific forms of travel and destinations as feminine and others as masculine. Writers and advertisers in popular magazines asserted that women could not travel autonomously, that they required mediators; thus, women could travel by airline or railroad, but could not drive themselves across the country. Popular discourse categorized destinations into a highly gendered division between the morally appropriate feminine beach and the morally dangerous masculine mountains while the image of "the pampered woman" conveyed the ease of air travel, implicitly gendering the method as feminine even as it substituted for men of the affluent classes. Moreover, discourse about travel constructed the road as a specifically male domain, and connected it to ideas of individual autonomy and control, physical danger, and masculine outdoor activities. The automobile, with its ability to transport the whole family as a single unit, existed as a masculine domain that gave "Dad" a feeling of direct control over the family's travels. Camping, taking place in the masculine mountains and connected to the male realm of the road, was also understood as masculine. Further, the popular discourse about travel relied on women's labor to mediate the tension between public spaces and private movement by recreating the home on the road.

These popular magazines adapted what appeared to be unchanging and traditional gender roles to promote a discipline of leisure. Popular culture during the 1940s and 1950s produced a discourse about travel that strengthened the hegemonic postwar domestic moral ideology and restrained individuals and society even as it claimed to provide freedom for self-expression and exploration. Gendered representations of space and transport served to naturalize and legitimate the annual family vacation that fostered the development of roadside chains like Travelodge, Howard Johnson's, and Holiday Inn that effectively institutionalized the private nature of the public space of the road.[7]

Notes

1   These representations of the privatization of the public spaces aboard trains appear to echo nineteenth-century ideas about women and railroads uncovered by Amy Richter in "At Home Aboard: The American Railroad and the Changing Ideal of Public Domesticity," paper presented at Gendered Landscapes Conference, The Pennsylvania State University, 31 May 1999.
2   In my research surveys, I have never found a similar image of a "pampered" male appearing during this period. Indeed, the use of the "pampered woman" image appears to have contradicted the desires of customers. According to a 1944 opinion survey of prospective airline passengers, female respondents overwhelmingly disapproved of Pullman-type sleeper service to Hawaii, while male respondents wanted it 2–1 (Matson Navigation Company 1944: A).
3   The Automobile Club had an obvious economic interest in lowering accident rates since it offered both low-cost automobile insurance and free towing for club members.
4   For more examples, see Newill and Newill (1946: 89). This reference to the dangers of urban driving is rather odd since it was often assumed in most cases that women did significant amounts of urban driving as a part of their domestic duties.
5   See for example *Holiday* (1946b: 105–107).
6   The quote also demonstrates the inexorable discursive connection between children and women.
7   Most surveys of postwar families and travel showed that more than half of American families took a vacation of at least a week and drove their automobile. For example, one 1950 study demonstrated that 62 percent of families surveyed took vacation trips with an average of 10.5 days per trip and that 79.5 percent traveled by private automobile. These results were consistent with most other surveys of vacation travel at the time. See Research Department (1950: 17, 33).

Bibliography

Borland, H. (1946) "See the Land," *Holiday*, September: 95–97.
Callahan, L. (1948) "Don't Take to the Woods," *Westways*, June: 10–11.
Fishbein, M. (1946) "It's Doctors Orders," *Holiday*, March: 51–52.
Hine, A. (1952) "The Best Way to See America," *Holiday*, July: 54–58.
Holiday (1946a) Budd Railroad Car advertisement, March: 97.
—— (1946b) "Tips to Travelers," March: 105–107.
—— (1947) Milwaukee Road advertisement, December: 127.
I Love Lucy (1956) "Off to Florida," James V. Kern director, 12 November.
Matson Navigation Company (1944) *Post-War Travel Survey*, San Francisco CA: Matson Navigation Company: A.

May, E.T. (1988) *Homeward Bound: American Families in the Cold War*, New York: Basic Books.

Newill, P. and Newill, P. (1946) "Carefree Car Travel," *Holiday*, April: 89–91.

Popenoe, L. (1948) "The Girls Know How," *Westways*, April: 13–14.

Research Department (1950) *The Vacation Travel Market of the United States: A Nationwide Survey*, Philadelphia PA: Curtis.

Sale, E. (1946) "40 Girls in Bermuda," *Holiday*, September: 51.

Shearer, L. (1947) "How to Cross the Country," *Holiday*, January: 52–53.

Sunset (1947) Canadian Railway advertisement, April: 12.

—— (1948) United Airlines advertisement, November: 27.

—— (1949) United Vacations advertisement, March: 3.

—— (1950) Southern Pacific advertisement, October: 6.

—— (1951) Pan American advertisement, February: inside cover.

—— (1952) United Vacations advertisement, February: 13.

Sunset Ideas for Family Camping (1957) Menlo Park: Lane.

Travel (1930) Rock Island Railroad advertisement, December: 5.

Westways (1946) April: cover.

—— (1948) July: cover.

Wilkins, A.V. (1949) "How to Travel with a Male," *Good Housekeeping*, December: 118–121.

7 A wilderness for men

The Adirondacks in the photographs of Seneca Ray Stoddard

Frank H. Goodyear, III

> On wings of thought swifter than the lightning's flash cleaving through space, we sweep away across the drowsy earth, over smoke-polluted cities, sun-scorched meadows, burning plain and highways with their flaunting skirts of sand, nor rest until the fragrant odor of wild flowers and the dewy breath of forest trees come like incense wafted to us from below.
>
> (Stoddard 1874: 1)

With these words, Seneca Ray Stoddard (1843–1917) welcomed readers to the world of the Adirondacks. In this opening sentence from the guide-book that would establish him as the expert on tourism in upstate New York during the last three decades of the nineteenth century, Stoddard sought to lift up readers from their familiar everyday existence and carry them as if by magic carpet into the Edenic surroundings of the North Woods. From above, readers would be able to note the dramatic changes in the landscape as they moved from an urban environment out into a primeval wilderness. Left behind was a world that Stoddard described as dirty, hot, and mundane. In its place one discovered a paradise of natural beauty and mystery.

Republished annually throughout his lifetime, Stoddard's different guide-books to the Adirondacks were not the only means by which he influenced travel to this region. His work as a landscape photographer beginning in 1870 was equally important in precipitating the transformation of the Adirondacks into a fabled tourist destination. Though other artists and writers also fixed their gaze on this subject, it was Stoddard who was most significant in mediating the American public's understanding of the Adiron-dacks during this period. By transporting audiences to sites he hoped to highlight and by shaping popular perceptions regarding the region as a whole, his writings and photography exhibited a keen sensitivity to both practical questions and larger cultural issues that would-be travelers held. Together with other developers – including hotel owners, railroad and steamship operators, and local commercial proprietors – Stoddard was instrumental in the reconfiguration of the Adirondacks as a marketable commodity. Later in his career, when timber and mining companies threatened tourism's position – and his own professional livelihood – his writings and photography were equally vital in limiting the extension of these new interests. He became, in effect, an arbiter of the moral value of exploiting or protecting the "wilderness."

In the rhetoric of Stoddard's guidebook, it was not simply the vision that he and others drew of the city, nor of the countryside, but rather the stark contrast between the two that attracted potential visitors and precipitated the tremendous growth in summer travel to the Adirondacks. By one estimate, the number of annual visitors grew from approximately three thousand in 1869 to a quarter of a million by the turn of the century (Strauss 1987: 281). And yet, though Stoddard and others reveled in those developments that enabled one to change worlds so conveniently and swiftly, it was his construction of the Adirondacks as a wilderness for men that made a trip to this region such a craze during this period. In privileging the North Woods as an elite male space, a place where women – not to mention, African-Americans, Native Americans, and even the region's local inhabitants – were pushed to the periphery both literally and figuratively, Stoddard's work was one manifestation of a broad effort to reconcile increasing cultural anxieties about masculinity and civilization at the end of the nineteenth century. Though these marginalized groups remained a physical presence in the region, often serving as the labor that built and supported tourism's infrastructure, their erasure from the iconography of the site supported developers' efforts at reconfiguring the Adirondacks' reputation as a wilderness reserved for those at the forefront of the emerging industrial capitalist economy.[1]

Stoddard's notion of the Adirondacks as a distinctly masculine space was predicated on a select set of perceptions regarding its rugged terrain and its supposed "primitive" nature. Here those men who worked in white-collar occupations were presented with the opportunity to assert their own physical strength and virility. Here they could connect their own professional work with the achievements of those heroic figures from America's past who had explored and settled the nation. During a period when fewer occasions existed for urban men to display publicly markers traditionally associated with their gender, many found wilderness as an antidote to the ills of civilization. Though developers cast the Adirondacks as a site where men could reclaim their primal manhood through exposure to unadulterated nature and through imitation of pre-industrial activities, it is important to recognize that there was nothing absolute about this designation. In fact, the Adirondack wilderness was at times reconfigured as a feminized body. Stoddard's later campaign against those resource extraction companies who threatened to "rape the land" he had helped to establish as a wilderness resort indicates that landscape, like gender itself, should be viewed as a historical, ideological process that is always shifting to meet certain ends.[2]

The Adirondacks were suited to this type of appropriation owing in part to the fact that the region was poorly understood by those outside of it. Before the Civil War, no written document or map existed that gave one a detailed rendering of the entire area. A number of surveyors and real estate developers, sportsmen, and artists had ventured through sections of the Adirondacks before Stoddard's first trip in 1870; however, their written and visual accounts tended to provide one with only a collection of loose, often contradictory, impressions. Visitors debated a number of questions, suggesting the region's uncertain status in the public's eye. For example, were the woods safe for human recreation, or was it, as one early observer suggested, an "impenetrable mass of natural chaos"? To what extent had the influences of "civilization" entered this region? Without a text that might present a more complete and accurate description, visitors to the North Woods remained hard-pressed to conceptualize accurately the exact nature of the region. As a result, people's perceptions remained fragmented, and the landscape largely a mystery, despite its close proximity to the heavily populated Eastern seaboard. This confusion acted to blunt its potential as a tourist destination.[3]

As the first commercial photographer to make a concentrated study of the Adirondacks, Stoddard was instrumental in transforming the region into a marketable tourist product.[4] Though writers and image-makers tended to complement one another, photography was ideally suited to provide tourists with usable information better than any other communicative medium. The publication in 1869 of William H.H. Murray's popular romance, *Adventures in the Wilderness; Or, Camp-Life in the Adirondacks*, is often credited as precipitating a new era of mass tourism in the Adirondacks; however, while books such as Murray's were important in relating the proper equipment to bring and in furthering notions about the region's romantic past, only photography gave potential visitors the idea they could see the site in the present.[5] It put them at ease about the type of "adventure" they could expect. Given the medium's supposed verisimilitude, photography appeared to provide a transparent window into the Adirondacks, lessening the sense of displacement many travelers experienced. In this way, Stoddard's photographs – together with his popular guidebooks – began the process of bringing together the region as a unified, consumable whole. Sold from his studio in Glens Falls, New York, and by merchants in the stores, hotels, steamships, and railroads that serviced the region, his series of photographic views helped to standardize the Adirondacks' representation. Although Stoddard was careful not to upset the mythic associations earlier chroniclers had established, he did recast them in such a way that they became understandable and manageable to his target audience.

Unlike many American tourist destinations, the Adirondacks have come to be understood not in terms of a single iconographic subject. There is no clearly recognizable geographical site that tourists commonly associate with the region. The fact that the Adirondacks stretch over an area that is larger than the state of Massachusetts contributes also to this lack of a clear identity. While guidebook writers have often compared certain mountains and lakes with more famous examples elsewhere, the Adirondacks do not contain the awe-inspiring waterfalls, canyons, and geysers that are found at other celebrated tourist destinations. Instead, during this period, developers promoted certain male-centered activities and ideas as central to the region's iconography. Stoddard's work in representing this landscape and the people who frequented it helped to fix these popular beliefs in the American public's imagination.

Most celebrated was the notion that the Adirondacks represented an unparalleled sportsman's paradise. Here one could fish for brook trout, hunt for moose and deer, canoe through miles of backcountry lakes and waterways, and camp out under the stars at night. In an age increasingly dominated by new technologies and work patterns, the Adirondacks provided the backdrop for an elite group of men to reenact what they imagined to be a morally pure, pre-industrial existence. Entertaining these men as they did, the woods also allowed them to escape the present. Visitors wrote frequently about the sensation of reliving their boyhood or recapturing a sense of youthfulness they had missed as boys.[6] While most sportsmen were fully conscious of this fallacy, they found evidence to suggest they were traveling through a space that was removed from time. For example, visitors often compared their guides to figures from a bygone era. Certain of these woodsmen became legendary, as they represented the historic link between the past and the present. Perhaps not surprisingly, some of Stoddard's most popular views were those portraits he created of such famous guides as Orson "Old Mountain" Phelps, Alvah Dunning, Mitchell Sabattis, and Bill Nye. In his guidebook, *The Adirondacks: Illustrated*, Stoddard wrote about and pictured all of these individuals in terms consistent with this popular image. Of Mitchell Sabattis, an Abenaki tribal member who worked for many years in the region, Stoddard wrote:

> [Born] a pure blood of the tribe of St. Francis, he early took to the woods as naturally as a duck to water . . . [He] has probably seen more of wood life than any other man in the wilderness, a fearless and successful hunter and is generously admitted by other guides to have the best knowledge of the woods than any man in the country . . . The old hunter is still hale and hearty, bidding fair, with his iron constitution, to guide for many a year to come.
>
> (Stoddard 1874: 99–100)

Though American settlers had by now displaced the Abenaki and various other Native groups who once had traveled through this region on hunting expeditions, Sabattis remained a bridge to that "primitive" past.

A second idea that held great currency during this period was the notion that outdoor adventure among the Adirondacks could act as an effective medicine – both physical and moral – for those city dwellers who were overwhelmed not only by certain diseases but also by the so-called "wear and tear" that the urban environment produced. Many visitors to the Adirondacks wrote about nature's restorative power; one even established a sanatorium for tuberculosis patients at Saranac Lake.[7] However, none was more influential in legitimating the idea that the "wilderness" was good for a man's health than Silas Weir Mitchell, a Philadelphia physician who specialized in nervous disorders. In a series of popular articles first published in *Lippincott's Magazine* and then collected in an edited volume, Mitchell explored the effects that the city had on men's health. "The cruel competition for the dollar, the new and exacting habits of business, the over-education and the overstraining of our young people, have brought about some great and growing evils," Mitchell reported in an article from 1869 (Mitchell 1869: 493). For those who suffered from stress and exhaustion, he prescribed the "Camp Cure":

> The nerve disorders which come of overwork, with worry, must surely multiply with the growth of cities and the keener competitions which such growth ensures ... The surest remedy for the ills of civilized life is to be found in some form of return to barbarism ... Civilization has hurt – barbarism shall heal. In a word, my tired man who cannot sleep, or who dreams stocks and dividends and awakens leg-heavy, and who has fifty other nameless symptoms, shall try a while the hospital of the stone-carver. He shall reverse the conditions of his life. Wont to live in a house, he shall sleep in a tent, or, despite his guide's advice, shall lie beneath "the moon's white benediction."
>
> (Mitchell 1874: 192–193)

Having gone on to explain how one might go about replicating this state of barbarism, Mitchell then recommended a number of places where the overworked city-dweller could locate this remedy. Leading his list were the Adirondacks. Although there were other ideas that became associated with the region, these two notions – the Adirondacks as sportsman's paradise, and the Adirondacks as the urban male's full service hospital – were most important during this period.

While many perceived the North Woods to be a domain ideally suited for men, the historic record reveals that women participated also on these male-fashioned hunting and fishing trips. As travel accounts written by men point out, though, few had much patience or respect for those groups who included women. Also, many of these same writers looked upon those women who enjoyed their time in the woods with suspicion, believing that only men possessed the constitution and the desire for this type of recreation. As Stoddard remarked in *The Adirondacks: Illustrated*, "unless a lady is perfectly at home in the saddle, she will be more apt to wish she was 'at home' in reality" (Stoddard 1874: 59). Similarly, evidence suggests that women often came to the Adirondacks seeking the same health benefits as their male companions. However, again, many discouraged their participation in this type of wilderness-centered therapy. Mitchell and other doctors warned that the region might not be suitable for women. Having also devoted attention to nervous disorders and mental depression among women, Mitchell prescribed a remedy for women that extended the ideology of "separate spheres" into the world of medicine. In a treatment known famously as the "Rest Cure," he directed women diagnosed with this condition to cease all activities of both a mental and physical nature. Instead, they were to remain isolated from their family, preferably in bed.[8] As these examples indicate, women's marginalization at sites such as the Adirondacks reflected popular ideas concerning gender, morality, and society, while simultaneously masking deeper anxieties surrounding American masculinity and male sexuality.

Another manifestation of this desire to keep the Adirondacks an exclusive male enclave was the growth of alternative recreational spaces for women. As New York City newspaper editor Thomas Bangs Thorpe argued, the Adirondacks were and should remain a space for men:

> We do not consider the wild woods a place for the fashionable ladies of the American style; they have, unfortunately, in their education, nothing that makes such places appreciated, and no capability for physical exercise that causes the attempt to be pleasantly possible.
>
> (Thorpe 1869: 565)

According to Thorpe and others, Saratoga Springs, though, was a destination that would perfectly accommodate married women whose husbands and sons had ventured off into the Adirondacks. With its famous hot springs and its opulent hotels and restaurants, Saratoga Springs, located halfway between Albany and Glens Falls, was described in the popular literature as the epitome of indulgence and idleness. According to prevailing ideas of acceptable female recreation, such a resort was, therefore, ideal for women. The following statement promoting travel to the North Woods from an account published in 1872 makes clear the well-defined distinction between these two sites:

> You monarchs of the business world, whose wealth can hardly be estimated, instead of dragging your weary bodies in the train of your pampered sons and daughters to Saratoga, Long Branch, and other earthly Pandemoniums, why not, O, why not, at one half of the expense, and one hundred fold more benefit, cut loose from fashion's moorings, fly off to these grand old forests, these sylvan lakes, these singing streams, and for once before you die enjoy one summer of

sweet content . . . Ten dollars a day at Saratoga for the necessities of life, and Satan only knows how much for luxuries. Come here for one-fifth the amount and treat yourself to something not only novel, but innocent and health-giving.

(Smith 1872: 94–95)

During the last three decades of the nineteenth century, evolving relationships between men and women in American society played a key role in reshaping the topographies of various tourist destinations throughout New York and the nation at large.

Though this essay is predominantly aimed at considering the manner in which gender has been inscribed on a specific landscape, it is also important to acknowledge that prevailing class and racial hierarchies also informed popular perceptions regarding the Adirondacks. The representation of the region's local inhabitants in visitors' travel journals is just one place where the contested nature of class relations comes into focus. Since the end of the American Revolution, the Adirondacks had been home to an array of fur trappers and subsistence farmers. During the first half of the nineteenth century, when mining and timbering operations were first established in the area, these individuals provided a much-needed labor supply. Although many writers excluded them altogether from their travel accounts, those who did observe them described their existence as nothing short of barbaric. The following excerpt from a travel account published in 1881 is representative of the manner in which these local "year-round" residents were portrayed:

To the year-round resident in the wilderness the world is bounded by Canada on the north, Plattsburg on the east, Boonville on the south, and Malone on the west. All that lies beyond this clearly defined territory is dim, shadowy, and uncertain. The end and aim of life is to "guide" in summer, and "log" in winter. Nowhere else on the face of the earth is it so easy to divide all people into classes at once so distinct and comprehensive.

(Cook 1881: 86)

Only visible at the periphery of tourists' travel accounts, these individuals likewise had no place in the iconography of this exclusive men's club, though they proved an invaluable source of cheap and available labor.

For tourist promoters, the historic presence of Native Americans and African-Americans in the Adirondacks also presented both a problem and an opportunity. While not tolerant of their physical inclusion, they did work to embed these groups into the emerging mythology of the region. In particular, notable landmarks were given names taken from native languages, and "ancient" tribal history became the subject of popular stories. When youth camps were started around the turn of the century, the activities in which young boys participated were often bastardized forms of traditional Native practices. "Playing Indian" was common for adults and children throughout the Adirondacks.[9] In the context of tourism, this type of appropriation presented visitors with the opportunity to remove themselves from the realities of their everyday world and provided them with an outlet to define their manhood. Yet, except for guides like Mitchell Sabattis, the actual tribal members who once claimed these lands were no longer a vital presence, having been removed from the area to distant reservations. These descriptions of a structured hierarchy, strict exclusions, and redefinitions are traces of the moral framework enacted within a specific landscape.

Though not present in great numbers, African-Americans had a historic connection to the Adirondacks too. This fact was largely due to the work of John Brown, the famed abolitionist who settled with his family in the Adirondack community of North Elba in 1849. As he left for Kansas in 1855, Brown only lived there for six years; however, during this time, North Elba became a regular stop on the Underground Railroad. Brown also helped many fugitive slaves establish homes in the community. After his death at Harper's Ferry in 1859, Brown's body was returned to North Elba for burial. Although his grave became a popular landmark in the Adirondacks, the African-Americans whom he had assisted almost completely disappeared out of view in the popular literature. Stoddard's remarks in his guidebook indicate that he had little respect for either Brown or African-Americans. About Brown, he wrote: "[his] presence [in Kansas] was marked by dissensions and bloodshed," and he "urged men on to murder in the name of freedom and [to] read his Bible all the time . . . A fanatic he undoubtedly was" (Stoddard 1874: 67). While writing about North Elba, he described "one of John Brown's pet lambs," who was now working as a carriage driver in the tourist trade, in equally disparaging language (Stoddard 1874: 70).[10] Such comments about African-Americans, though, are rare in both guidebooks and travel journals to the Adirondacks, an indication of this community's marginal status in the region.

These themes of exclusion and elitism play a central role in Stoddard's landscape photographs. An examination of the subjects he selected and the manner in which he composed and marketed his views reveals his participation in reconfiguring the Adirondacks to reach the urban male audience that he and other developers targeted. His desire to fashion a career as a landscape photographer in the service of the burgeoning tourist trade also reflects a personal desire to move beyond what he perceived as the effeminate business of portrait photography. Having grown uninterested in such work not long after opening his first commercial studio, Stoddard chose to focus his energies on photographing a subject he considered grander in reputation and therefore more fitting to a young man with artistic ambitions (Adler 1997: 59–64). Yet, though he constructed views that figured the Adirondacks as a landscape far removed from the realities of the city, he rarely photographed a subject in which mankind was totally absent. As the following images illustrate, Stoddard frequently inserted both individuals and signs of the dominant culture's presence into these views to instill moral significance into his chosen landscapes.

It was not only the inclusion of individuals into his landscape photographs but also the poses they struck and the activities in which they were engaged that acted to redirect the meanings associated with wilderness. By regularly photographing men involved in such pursuits as canoeing, fishing, hunting, and camping, Stoddard represented the Adirondack landscape quite literally as the nostalgic playground of the elite male tourist. In *Indian Head, Ausable Pond* (Figure 7.1), for example, he pictured a seemingly uninhabited setting with a lone canoeist in the foreground. This man has

Figure 7.1 Indian Head, Ausable Pond, Seneca Ray Stoddard, *c.*1880

Photograph from the collection of The Adirondack Museum, Blue Mountain Lake, NY

Figure 7.2 Game in the Adirondacks, Seneca Ray Stoddard, 1889

Photograph from the collection of The Adirondack Museum, Blue Mountain Lake, NY

stopped rowing for the moment and has focused his attention on a fish he has just caught. Stoddard figured him in a scene that seems far removed from the "wear and tear" associated with the urban environment. And yet, from the photograph's title, the viewer learns that this man is not completely alone. Indeed, from this perspective, one is able to make out the outline of a human head from an outcropping of rock in the photograph's background. Though Stoddard did not discover the "Indian Head," his photographs were responsible for popularizing this landmark. Sold in a variety of different sizes and formats at locations throughout the region, this view helped to emphasize to the consuming public the supposed "primitive" nature of this landscape and the historic connection that Native peoples had to this place.[11] The Indian having now been turned to stone, it is a white man in a canoe who now plies these "ancient" waters. This and other photographs suggested to tourists that they had the moral right to define themselves as the heroic occupants of these lands.

Whereas Stoddard used such figures as the canoeist on Ausable Pond to complement the larger landscape subject, he created other views in which the activities of these individuals were more pronounced than the scenery itself. Such photographs as *Game in the Adirondacks* (Figure 7.2), for example, features four men playing cards around a campfire. Stoddard created this night-time photograph with the aid of flash powder, a technique with which he experimented throughout his career. He also used gouache to highlight areas in the negative that were inadequately articulated. His immediate subject is a card game being played around the campfire. Four rustically dressed men sit on logs around an overturned wicker basket, and a crude lean-to serves as a backdrop to the scene. Though Stoddard has illuminated this scene artificially, a lantern hangs from a pole on the lean-to. This carefully composed scene further points to the simple pleasures and male companionship that one could find outdoors in the North Woods. At the same time, the photograph's title engages the viewer in a not-so-subtle, yet revealing joke. Indeed, the "game" that most visitors associate with the Adirondacks are those animals that men hunt for sport. By using this title to describe a group of men playing cards, Stoddard purposely speaks to a select male audience for whom such activities are meaningful. Following a popular trend for such campfire scenes, he produced a number of views similar to *Game in the Adirondacks*. More iconographic than any natural landmark in the Adirondacks, these romantic vignettes were influential in redefining the notion of wilderness as the sportsman's dominion.

Although these elite tourists came to the Adirondacks under the premise of replicating a pre-industrial existence, they also wanted to know the area was packaged in such a way that it was able to be understood and experienced in relative comfort and safety. For individuals about to spend the time and money to make such a visit, this landscape needed to contain signs that developers were looking after its material progress. In leaving behind morally suspect "civilization," Adirondack tourists believed they were only getting rid of those negative qualities concerning life in the American city. The heat, the dirt, and the crowded streets were what visitors hoped to escape – not to mention their jobs, their families, and their individual concerns. Though many were prepared to "rough it" by sleeping under a thatched lean-to, others still longed for many of the material comforts to which they had long since grown accustomed. Through his photographs, Stoddard not only confirmed that one did not have to sacrifice personal comfort but also revealed that the Adirondacks possessed institutions which were downright luxurious. Towards this end, he photographed many of the hotels that were built during this period. In photographs such as *Hotel Ampersand at Saranac Lake* (Figure 7.3), Stoddard hoped to convey the rustic elegance that had grown up in the Adirondacks. As he had done in the photograph *Indian Head, Ausable Pond*, he placed in the view's foreground the familiar motif of a man in a canoe. Yet, in this image, the five-storied Hotel Ampersand rises up in the background, rather than the "Indian Head," signifying the material prosperity of those who now dominated the region.

Naturalizing this vision for the Adirondack landscape was not a casual enterprise for Stoddard. Not only did he have to carry great quantities of heavy equipment into the field, but he also had only a limited window

Figure 7.3 Hotel Ampersand, Saranac Lake, Seneca Ray Stoddard, 1891
Photograph from the collection of The Adirondack Museum, Blue Mountain Lake, NY

of opportunity to create views. In an article from an 1877 issue of *The Philadelphia Photographer*, the leading professional journal of its day, Stoddard described the labor that went into such landscape views: "I do not attempt to work alone; have found that the hours of suitable weather in the course of the year were too few to waste." On many occasions Stoddard brought along his brother-in-law Charles Oblenis to man the dark-box, while he attended to the camera itself. Often he also "secured the services of some boy, sometimes two, . . . and with extra inducement for quick time set him to running with plateholders between the camera and the dark-box" (Stoddard 1877: 147). The photograph of Stoddard creating views from atop an impermanent wooden platform that he built in the middle of a lake suggests the great lengths he went to in constructing his photographs (Figure 7.4). In no way were the images he produced informal snapshots; nor was Stoddard only recently discovering this landscape. Though many of his photographs convey the sense that he was encountering these different spaces for the first time, this feeling was one he achieved through much work and the assistance of the latest photographic innovations.

Figure 7.4 Stoddard's Photographic Platform, Seneca Ray Stoddard, *c.*1880

Photograph from the collection of the State Library of New York, Albany, NY

Despite Stoddard's careful efforts to select the extent of civilization's impact on the landscape, the history of the region makes clear that industrial capitalism had long underlain this "wilderness." The first iron works having been established in the Adirondacks in 1809, the mining and lumbering industry rose to prominence concurrently with the tourist trade. For most of the nineteenth century, those who wrote about the region celebrated the idea that the land could provide for both the tourist and the industrialist. Given the tremendous size of the area, many believed that mining and timbering would not spoil its beauty. In fact, some argued that the region's industrial development was essential to the continued growth of the tourist trade. An editorial from *The New York Times* in 1864 made this argument explicit. Arguing on behalf of a plan to build a railroad line through the Adirondacks, the writer explained:

> The Adirondack region will become a suburb of New York. The furnaces of our capitalists will line its valleys, and create new fortunes to swell the aggregate of our wealth; while the hunting-lodges of our citizens will adorn its more remote mountain-sides and the wooded islands of its delightful lakes. It will become to our whole community, on an ample scale, what the Central Park now is on a limited one ... In spite of all the din and dust of furnaces and foundries, the

Adirondacks, thus husbanded, will furnish abundant seclusion for all time to come; and will admirably realize the true union which should always exist between utility and enjoyment.

("Adirondack" 1864: 4)

Although the vision of a railroad through the Adirondacks to the St Lawrence River was never realized, industrial developers continued to expand their enterprises in the region.

However, the peaceful co-existence of the tourism, mining, and timber industries did not last. As the network of development covered increasing amounts of land towards the end of the century, conflicts arose over the proper manner in which to utilize the area's resources. Given Stoddard's investment in the tourist trade, the promoter was not slow in involving himself in this debate as an outspoken supporter of the burgeoning conservation movement. Having worked to transform the Adirondacks into a marketable tourist product, Stoddard took advantage of his privileged position as the region's leading commercial photographer to ensure that industry's future. Photography's ability to select and exclude had been influential in constructing a specific vision of tourism in the Adirondacks; on behalf of the conservation movement, he again used photography to shape the evidence in this debate. In doing so, he moved away from his earlier idea of the Adirondacks as a rugged, masculine terrain and came to represent this same landscape as a delicate world that required vigilant protection from the abuses of outside interests. In the Adirondacks and at similar "wilderness" sites around the nation, conservation was increasingly couched as a patriarchal effort to safeguard a feminized object of desire.

Working as the head of the New York State Survey's Photographic Division beginning in 1878, Stoddard created photographs that suggested to viewers that the continued extraction of natural resources would spell future trouble. Having read George Marsh's groundbreaking work on resource management, *Man and Nature*, he understood the long-term consequences that resulted from the destruction of certain ecosystems. In particular, he was struck by Marsh's observation that deforestation led to the eventual drying up of watersheds. As the headwaters of the Hudson River were located in the Adirondacks, Stoddard and others grew concerned that the timber industry represented a threat not only to the local area, but also to everyone downstream. The future of New York City's water supply was at stake if lumbermen continued to clear cut the North Woods. Furthermore, the desiccation of the landscape heightened the possibility of wild fires, a horror that portions of the state had experienced in the past.

An article written by Stoddard in 1885 for *Outing*, a popular magazine dedicated to outdoor recreation, reveals his thoughts concerning the Adirondacks' precarious health. Entitled "The Head-waters of the Hudson," the essay described in romantic terms the location where the Hudson River began and then ruminated on its uncertain future. By his account, the region was under attack by those who longed to extract greater amounts of its natural resources. As he explained in the following description of Lake Henderson and Indian Pass, the Adirondacks were meant to be left untouched:

> Of old it [Lake Henderson] furnished power for the iron-works, and, like its sister lakes that have been dammed for man's use, has sur- rounded itself with a hideous border of dead trees. Around it are steep mountains coming close to the water's edge; from its head we look up through grand old Indian Pass . . . words can convey no adequate idea of this wild pass; pen cannot describe it; the pencil only faintly suggests its grandeur. It is little different today from what it was a thousand years ago, or will be a thousand years hence, if the forests are not destroyed in sheer wantonness; for, save its wild beauty, there is little to attract man, no soil to till, no mines to drill, no timber that will pay for its own destruction.
>
> (Stoddard 1885: 61)

Like a protective father, Stoddard drew upon moral stereotypes and demon- ized anyone who attempted to sully this supposedly virgin landscape. At the same time, he equated lands damaged by resource extraction with despoiled women. To Stoddard, such industries had no place here.

During his work with the State Survey, Stoddard spent most of his time creating images that would assist others in writing about and mapping the Adirondacks. Although most of these views revealed the region's beauty, he also assembled a series of photographs that illustrated the damage being done by the timber industry. In *Drowned Lands of the Lower Raquette, Adiron- dacks* (Figure 7.5), for example, he presented a bleak picture of a once-thick forest now wasted on account of a lumber company's greed. His article from *Outing* described what had happened to this land:

> It [the dammed waters of the Raquette River] covers the lowlands over, and every green thing perishes, and then, when the summer is at its height, and the sun shines hottest through the leafless trees, the water is drawn down, forsooth, to furnish some distant saw-mill with power, leaving the dead, exposed, decaying vegetable matter and stagnant pools all slimy and festering in the heat. The beautiful valley, once fair and sweet as Eden, has become a foul, malaria-breeding pit.
>
> (Stoddard 1885: 63)

In describing the need to protect "fair and sweet" nature from outside threats, Stoddard used traditional notions of womanhood to invest the con- servation movement with a larger cultural significance. This photograph and others showing similar destruction were the first visual documents that

Figure 7.5 Drowned Lands of the Lower Racquette, Adirondacks, Seneca Ray Stoddard, 1888

Photograph from the collection of The Adirondack Museum, Blue Mountain Lake, NY

drew attention to this environmental catastrophe. As with his earlier views, he took great care in constructing this series of photographs for maximum moral impact. In this example, note how he used a human figure in the foreground scene to heighten the sense of melancholy associated with this place. Looking out at the wasted landscape, this individual acts as the representative victim of this lumber company's misdeeds. Also leading the viewer's gaze through the picture plane is a stagnant water pool that begins in the right foreground and extends diagonally into the distance. Unlike his views of unspoiled nature, this photograph highlights a body of water that is not restorative, but instead potentially deadly. These photographs transported his audience to the site itself, and created the impression that the region deserved to be set aside as a park. Although it would take a series of steps over a period that lasted more than two decades, the region was finally declared a state park in May 1892.

In writing about the phenomenon of tourism, historian Dean MacCannell has observed how "tourists may attempt to discover or reconstruct a cultural heritage or a social identity" (MacCannell 1976: 13). To the many men who went into the North Woods on holiday, this landscape presented them with an opportunity both to reclaim something real or imagined from the past and to reposition themselves in the larger culture. Whether creating a marketable commodity or assisting the conservation movement, Stoddard's work in the Adirondacks prepared and preserved the region as a place that privileged this exclusive clientele. His photographs naturalized this vision for the land at a time when other competing groups and institutions often contested this new usage. Yet, without a communicative medium like photography to counter Stoddard and other developers, women and other marginalized groups were handicapped in pursuing this resistance. Desirous of controlling the view, Stoddard helped to precipitate wide-ranging changes that affected both the land itself and the American public's understanding of wilderness.

Notes

1 In *Manliness & Civilization: A Cultural History of Gender and Race in the United States, 1880–1917* (1995), Gail Bederman writes articulately about the sources of the turn-of-the-century crisis in American masculinity and the varied responses this anxiety provoked.

2 In addition to Bederman's *Manliness & Civilization*, see also Carolyn Merchant, *The Death of Nature: Women, Ecology, and the Scientific Revolution* (1980); and Annette Kolodny, *The Lay of the Land: Metaphor as Experience and History in American Life and Letters* (1975).

3 The Adirondacks' borders were never defined until the New York State Legislature established them in 1892 at the time of the area's designation as a state park. Because a majority of the lands within the Park were privately owned, much confusion over its borders continued even after the official pronouncement. The history of the Adirondacks has been chronicled by various individuals. In particular, see Frank Graham, *The Adirondack Park: A Political History* (1978); Paul Schneider, *The Adirondacks: A History of America's First Wilderness* (1997); and Philip Terrie, *Forever Wild: Environmental Aesthetics and the Adirondack Wilderness Preserve* (1985). The cultural history of the region during the period of its settlement in the nineteenth century is best outlined in Philip Terrie, *Contested Terrain: A New History of Nature and People in the Adirondacks* (1997).

4 A few individuals had created photographs in the Adirondacks before Stoddard. For example, William James Stillman, the Boston artist who founded the first American art magazine, *The Crayon*, brought a camera with him on his trip into the Adirondacks in the summer of 1858. However, the goal of Stillman and other early practitioners was not commercial sale but personal experimentation with the new visual medium.

5 First to proclaim the importance of Murray's book was fellow Bostonian and famed abolitionist Wendell Phillips. Its publication "kindled a thousand camp fires and taught a thousand pens how to write of nature," according to Phillips. Graham used this passage and other evidence to suggest that "the opening of the Adirondacks to a great flood of men, women, and children can be dated" to the publication of Murray's book (Graham 1978: 25–26). However, while the book did enjoy commercial success at first, largely as a result of the fact that no other book of its time provided as much detailed information about how to negotiate a trip to and through the Adirondacks, it was not the first popular guidebook to the area. Nor did Murray's book remain in print long. Soon after its publication in 1869, disputes about some of Murray's conclusions erupted in the

popular press. These issues and his controversial resignation as the minister of Boston's Park Street Church in 1874 contributed to the decline in the book's importance. The appearance of Stoddard's *The Adirondacks: Illustrated* in 1874 and Edwin Wallace's *Descriptive Guide to the Adirondacks* in 1872 ushered in the era of the standardized annual guidebook to the area and soon left Murray's book an outdated collection of romantic stories.

6　Ralph Waldo Emerson was one of many writers during this period who commented on nature's ability to transform men into boys again. See, for example, his poem, "The Adirondacs," where he wrote about a party of grown men sleeping under the stars: "They fancied the light air / That circled freshly in their forest dress / Made them to boys again. Happier that they / Slipped off their pack of duties, leagues behind, / At the first mounting of the giant stairs" (Emerson 1867: 45–46).

7　Dr Edward L. Trudeau first visited the Adirondacks in 1873 as a tuberculosis patient. His recovery was so impressive that he began recommending his own patients to do as he had done. In 1884 he built his first sanatorium in order to cater to this new class of Adirondack visitors (Graham 1978: 45–52).

8　The feminist reformer Charlotte Perkins Gilman endured the "Rest Cure," and later wrote about this period of suffering in her short story, "The Yellow Wallpaper." Fourteen years after he published his work on the "Camp Cure," Mitchell acknowledged the fact that women too could benefit from the "Camp Cure." "Wishing now to correct my error of omission," Mitchell dedicated a chapter of his popular medical journal, *Doctor and Patient*, to the subject of "Out-Door and Camp-Life for Women" (Mitchell 1888: 155–177).

9　As Philip Deloria has demonstrated in *Playing Indian* (1998), the non-Native appropriation of Native American history and culture has been a centuries-old American tradition.

10　Jeanne Adler's biography of Stoddard concludes that his "views on race reveal an intolerance" about not only African-Americans but also Irish and Italian immigrants and Jewish-Americans (Adler 1997: 88).

11　Although some have credited Stoddard with discovering the "Indian Head," many writers had described it before his first trip into the Adirondacks. Alfred Street – in his 1869 travel account *The Indian Pass* – noted its presence and recounted a local legend about the specific Native American whose head was depicted there. The story ends by stating that the tribe to which he belonged had "vanished from the region" (Street 1869: 115).

References

"Adirondack" (1864) *The New York Times*, 13 (August 9): 4.

Adler, J. (1997) *Early Days in the Adirondacks: The Photographs of Seneca Ray Stoddard*, New York: Abrams.

Bederman, G. (1995) *Manliness & Civilization: A Cultural History of Gender and Race in the United States, 1880–1917*, Chicago: University of Chicago Press.

Cook, M. (1881) *The Wilderness Cure*, New York: William Wood.

Deloria, P. (1998) *Playing Indian*, New Haven CT: Yale University Press.

Emerson, R.W. (1867) *May-Day and Other Pieces*, London: George Routledge.

Graham, F. (1978) *The Adirondack Park: A Political History*, New York: Knopf.

Kolodny, A. (1975) *The Lay of the Land: Metaphor as Experience and History in American Life and Letters*, Chapel Hill NC: University of North Carolina Press.

MacCannell, D. (1976) *The Tourist: A New Theory of the Leisure Class*, New York: Schocken Books.

Merchant, C. (1980) *The Death of Nature: Women, Ecology, and the Scientific Revolution*, New York: Harper & Row.

Mitchell, S.W. (1869) "Wear and Tear," *Lippincott's Magazine*, 4 (November): 493–502.

—— (1874) "Camp Cure," *Lippincott's Magazine*, 14 (August): 192–200.

—— (1888) *Doctor and Patient*, Philadelphia PA: J.B. Lippincott.

Schneider, P. (1997) *The Adirondacks: A History of America's First Wilderness*, New York: Henry Holt.

Smith, H.P. (1872) *The Modern Babes in the Wood; or, Summerings in the Wilderness*, Hartford CT: Columbian Book.

Stoddard, S.R. (1874) *The Adirondacks: Illustrated*, Albany NY: Weed, Parsons & Co.

—— (1877) "Landscape and Architectural Photography," *The Philadelphia Photographer*, 14 (May): 146–148.

—— (1885) "The Head-waters of the Hudson," *Outing*, 7 (October): 58–63.

Strauss, D. (1987) "Toward a Consumer Culture: 'Adirondack Murray' and the Wilderness Vacation," *American Quarterly*, 39 (Summer): 270–286.

Street, A. (1869) *The Indian Pass*, New York: Hurd & Houghton.

Terrie, P. (1985) *Forever Wild: Environmental Aesthetics and the Adirondack Wilderness Preserve*, Philadelphia PA: Temple University Press.

—— (1997) *Contested Terrain: A New History of Nature and People in the Adirondacks*, Syracuse NY: Syracuse University Press.

Thorpe, T.B. (1869) "The Abuses of the Backwoods," *Appleton's Journal of Popular Literature, Science, and Art*, 2 (December 18): 564–565.

Wallace, E. (1872) *Descriptive Guide to the Adirondacks*, Syracuse NY: Edwin Wallace.

Part III

Memories of home

8 Mapping the Amazon's salon

Symbolic landscapes and topographies of identity in Natalie Clifford Barney's literary salon

Sheila Crane

> Yet we must take courage and examine this house. So vague an inheritance has its responsibilities. We must not appear to know. Let us ask assistance from passages, from rooms an explanation. Even the memories they contain must become ours.
>
> (Natalie Clifford Barney)

In the early decades of the twentieth century, writer, aphorist, and salon host Natalie Clifford Barney was a prominent figure among both expatriate American and French literary communities in Paris. Her notoriety was initially assured by her association with the aging doyen of the French literary scene, Remy de Gourmont, whose correspondence to Barney was serialized in the literary review *Mercure de France* and subsequently published as *Lettres à l'Amazone* (*Letters to the Amazon*) in 1914. Although Barney had long been nicknamed "the Amazon" by her intimates, Gourmont's publications solidified her literary identity in these terms even as they helped to further her own writing career. More than for her own writings, however, Barney was renowned for her Parisian literary salon where Anglophone and Francophone luminaries such as Gertrude Stein, Robert de Montesquiou, Isadora Duncan, Radclyffe Hall, André Gide, Ezra Pound, and Dolly Wilde could often be found in attendance.

Even before establishing a formal literary salon, Barney had regularly staged performances and dramatic readings with her friends in the extensive walled gardens surrounding her home in Neuilly on the outskirts of Paris. In 1909, Barney moved from Neuilly to Paris, installing herself at 20 rue Jacob, around the corner from the abbey church Saint-Germain-des-Prés in the Latin quarter. Barney thus repositioned herself and her residence in the heart of the established literary and artistic topography of the Left Bank. In this setting, she invited members of Parisian high society along with writers, scholars, artists, and friends to her weekly Friday late afternoon gatherings. Barney's salon was most active in the 1910s and

1920s, although it continued to meet during the following decade until the beginning of the Second World War (Wickes 1976; Causse 1980; Benstock 1986; Rodriguez 2002). Barney's house and surrounding gardens, where a miniature Greek temple known as the Temple de l'Amitié (Temple of Friendship) stood, comprised the landscape within and against which Barney framed both herself and the community she gathered there.

Twenty years after her move to 20 rue Jacob, Barney published a drawing of her salon as the frontispiece to her 1929 collection of literary portraits, *Aventures de l'esprit* (*Adventures of the Mind*)[1] (Figure 8.1). The drawing featured the names of guests who attended her salon, several of whom were also described in the essays that followed. Barney's drawing has generally been dismissed as an unsophisticated sketch of her salon that is only remarkable insofar as it provides a useful record of visitors to her salon (Gatton 1988: 52; Chalon 1992: 186–7; Elliott and Wallace 1994: 158). I argue instead that through its mapping of both the physical structure of the salon and relationships among people within these spaces, the drawing provides much more than merely a list of names. Instead, it is a striking depiction of the symbolic landscape of Barney's salon as it was both imagined and experienced.

In this chapter I examine the relationship of Barney's drawing to a range of artifacts – including architectural structures, gardens, portraits, and other significant objects within these spaces – that comprised the landscape of her salon and to the performative salon ritual staged within these spaces. I am particularly interested in excavating the process through which Barney re-appropriated and reinvented the existing landscape of 20 rue Jacob over the course of her almost sixty-year tenure there.[2] The reiteration of the salon ritual was central not only to Barney's own self-fashioning; it also became a means of proposing and solidifying a community organized in relation to the salon's host and to the spaces of her home. 20 rue Jacob was not merely the location of her residence and literary salon but a landscape through which Barney self-consciously envisioned an alternative ethic, challenging normative literary institutions, social practices, gender definitions, and affective relationships.

Entitled "Le Salon de l'Amazone" ("The Amazon's Salon"), the drawing that accompanied Natalie Barney's 1929 collection of literary portraits seems to have been created by the author herself (Wauthier 1963: 15; Lottman 1969: 2; Chalon 1992: 186). Most of the names of salon guests are contained within the outlines of a roughly rectangular area that dominates the drawing, although a few appear to spill out of the openings representing the entrance and exit doors along the left edge and into the garden at the bottom. In the upper portion of the drawing, the outlined façade of the structure known as the Temple of Friendship that stood in a second garden behind the main house is flanked on either side by a row of names. A meandering line, with Barney's literary moniker "the Amazon"

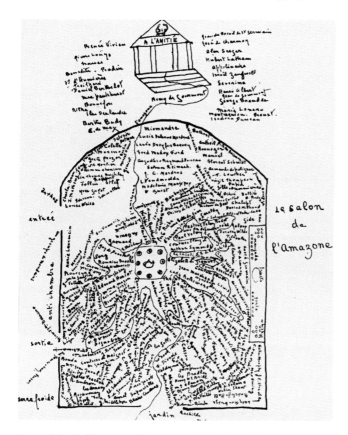

Figure 8.1 "Le Salon de l'Amazone," Natalie Clifford Barney, frontispiece to
     Barney, 1929

Photograph: Bancroft Library, University of California, Berkeley

inscribed at various points along it, winds its way through the mass of
names and up the steps of the Temple of Friendship.

In the drawing, Barney represented the boundaries of what appeared to
be the ground-floor of her two-story, seventeenth-century *pavillon*, a small,
semi-detached dwelling that stood at the far end of the cobblestone courtyard behind the four-story apartment building fronting the street at 20 rue
Jacob (Figure 8.2). The abbreviated rectangular outlines within which the
names are arranged suggest the diagrammatic conventions of an architectural plan and thus imply that Barney was interested in conveying a credible
likeness of the actual reception space of her salon on the ground floor of
her residence.[3] Although the main sitting room comprised the structure's
formal reception space, during salon meetings guests mainly congregated
in the adjacent dining room that was dominated by an octagonal table.
While neither the actual plan of these two rooms nor the location of the

Figure 8.2 Barney's *pavillon* at 20 rue Jacob in Paris, *c.*1920

Photograph: Caisse Nationale des Monuments Historiques et des Sites, Paris

entry hall and kitchen that also stood on the ground floor is depicted in Barney's drawing, certain aspects of the drawing are nevertheless faithful to and expressive of specific architectural details. For example, the rounded line at the top of the drawing denotes the actual contours of the Rotunda, a semi-circular alcove in the dining room that formed the backdrop for readings and performances delivered at many salon gatherings. Rather than attempting to communicate the exact form or the entire landscape of 20 rue Jacob in her drawing, Barney focused on select elements that were in turn significantly revised in her depiction of them. In this way, she fused elements of the two rooms where her salon was held – the curve of the Rotunda and the table in the dining room with the main entrance door to the sitting room – into a single, unified space.

Given this representational strategy, Barney's drawing must be understood as less a documentary rendering than an experiential mapping of her salon. Even though the ground floor was divided into two separate rooms, the sitting room and the adjacent dining room were perceived by salongoers as an uninterrupted space. In the eyes of her guests, this spatial continuity was further emphasized by the fact that both rooms were covered with the same fading red wall coverings, crowded with a similarly haphazard collection of chairs along their walls, and reflected back into one another in the mirrors they each contained. Descriptions by Barney's visitors repeatedly mention the difficulty they had discerning the actual dimensions of these rooms, due in part to the shadowy light filtered by the trees in Barney's two overgrown gardens and the ivy-covered windows of her *pavillon* (Wickes 1975: 110; Cassou 1962: 34). Thus while the diagrammatic form of her drawing does not delineate the exact structure of these rooms, it does convey a sense of how they were used during salon gatherings.

While the interior space that Barney delineated in the drawing was almost entirely filled with names, a few key objects were also included. Along the right edge of the notional room, beverages and fruits were carefully arranged atop a narrow buffet table. At the center of the drawing, an eight-sided form punctuated by the outlines of eight plates represented Barney's famous octagonal table. Whereas these two elements might seem at first to be inconsequential details, the choice to include them was significant. Together, the two tables referenced the unique traditions of 20 rue Jacob; in contrast to salons held by her contemporaries, guests at Barney's Friday gatherings were always invited to sit down and eat (Wickes 1975). As the central organizing element within the drawing, the octagonal table effectively formed the structural heart of the salon. At the center of the table, a teapot was delineated in profile instead of as viewed from above. Here Barney betrayed the spatial logic of her drawing in order to signal this object's importance. The drawing thus illustrates what Janet Flanner later remembered, that at 20 rue Jacob "we all clustered around the teapot" (quoted in Wickes 1975: 130). In this way, select objects that held particular importance to Barney and her salon were transformed in the drawing into iconic emblems.

Given its location at the top of the drawing, the Temple of Friendship clearly also played a critical role. Hidden behind Barney's house, this striking architectural folly was a remnant of what had once been much more sprawling gardens (Champion 1932; Melicourt 1975). Although the exact origins of Barney's temple are unknown, similar structures were constructed in England and France – beginning in the early eighteenth century – that were often dedicated to specific relationships or more generally understood as sites of amorous liaisons (Macon 1908; Dams and Zega 1995). For example, the Temple of Love at Versailles that Marie-Antoinette commissioned for the gardens surrounding the Petit Trianon became the centerpiece around which she orchestrated her *fêtes*, masked balls, and outdoor entertainments. In pamphlets criticizing Marie-Antoinette's libertine practices, including her supposed affairs with women, the Petit Trianon and extensive gardens where the Temple of Love stood were explicitly characterized as dangerously libidinous spaces (Castle 1992). Whether or not Versailles was a direct reference for Barney, she was invested in the broader historical tradition in which structures like her garden folly were understood as sites of and symbols for friendship, love, and erotic attachments.

At 20 rue Jacob, the front façade of Barney's small Greek temple featured four Doric columns supporting an entablature inscribed with the phrase "A l'Amitié" ("To Friendship") and surmounted by a garland emblazoned with the letters DLV (Figure 8.3). Although salon gatherings were never held inside the temple, its doors were often left open so that guests could take a brief tour and pay homage to the "cult" of friendship that the building evoked (Causse 1980: 133). Such associations were further solidified by the fact that Barney used the small room inside the temple for dinner parties limited to her most intimate friends. In her drawing of "The Amazon's Salon," Barney reduced the temple to its basic elements: the four columns, the inscriptions, and the stairs leading to the entrance. Given that this façade could be seen through a door near the Rotunda that opened

Figure 8.3 Temple de l'Amitié (Temple of Friendship) in Barney's rear garden at 20 rue Jacob in Paris, *c.*1920

Photograph: Caisse Nationale des Monuments Historiques et des Sites, Paris

onto the rear garden, the drawing represented a view of the temple's elevation as it would have been seen from inside the dining room. By invoking
the visual relationship of the garden folly to the interior reception space
of her salon, Barney's drawing thereby emphasized the temple's centrality
to the experience of her Friday gatherings.

An anecdote Barney related in *Adventures of the Mind* suggests that Barney
herself was aware of and actively invested in such a process of isolating
particular objects within her immediate environment in order to infuse
them with heightened meaning. Her description of the art historian Bernard
Berenson's first meeting with Pierre Louÿs, a formal introduction that
Barney herself orchestrated, is particularly revealing in this light. According to Barney, upon first seeing the interior of Louÿs's house, Berenson
expressed his complete bewilderment in the face of what he saw as Louÿs's
absolute lack of taste in the selection of paintings and objects displayed
there. Looking back on this incident, Barney explained that the absolute
incommensurability between Berenson's aesthetic evaluations and those
of Louÿs was, to her mind, perfectly logical. Whereas Berenson was
interested in exercising a purely visual aesthetics, Louÿs was:

> like all men of letters . . . [who] have a whole other understanding of
> what they have before their eyes than other people with underdevel-
> oped taste. They proceed less by aesthetic evaluations than by symbols.
> Their relationship with external objects become symbols, almost
> mystical ones.
>
> (Barney 1992: 43)

I would argue that, as a "woman of letters," Barney certainly counted
herself among the proponents of such a practice of willfully attaching symbolic meaning to selected objects, people, and sites within her surrounding
environment.

By isolating critical elements within the depicted space of her salon and
in relationship to one another, Barney created a remarkable diagram of
her own subjective interpretation of the variegated topography of 20 rue
Jacob. By highlighting key features of its architectural landscape (such as
the tea table and the temple), Barney figured these elements as symbolic
of the salon itself, its temporalities, and its ritualized interactions. Thus, in
spite of the rough quality of its outlines, the drawing represents a carefully
constructed symbolic landscape.

While the drawing focused on the relationship between interior rooms
and exterior gardens at 20 rue Jacob, it did not indicate how Barney's
house, temple, and gardens were organized in connection with the surrounding urban landscape. The composition instead emphasized the
boundaries of Barney's salon and the self-contained nature of its architectural spaces that were effectively hidden at the end of a private courtyard
and emphatically shielded out of view from the rue Jacob. Nevertheless,

the prominent literary and social figures named in the drawing evidence Barney's dramatic transformation of her notionally private dwelling into an important site within the broader topography of the Parisian literary scene. That is, even as Barney detached her home from its actual physical surroundings in both the drawing and the private ritual of the salon, she effectively reorganized the Parisian literary landscape within the space of her dining room.

The profusion of names scrawled in the drawing collectively represents a group portrait of Barney's salon guests. Oriented in many different directions and written with lesser or greater degrees of legibility, the names seem haphazardly organized, as if written at random. However, upon closer inspection, it is clear that Barney did create deliberate groupings among the mass of visitors represented in the drawing. In the lower right-hand corner of the salon interior, a cluster of Anglophone literary figures – including Gertrude Stein, Alice B. Toklas, Djuna Barnes, Mina Loy, Ezra Pound, Carl Van Vechten, Janet Flanner, J.E. Strachey, Lady Una Troubridge, and Radclyffe Hall – huddle together between the end of the buffet table and the door to the garden. In addition, several of the names inscribed in a relatively readable manner beneath the rounded curve representing the Rotunda include people who gave memorable performances or readings in this alcove at one of Barney's Friday afternoon gatherings: Colette and Marguerite Morena, who performed scenes from Colette's play *La Vagabonde*; Wanda Landowska, who played harpsichord for a performance of the eighteenth-century piece *Le ton de Paris*; George Antheuil, whose *Ballet méchanique* composed for player piano, car horns, and airplane propellers had its debut here in 1926; and Paul Valéry, who, in 1927, sang two of Gertrude Stein's pieces he had set to music (Barney 1929; Barney 1960; Antheuil 1962; Chalon 1992).

Several of Barney's closest circle occupy seats of honor around the octagonal tea table, which was not only, as I have suggested, a central organizing structure of the salon event but also the site of smaller dinner parties restricted to her intimates. Here, the unusual legibility of the names of the painter Romaine Brooks, with whom Barney shared an intimate relationship for over forty years, and the novelist and political essayist Elizabeth de Gramont, one of Barney's lovers and dearest friends, distinguishes them from the rest of the crowd and clearly indicates that both Brooks and Gramont held privileged positions within the salon gathering as imagined by Barney. When read together with the names flanking it, the tea table establishes a hierarchy of relationships between individuals depicted within this space even as it functions as a symbol of the salon itself.

While several of Barney's most intimate friends and lovers were given prominent positions, her drawing was far from restricted to them. Instead, the drawing brought together all those who attended her Friday afternoon gatherings in the same imaginary moment. Rather than representing a specific salon meeting, Barney depicted its long-term history in a single, iconic image. Through the synchronic mapping of both this place and the community gathered there, the drawing attempted to establish Barney's salon as a momentous literary and artistic site. In fact, precisely by including the names of even those people who may have attended only once, as William Carlos Williams claimed to have done, Barney represented her salon as *the* meeting place for all of the most illustrious literary personalities in Paris.

While the names of recognizable individuals dominate the image, they are intermingled with descriptions of anecdotal characters, including "a group of mondains," "a beauty of the day," and "one of the disenchanted women." The presence of such peripheral figures suggests that Barney's salon had such a high profile that it had even become a destination for anonymous individuals hoping to advance their social or artistic status. In addition, while the named individuals and the roles they played in her salon necessarily changed over time, by introducing these generic types into the salon landscape, Barney underscored the timelessness of the gathering. In this way, the drawing projects and solidifies the importance of Barney's salon and her own position as writer within the contemporary literary scene by way of both the sheer numbers of guests enumerated here and, perhaps most importantly, through its composite assemblage of well-known artistic and literary figures.

Unlike the collection of names contained within the outlines of the salon interior, those flanking the Temple of Friendship at the top of the drawing are arranged in two orderly rows. In this register, Barney inscribed the names of those people associated with her salon who had died by the time this map was created for the 1929 publication of *Adventures of the Mind*, including such luminaries as Pierre Louÿs, Anatole France, Rainer Maria Rilke, Guillaume Apollinaire, and Marcel Proust. Here Renée Vivien, a poet and one of Barney's lovers who committed suicide in 1909, holds a primary position. Given his role in establishing the prominence of Barney and her salon, it is also not surprising that Remy Gourmont's name appears near that of "the Amazon" on the steps of the Temple of Friendship. Collectively, the names surrounding this structure comprise Barney's own personalized pantheon of literary influences, the veritable muses of her salon. Hovering above the interior space, the list of deceased habitués contrasts sharply with the synchronic representation of salon attendees in the lower portion of the drawing in part because the former is organized in reference to a single and actual temporal moment.

The separation of these two zones distinguishes the Temple of Friendship as a monument to those who had died at the same time as it positions Barney as the link between these two registers and the sole person who circulates between them. In fact, in at least one salon meeting, the Temple actively functioned as a memorializing space. In July 1915, a commemorative program in honor of Renée Vivien began with a reading of one of her poems in front of the temple before moving inside the main *pavillon*

(Goujon 1983). Barney's desire that the Temple of Friendship function as a space set apart from everyday life was not restricted to the drawing, but also informed the one major architectural renovation she made during her tenure at 20 rue Jacob. Although her house and gardens were set back and protected from the main street, Barney erected a garage at the end of the rear alley that further restricted access to the temple (Chapon *et al.* 1976: 6). With this addition, the space of the garden and its folly was even more clearly distinguished from that of the surrounding urban landscape. In this way, the Temple of Friendship was carefully framed within the topography of the salon as a memorial site and symbolic link between the "living" salon and its deceased members, or between the current and past communities that Barney gathered around herself at 20 rue Jacob.

As much as the drawing represented a collectivity, all of these individuals were organized in relation to their host, Natalie Barney. In his reminiscences, Samuel Putnam observed that at 20 rue Jacob, "it was obvious that the 'Amazon' was the center around whom all revolved" (Putnam 1962: 141). At her weekly gatherings, Barney carefully orchestrated her self-presentation and movements for maximum effect. Dressed in her usual Friday attire of a white Vionnet dress with large sleeves and often with the added flourish of an ermine cape, Barney would descend from her upstairs bedroom after the first guests had arrived and seat herself across from the entrance or circulate among her guests (Wickes 1975; Orenstein 1979; Barney 1929). Barney's actual physical presence was accentuated and reiterated in the portraits displayed at 20 rue Jacob, including two paintings of Barney by Carolus-Duran that hung in the entryway, one of her as an adolescent wearing a green cape and one as a young woman with long, flowing blonde hair.

The most prominently displayed portrait of Barney was *The Portrait of the Amazon*, painted by Romaine Brooks in 1920 (Figure 8.4). From its position in the dining room next to the Rotunda, this image of Barney presided over the readings and performances that took place there. In the painting, the sitter is framed by a garden that is summarily rendered in muted tones and visible through a window in the background. The relationship between figure, interior space, and exterior garden constructed in the portrait was echoed in its placement within Barney's salon, as it was mounted directly over a window looking out into the rear garden. Thus, the portrait and its display emphasized the connection between the sitter and the landscape of 20 rue Jacob that was at the heart of Barney's own identity.

In the portrait Barney is represented seated in front of a window and next to a small table where a folded sheet of paper is anchored beneath a small horse statuette. Following long-standing conventions of portraiture, Brooks depicted her sitter with attributes that function as signs of her literary profession and personal identity; the horse is intended to index her reputation as "the Amazon" and the paper works to further emphasize her status as writer. For Barney, her moniker "the Amazon" referred

Figure 8.4 Portrait of the Amazon, or *Portrait of Natalie Clifford Barney*, Romaine Brooks, 1929, oil on canvas, Musée Carnavalet

Photograph: Photothèque des Musées de la Ville de Paris

to her avid penchant for horseback riding and to the costume of riding jacket and feathered hat she wore in the 1910s when visiting friends.[4] At the same time, this appellation indexed her self-identification with the historical tradition of Amazons as models of actively independent and free-thinking women who rejected what Barney saw as the traditional and limiting gender roles of wife and mother. As Barney explained in an unpublished version of her *Pensées de l'Amazone*, "Having realized in myself this extreme paradox of being the most feminine of women and the most perverted, the men with which you content yourselves as husbands, as sons or lovers, were hopelessly inadequate" (Causse 1980: 36–7). In *Adventures of the Mind*, Barney described herself more generally as an urban Amazon who hunted sympathetic and challenging minds (Barney

1929: 20). Her salon might thus be understood as an assemblage of captured "prey" gathered around their modern-day intellectual huntress.

The portrait itself was a token of exchange and, like many objects within Barney's salon, a testament to particular ties of *amitié*, in this case her relationship with her lover Romaine Brooks. While in the painting Brooks highlighted references to Barney and her salon in both its spatial composition and the objects it included, these elements would be most meaningful to those who were already familiar with Barney or at least with her reputation. Although in the portrait Barney was enveloped by a thick fur coat rather than wearing her riding costume, the horse statuette was intended to index her literary and gendered identity as an "Amazon." Given that these references might not have been understood by all viewers, the portrait worked to consolidate group identity among Barney's salon guests for whom those clues held special meaning, particularly as framed within the symbolic landscape of 20 rue Jacob.

While the portrait reiterated Barney's performative role as the presiding force of the salon, in her own drawing Barney cast herself even more overtly as the structuring device of the salon and the very center of its meaning. At various points along the line that weaves its way from the space of the garden at the bottom of the drawing, around the central tea table, and up the steps of the Temple of Friendship, Barney wrote "*l'amazone*" as a means of visually representing her movement through this landscape. Since "the Amazon" is the only name that is repeated in the image, such implied motion vividly contrasts with the static, singular names of the individual visitors. Barney's focus on her own position within the salon was evident even in the process of the drawing's construction. That is, as Barney wrote each individual's name in the drawing, she emphasized her own self-inscription because each "signature" was literally rendered in her own hand. In this respect, the creation of the drawing recalled the structure of the book it preceded, which comprised a series of written descriptions of important literary figures through Barney's own account of them, their correspondence with one another, and their shared experiences. In both the drawing and the book's text, each of her guests was depicted in relationship to their host and thus in some sense as an extension of Barney's own identity.

Barney herself seems to have been aware of this process, at least in relation to the writing of her book. As she remarked near its conclusion, "In leaning over multiple faces have we discovered only our own image?" (Barney 1992: 198). Like the multiplication of Barney's own painted image on the walls of her salon, the repetition of her moniker "the Amazon" along the meandering line in the drawing depicted multiple sites for encounters between Barney and her guests.[5] In organizing her weekly gatherings, it was Barney who actively brought people together, in the same way that in the drawing the line of her suggested movement organized the individuals assembled in these spaces and united them into an imagined community. The drawing thus functioned neither simply as a group portrait nor merely as a self-portrait of a single individual. Instead, Barney figured herself in relationship to others, constructing her salon and her own identity in terms of the people she gathered around her.

At the same time that Barney's own self-construction was consolidated through others, she also worked to shape the group identity of her guests. Through the ritualized performances of her salon, Barney hoped to enact a new vision of community following the philosophy of friendship she articulated in her writings. At 20 rue Jacob, the Temple of Friendship formed the locus of this projected moral landscape. Delineated at the top of the drawing, the Temple of Friendship appeared to float above the salon interior. The iconic form of its outlined façade was thus pictured as the architectural figurehead of Barney's salon and the generative structure for the imagined community gathered beneath it. By isolating this structure from its surroundings, Barney reframed the Temple of Friendship as a symbolic monument onto which her own ideas about the past and the nature of friendship could be projected and, at times, literally staged. Its architectural form, a Greco-Roman temple constructed in miniature, was central to this process since the model of *amitié* that Barney was interested in salvaging was based on her rereading of a specific history of antiquity, that of Sappho and the women poets who surrounded her.

In 1904, Barney and the poet Renée Vivien made a pilgrimage to Mytilène on the island of Lesbos with the hopes that they might reestablish just such a community there (Jay 1988; Goujon 1978). Disillusioned with modern-day Greece and the dearth of existing traces of Sappho's legacy, Barney soon returned to France and embraced the sprawling, enclosed gardens surrounding her home in Neuilly as an ultimately more controllable space in which to stage a blend of literary and amorous performances inspired by her idealized vision of Sappho.[6] Even as they gestured towards antiquity, Barney's *fêtes pastorales* simultaneously exploited the eighteenth-century tradition of the garden as a picturesque stage for romantic encounters and a libertine space in which social conventions might be more easily transgressed.

In 1906, Barney and several of her friends staged Barney's play "Equivoque" ("Ambiguity") recounting the end of Sappho's life (Barney 1910). In this piece, Barney refocused the myth of Sappho on homoerotic desire. Instead of throwing herself into the sea out of jealousy because her lover, the male poet Phaon, was going to marry his female lover Timas, in Barney's version Sappho commits suicide because she is really in love with Timas.[7] The small circular colonnade that stood in Barney's garden in Neuilly served as an appropriately classicizing backdrop for this and other performances. At the same time, the production relied on the broader architectural framing of the landscape whose walls created a space physically and notionally removed from the city and its social norms, as in the rear garden at 20 rue Jacob.

In Paris, instead of restaging the life of Sappho as she had in Neuilly, Barney attempted to create a living community of modern, creative women in the real topography of her salon and garden. In an essay entitled "The Trial of Sappho: Fragments and Testimonies" included in her 1920 *Pensées d'une Amazone* (*Thoughts of an Amazon*), Barney cited a late nineteenth-century description of the poet's house as a site where women gathered to share literary creations and erotic relationships with one another: "Sappho gathered in her house, that she called the House of the Muse's Female Friends, beautiful young women friends with whom she sang and to whom she attached the exalted love of a hot-blooded Southerner" (Christ quoted in Barney 1920: xxix). Although Barney's salon was not restricted to women, she reimagined the spaces of her own home in part in reference to Sappho's House of the Muses and the creative friendships it fostered. Thanks in part to its inscription, the Temple of Friendship at 20 rue Jacob became the architectural embodiment of Barney's ideal of *amitié*, which she described as "a pact above passions, the only indissoluble marriage" (Barney 1939: 164). For Barney, *amitié* was a flexible term combining nineteenth-century ideals of friendship and sapphic eroticism which were understood together as the necessary basis for productive artistic creativity.

Barney believed friendship was an intersubjective process, shared and constructed between two people through their creative, intellectual, and (potentially) erotic exchanges. Ideally, however, this original diadic structure was not a stable end in and of itself, but should instead be oriented towards and eventually absorbed into a broader community. As she later explained, "And didn't the great Sappho live in harmony, not with one but with several of her female friends who, in succession, felt those sweet rivalries that were more a subject of inspiration than of discordance" (Barney 1963: 173). Following her reinterpretation of Sappho's life, Barney rejected conventions of fidelity and monogamy and emphasized instead the importance of constancy, passion, intellectual engagement, and creative exchange to her ideal model of social, artistic, and erotic relationships (Barney 1910, 1939, 1963). Barney thereby envisioned a model for a new form of modern sociability constructed through creative exchange and fluid erotic bonds in which female homoeroticism occupied a privileged postion.

Barney's ethics of friendship were intended as a critique of both contemporary social practices and institutional structures. In response to the masculine bias of the French Academy, which only accorded membership to male writers and artists, Barney created her own Academy of Women, composed of select Francophone and Anglophone writers, poets, and artists who were active in her salon (Barney 1929). Over the first six months of 1927, each woman invited to join Barney's Academy was honored at her own induction ceremony during a Friday afternoon gathering at 20 rue Jacob.[8] Although many other performances and gatherings had taken place in front of the Temple of Friendship, Barney called the Academy of Women "one of its most sacred missions" (Barney 1992: 129). As the literal backdrop for these events, the Temple of Friendship and the garden that surrounded it framed a site where such an ideal of female creativity and homosociality might not merely be imagined but finally actualized.

Barney's miniature Greco-Roman temple offered an appropriate architectural support, one that could accommodate both the historical references and the contemporary investments of Barney's ethics of friendship. Over time, the Temple of Friendship became so thoroughly intertwined with Barney's own identity and her philosophy of creative and affective relationships that Barney herself insisted that "it must not be by chance that I have in my garden a little Temple of Friendship" (Barney 1939: 210). Her retrospective assertion palpably demonstrates the thoroughness with which Barney appropriated existing structures to her own ends. Through the performative ritual of the salon, Barney's own writings, and her drawing, 20 rue Jacob was thus transformed over time into a deeply symbolic moral landscape.

Acknowledgments

I thank Katherine Fischer Taylor and Angela Rosenthal for reading and critically responding to earlier versions of this text. My deepest debt of gratitude goes to Sarah Betzer for reading numerous drafts of this chapter and, most of all, for her unflagging support of this project.

Notes

1 Here I follow J.S. Gatton's translation of *Aventures de l'esprit* into *Adventures of the Mind* (Barney 1992). However, it is important to note that the French word "l'esprit" does not refer merely to the mind but to a studied combination of wit, intellect, and spirit. All other translations, unless otherwise noted, are by the author.
2 Having always rented the *pavillon* at 20 rue Jacob, the owners forced Barney to move out in 1968, four years before her death (Jay, in Barney 1992: 2).
3 The point is not that architectural drawings are accurate representations of space, but that they generally follow certain conventions, such as their orientation on the pages which allow the viewer to understand the representation following horizontal and vertical axes. That is, in Barney's drawing, the map is read from right to left and the figure at the top of the page is understood to be either above or beyond the main rectangular area. Of course, the inconsistencies are also important here, as in, for example, her combination of architectural plan with profile views, an aspect of her drawing to which I will return later.
4 Magdeleine Wauthier described Barney's "Amazon" costume as follows: "What a surprise it was, the day she came, without warning, to see me. She was dressed, like a 'Montpensier,' as a seventeenth-century amazon, in a long, fitted jacket with black facing and on top of her very blond hair, a felt hat with ostrich feathers like a 'Frondeuse,' enemies of the established order and of Mazerin" (Wauthier 1963: 12). The Duchesse of Montpensier, known as "la Grande Amazone" (the Great Amazon), took part in the opposition movement ("la Fronde") against the royal order in the mid-seventeenth century, and in 1652

became famous for having fired a cannon on top of the Bastille against royal troops.

5   The inscription of Barney at various sites in the drawing as a means of emphasizing her role at the center of the salon ritual recalled the costume she wore to a ball she gave at 20 rue Jacob in 1911 or 1912 that literally allowed her to disappear and reappear at will in the darkness of her garden: "I myself wore a costume of a Japanese firefly catcher, with an electric battery hidden in the bottom of a basket on one arm, from where I freely manipulated a group of small lightbulbs, very closely imitating a firefly's glow, that twinkled even up around my hair. This helped my guests to distinguish me, even in the depths of the garden, faintly lit by Chinese lanterns suspended from the trees" (Barney, 1960: 108–9).

6   While most of the performances in Barney's garden were given to an audience of Barney's female and male friends, on at least one infamous occasion, Mata Hari performed Javanese dances in Barney's garden to an exclusively female audience. Colette recounted both Mata Hari's performance and her own participation in a reading of Pierre Louÿs's "Dialogue au soleil couchant" in her book *Mes Apprentissages* (quoted in Barney 1960: 191–2).

7   Barney included an additional twist to the conventional narrative of Sappho's suicide. At the end of the play, upon discovering that Sappho has thrown herself off a cliff into the sea, Timas is inconsolable, and heads for this cliff to rejoin her true love in the sea.

8   Barney's Academy of Women was comprised of Lucie Delarue-Mardrus, Anna Wickham, Colette, Rachilde (the pseudonym of Marguerite Eymery Valette), Aurel (the pseudonym of Antoinette Gabrielle Mortier de Faucamberge), Mina Loy, Elizabeth de Gramont, Djuna Barnes, Gertrude Stein, Romaine Brooks, Renée Vivien, and Marie Lenéru.

References

Antheuil, G. (1962) "Parisian Apex," *Adam International Review*, 29: 145–6.

Barney, N.C. (1910) *Actes et entr'actes*, Paris: E. Sansot.

—— (1920) *Pensées d'une Amazone*, Paris: Emile-Paul Frères.

—— (1929) *Aventures de l'esprit*, Paris: Émile-Paul Frères.

—— (1939) *Nouvelles pensées de l'Amazone*, Paris: Mercure de France.

—— (1960) *Souvenirs indiscrets*, Paris: Flammarion.

—— (1963) *Traits et portraits*, Paris: Mercure de France.

—— (1992) *Adventures of the Mind*, trans. J.S. Gatton, New York: New York University Press.

Benstock, S. (1986) *Women of the Left Bank: Paris, 1900–1940*, Austin TX: University of Texas Press.

Cassou, J. (1962) "L'Amazone amie," *Adam International Review*, 29.

Castle, T. (1992) *The Apparitional Lesbian: Female Homosexuality and Modern Culture*, New York: Columbia University Press.

Causse, M. (1980) *Berthe, ou un demi-siècle auprès de l'Amazone*, Paris: Éditions Tierce.

Chalon, J. (1992) *Chère Natalie Barney: Portrait d'une séductrice*, Paris: Flammarion.

Champion, P. (1932) "Temple à l'Amitié," in *Mon Vieux Quartier*, Paris: Éditions Bernard Grasset.

Chapon, F. *et al.* (1976) *Autour de Natalie Barney*, Paris: Université de Paris.

Dams, B. and Zega, A. (1995) "Betz: Le Temple à l'amitié," in *La Folie de Bâtir: Pavillons d'agrément et folies sous l'Ancien Régime*, Paris: Flammarion.

Elliott, B. and Wallace, J. (1994) *Women Artists and Writers: Modernist (Im)positionings*, New York: Routledge.

Gatton, J.S. (1988) "Natalie Clifford Barney: Literary Amazon of the Left Bank," *Kentucky Review*, 8: 47–55.

Goujon, J.-P. (1978) *Renée Vivien à Mytilène*, Reims: À l'Écart.

—— (ed.) (1983) *Correspondances croisées: Pierre Louÿs, Natalie Clifford Barney, Renée Vivien*, Muizon: À l'Écart.

Gourmont, R. (1914) *Lettres à l'Amazone*, Paris: Crès.

Jay, K. (1988) *The Amazon and the Page: Natalie Clifford Barney and Renée Vivien*, Bloomington IN: Indiana University Press.

Lottman, H.R. (1969) *New York Times Book Review*, 28 September: 2, 46–7.

Macon, G. (1908) *Les Jardins de Betz, description inédite*, Senlis: Imprimerie Eugène Dufresne.

Melicourt, M. (1975) "Le Temple de l'Amitié à Paris," *Vieilles Maisons Françaises*, 66: 28–9.

Orenstein, G.F. (1979) "The Salon of Natalie Clifford Barney: An Interview with Berthe Cleyrergue," *Signs*, 4: 484–96.

Putnam, S. (1962) "The Incomparable Rue Jacob," *Adam International Review*, 29.

Rodriguez, S. (2002) *Wild Heart, A Life: Natalie Barney's Journey from Victorian America to Belle Époque Paris*, New York: Ecco.

Wauthier, M. (1963) "Portrait de Natalie Barney," in N.C. Barney, *Traits et portraits*, Paris: Mercure de France.

Wickes, G. (ed.) (1975) "A Natalie Barney Garland," *The Paris Review*: 86–134.

—— (1976) *Amazon of Letters: The Life and Loves of Natalie Barney*, New York: G.P. Putnam.

9 Pincushions, dormitory kitchens, and seed gardens

Gender identity and spiritual place at the West Union Shaker village

Christine Gorby

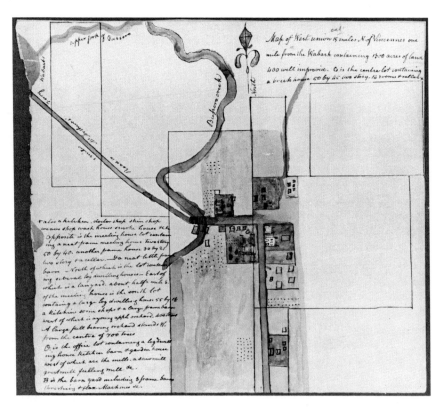

Figure 9.1 Reconstructed drawing showing the building types and general layout of the entire West Union Shaker village, based on the historic surveyor's map dated between 1824 and 1827

The understanding of how religious beliefs and practices structure the cultural landscape is a hidden story within the development of American history. In a similar vein only 5 percent of national, state, and local landmarks in America focus on any aspect of women's history at all (Hayden 1995: 54).[1] The formation of the sacred within gendered landscapes, in particular, has remained even more concealed from public awareness, thereby limiting insight into women's own unique spatial environments. Yet to uncover the hidden histories of gendered sacred spaces at historic places involves a process beyond self-education toward self-knowledge. For a visitor the development of self-knowledge involves questioning historical accounts and interpretations at every reconstructed site. Visitors must not only be more readily able to make connections with the individual histories and memories evolving from their own past but also with those that are unknown. "Places trigger memories for insiders, who have shared a common past, and at the same time places often can represent shared pasts to outsiders who might be interested in knowing about them in the present" (Hayden 1995: 46). The opportunities for achieving such "self-knowledge" on the part of visitors become limited when certain sisters' histories are excluded to retain the more familiar utopian aspects of the Shaker Society.

Not allowing for contradictory readings of the Society has an impact when considering larger issues such as how past religious beliefs have roots in contemporary culture. Few sites today are committed to telling the history of religion. This is especially significant in regard to women's connection with the spiritual. The Shaker religion is important culturally in many respects, but it is most unusual for incorporating a female into the deity (Setta 1979: 175). While no other religion in America can claim such a distinction, it is still important to remember not only the Society's progressive attitudes but also its shortcomings in the treatment of women. The Shaker religion is also of consequence to contemporary American culture because perhaps no other religious group has more literally tied religious meaning with the landscape as a source of spiritual self-renewal. At a time when there is growing rejection of the mass-media society and renewed interest in the spiritual, historic Shaker sites have much to offer. Also, in a country where there are "few traces to link pilgrims back to the long history of their religion" (Bharwajand and Rinschede 1990: 23), it is meaningful for women and others to be able to discover the sacred by locating themselves where the sacred has been experienced.

Historic traces

In 1810, following a great 50-year period of religious revivalism in the mid-west colonies (Woloch 1992: 129), the Shaker village of West Union, Indiana, was founded. This happened 34 years after the creation of the first Shaker village near Albany, New York, (1776) and 36 years (1774) after Ann Lee, the founding leader of the Shakers, arrived in America.

Established in the Northwest Territory, Busro (later renamed West Union) was the most western community ever developed by the Shakers. The village was dissolved, however, in 1827 after only seventeen years of development, an unusual decision by the optimistic, hardworking Shakers who rarely gave up on any endeavor.

Numerous setbacks caused the abandonment of the West Union village, including problems stemming from its location in the swampy surrounding lowlands that caused incessant malaria in the village. Deaths from "the fever" were the primary cause in the drop of community membership from a high of 300 at the settlement's beginning to only 140 members at the end. Conflicts with people from other settlements near the village also caused the decline of the community. Sited as they were between a major military post in Vincennes, Indiana, and numerous Native American villages to the north, the pacifist Shakers, despite attempts to befriend both sides, became caught between the invasions and cross-attacks. Hostilities between the military and Native Americans culminated in The Battle of Tippecanoe, the first battle in The War of 1812, signaling the end of any cooperation (Dowd 1992: 183). During this period the West Union village was even taken over by the military and the Shakers were forced to leave their own property. Because the life of this Shaker village was so brief and because their physical buildings remain, historians have given little in-depth attention to this site. It is also not generally within the thinking of most Americans to reconstruct or reconsider their failures.

The planning of Shaker villages stems from a religious belief in function and usefulness through orderliness, most clearly evident in the overriding sacred spatial geometry of Shaker villages, the orthogonal or "four square." This geometry originated with a central belief in a dual god (Holy Mother Wisdom and Almighty God the Father) and a dual messiah (Jesus Christ and Ann Lee). The "four square" metaphor was also carried through to the Shaker ministry of two men (elders) and two women (eldresses) who governed each village. The translation of this sacred doctrine into the physical landscape can be seen everywhere in Shaker landscapes from their architecture which is built orthogonally with entry paths, sidewalks and main thoroughfares, fencing and trees at right angles with fields and barns, and cultivated fields and pastures organized in endless geometric square forms. Emphasis on the orthogonal meant that the spatial order between the architecture and landscape was always maintained, a concept further emphasized in Shaker village maps.[2] While there was a desire for uniformity in the total reading of the village, the Shakers viewed the community as a temporal landscape that should always be evolving. This is evidenced by the ease with which Shakers would change the use of a field or building – even moving a structure if necessary. The emphasis on an evolving community structure parallels the way in which the Shakers also viewed the development of their own faith – not as a static but as a continual spiritual process.

While the orthogonal landscape was symbolic of the dual nature of God, spatially the grid "neutralized" the limitless prairie territory through the imposition of an economically efficient and rational form (Sassen 1996: 144). It formally linked the three smaller groups of members or "families" that were in different locations within the vast grid site.[3] This had the negative effect of nullifying each group's individuality by the development of bland, often repetitious construction. But it also maintained the idea of community because the "families" were spatially and socially linked by a geometric system of enclosure. The Millennial Laws of 1821, the rigorous set of rules and rituals which the Shakers followed in work, life, and prayer, did not encourage the "family" groups to be active sites of difference. The grid did, though, become an expression of power and precision onto the gendered landscape. On the surface, equality through uniformity was the centerpiece of Shaker thought and expression within their communal order.

Mapping

On the West Union site today, only surface fragments of the original village remain and limited archaeological work has been completed in the "Center Family" area (Janzen 1992: 7, 12, 20, 22). A drawing of the Shaker's original 2,600-acre village (Conlin 1963: 57) overlaid with the present-day configuration of the landscape helps to visualize the immense scale of their overall territory (see Figure 9.2). The only surviving visual historic documentation of the village is a hand-drawn surveyor's map made some time between 1824 and 1827 near the end of the community (see Figure 9.3). Like many other plans of Shaker villages, it shows the West Union orthogonal layout with a primary north-to-south axis. The two most important spiritual buildings to the community, the meetinghouse and the dormitory, are located along this axis and are shown in elevation, emphasizing their significance spiritually and communally to the members. Sisters and brethren worshiped together in the meetinghouse. Meals were taken in the dormitory and sisters and brethren slept in separate quarters in accordance with their spiritual belief in celibacy.

Like most Shaker maps, the West Union surveyor's map is not constructed topologically; that is, it does not represent actual distance or scale. Nor does the map establish particular viewpoints; the entire community is pictured to reinforce the communal nature of the religious order. Shaker village maps in general were much more artistic and detailed than the West Union plan (Emlen 1987: 4). Select male brethren made them as planning and communication tools of their own and as a way to promote uniformity in the built environment and kinship ties in distant Shaker villages (Emlen 1987: 3). Sisters were not involved in map-making because the central ministry forbade them from studying such necessary skills as architecture, horticulture, mill complexes, granite-working, surveying, or building and farm trades. While sisters had no active role in the physical

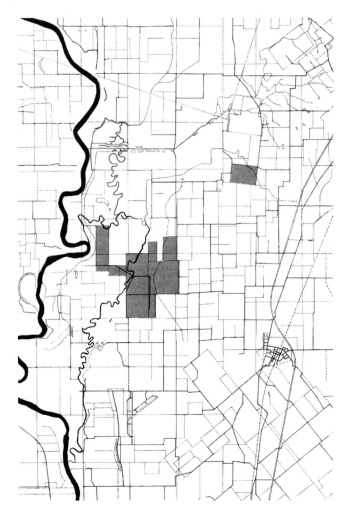

Figure 9.2 Drawing overlay of the original West Union Shaker village with the
present day landscape, showing the extent of the Shakers' holdings in
1827 (redrawn by the author)

Note: Of the Shakers' 2,600-acre property, 1,300 were located in Illinois Territory adjacent
to the Shakers' holdings in Indiana. Because the Illinois land parcel was relatively undevel-
oped it is not shown on this map.

structuring of their landscapes, a few maps do document a very limited
number of their daily activities indicated by drying racks, wash houses, and
kitchens.

The most complete set of Western Shaker village maps were made by
Isaac Young, a Shaker from New Lebanon, New York, who traveled to
the Ohio–Kentucky region in 1834. What is interesting about his maps,

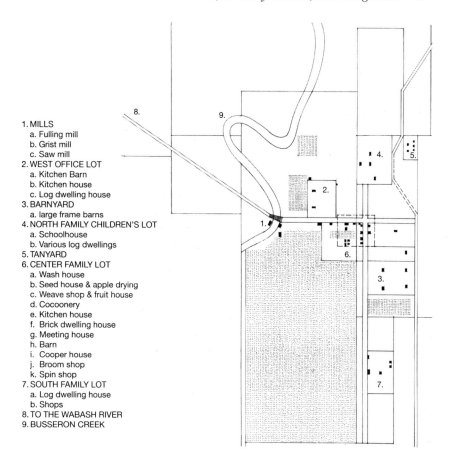

1. MILLS
    a. Fulling mill
    b. Grist mill
    c. Saw mill
2. WEST OFFICE LOT
    a. Kitchen Barn
    b. Kitchen house
    c. Log dwelling house
3. BARNYARD
    a. large frame barns
4. NORTH FAMILY CHILDREN'S LOT
    a. Schoolhouse
    b. Various log dwellings
5. TANYARD
6. CENTER FAMILY LOT
    a. Wash house
    b. Seed house & apple drying
    c. Weave shop & fruit house
    d. Cocoonery
    e. Kitchen house
    f. Brick dwelling house
    g. Meeting house
    h. Barn
    i. Cooper house
    j. Broom shop
    k. Spin shop
7. SOUTH FAMILY LOT
    a. Log dwelling house
    b. Shops
8. TO THE WABASH RIVER
9. BUSSERON CREEK

Figure 9.3 Composite drawing of the archaeological findings of the "Center Family" area (redrawn by the author)

though, is not what he documented but what he chose not to illustrate. By comparing four of the primary village maps in Issac Young's journal (Pleasant Hill, Watervliet, Whitewater, and South Union) it is evident that Young consistently names brothers' work sites but does not identify those of the sisters. A good, typical illustration of this type of omission is the Shaker village map of Pleasant Hill, Kentucky, to which many Shaker members from the West Union, Indiana, village went to live after the dissolution of their own community. Young, for example, identifies twelve types of work places for brethren on his map (grist mill, oil mill, saw mill, cooper shop, blacksmith's shop, tan house, etc. – fourteen buildings in total) but only lists two work sites for sisters (two wash houses and eight other structures he only classifies as "shops"). While Shaker sisters had less choice about the work they did, in a typical Shaker village site there would still be numerous specific buildings where their productive skills were put to use. These might have included, for example, a spin shop, seed house, herb shop, weave shop, or fruit house. Young's representations are a good reminder that maps, as instruments of knowledge and authority, can be used to conceal as much as reveal. Naming, then, was one way that Young subtly affirmed the specific contributions that brethren and not sisters were making to their communal order.

While conventional cartography is one source for reclaiming gender roles, mapping the cultural landscape within a "lived geographical place" (Nash 1994: 228) is another way to conceptualize women's space. New readings of the "lived" cultural landscape have emerged by negotiating such divergent fields as social history, architectural preservation, environmentalism, and public art – areas that also draw from shared collective memories (Hayden 1995: 45). As space has become reinterpreted as a social product (Yaeger 1996: 5) and not just a place for the re-enactment of historical events, cultural studies have "reinvent[ed] the story it sets out to critique . . ." (Yaeger 1996: 15). From anthropology "culture" has been interpreted not only as a dynamic place always being "created, contested, and recreated" but as a practice contextualizing space within a mass-media society (Warren 1997: 175). Because the cultural landscape, then, is viewed as a continually active rather than passive place, it can be reinterpreted as a "storehouse of culture and history rather than as a scientific problem" (Hayden 1995: 66).

Emphasis on reconstructions as utopian

As feminist historian Marjorie Proctor-Smith has stated, even though a woman, Ann Lee, founded the Shaker order, the organization became largely institutionalized and controlled by men (Procter-Smith 1991: 8). Being more cognizant of how the outside world viewed them, the Shakers turned to male leadership to be more acceptable to the outside world and the potential new members they were trying to attract (Procter-Smith 1991: 10).[4] Even though women had access to leadership, a unique aspect in nineteenth-century communities (Procter-Smith 1985: 19), the male ministry assumed the real power in the community. A select group of male brethren controlled most of the political, economic, and social aspects of the Society through their letter writing, spiritual texts, allocation of funds, and structuring of the physical landscape. Sisters, then, were positioned both inside and outside the power and authority of the community. While female gender identity at the turn of the nineteenth century was beginning to be negotiated and contested in the outside world, within the private sphere of the Shaker community, select male brethren continued to control women's individuality. The following excerpt from a Shaker newsletter written by a Shaker sister describes the indoctrinated ways in which "equality" between male and female was characterized:

> Woman is not man's equal in physical strength; neither as a general rule, is she his equal in logic and the sterner qualities of the mind; but she possesses some properties which he does not; and combine the faculties of both, and when redeemed from the blight of sin, they will make one perfect whole, without schism, capable of honoring God and beautifying the earth.
>
> It is for man's interest that woman should find and fill her proper sphere, and be something higher, purer, and better than a slave to man's passions, and to make a way for her to become a co-worker with him in elevating the race. How much her influence is needed in all ranks of society! Let woman use her intelligence to find out her proper sphere of action, and in what consists her adorning, and she will soon cease to desire to be a mere thing of outward beauty, or an idol to be worshipped.
>
> (Antoinette 1871: 20).

"Equality" between men and women, then, was contextualized to fit within the prevailing attitudes toward women in the outside world during the early nineteenth century – woman weaker in strength and mind, evil woman in need of purification, and woman as object of beauty. Parity went only so far as a woman herself was capable of achieving given her endowed "limitations."

In addition to inequalities of leadership, limitations of labor and economic opportunities for Shaker women shaped the gendered landscape, sometimes in ironic ways. Although the Shaker religious order is viewed by modern-day tourists as "utopian" in every respect, work for both men and women followed the standard divisions of labor commonly found in the outside world during the nineteenth century (Woloch 1992: 224).[5] With the exception of factory work, Shaker women long before had similar opportunities to outside women in workshops and as students or teachers. What they produced, studied, or taught was dictated to them, however. Also, because only one teacher was needed within each Shaker village, those who could aspire to this position were limited.

Although it was pragmatic for a celibate society to divide work between the sexes, the need to enforce sexual abstinence in every aspect of the society created extreme spatial separation. The separation of work opportunities had the effect of locating males and females in distinct spatial relationships within the landscape. At the West Union site brethren could work in three types of mills, a tan yard, cultivated fields and pastures, or the orchards. A monthly rotational system meant brethren might be sawing and planing walnut and cedar lumber one month and caring for livestock and sheep the next (Sebree 1956: 11). A variety of craftsmen jobs such as "tayloring," "smithing," cabinet making, carpentry, or broom making were also available to male members and rotated. Sisters had less choice. Much of their work involved domestic chores and sewing-related jobs such as cleaning, sweeping, and cooking. These jobs were allocated to sisters in addition to their work in their "other" occupations. Women would arise before the brethren early in the morning to begin such work. Some of the goods they produced for sale in their daily work included baskets, mops, pincushions, gloves, and seed bags. Foodstuffs for their own community and for sale included applesauce, maple sugar, dried apples, and jellies, and other items. The drawing reconstruction of the Center Family, a portion of the overall West Union village site, shows how sisters were relegated to traditional kinds of household work in the dormitory kitchen, spin shop, and laundry (see Figure 9.4). The drawing is based on historic published Shaker daybooks chronicling construction and land improvements over time in this village, recent archaeological findings, and western Shaker building practices. While the cocoonery represents a very unique industrial specialization not typically found in Shaker settlements, the choice of production activities sisters had in general was extremely limited and the nature of the activities was especially repetitive and task-oriented. Based

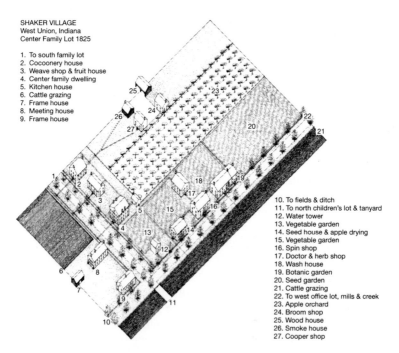

SHAKER VILLAGE
West Union, Indiana
Center Family Lot 1825

1. To south family lot
2. Cocoonery house
3. Weave shop & fruit house
4. Center family dwelling
5. Kitchen house
6. Cattle grazing
7. Frame house
8. Meeting house
9. Frame house

10. To fields & ditch
11. To north children's lot & tanyard
12. Water tower
13. Vegetable garden
14. Seed house & apple drying
15. Vegetable garden
16. Spin shop
17. Doctor & herb shop
18. Wash house
19. Botanic garden
20. Seed garden
21. Cattle grazing
22. To west office lot, mills & creek
23. Apple orchard
24. Broom shop
25. Wood house
26. Smoke house
27. Cooper shop

Figure. 9.4 Drawing reconstruction of the Center Family area of the West Union site showing the traditional domestic work to which women were relegated (drawn by Gorby)

on my personal experience visiting several historic Shaker sites, because labels or tour guides do not communicate otherwise, there is an impression to the outside visitor that the decision to engage in such jobs was a "rational choice" by Shaker women (Hanson and Pratt 1995: 4) and that they had the same opportunities for job variety as male brethren. While one modern visitor pamphlet from the Pleasant Hill village in Kentucky, for example, does mention the availability of job rotation for sisters, it does not describe the very limited range of choice available to them in contrast to male members (*A Look Inside a Shaker Kitchen* 1996: 2).

The experience of woman's landscapes at historic sites today also fails to reflect adequately the mass quantities of goods grown and produced by Shaker women for use in the community and for sale in the outside world. This prevents an understanding of the vast economic contribution sisters made to the society as a whole. Instead, a very genteel approach to women's economic production is recreated today. A visitor might see, for example, a small herb garden, a few candles dipped, or fragments of cloth woven. Every Shaker village was a complex production site with a diverse combination of industries, supplies, and markets. Some contributions made by women can be quantified through the economic value of products they made such as baskets, maple sugar, or barrels of applesauce. For example, in 1815 at West Union two hundred bushels of peaches and four hundred bushels of apples alone were harvested (Sheffler 1968: 68). Other domestic chores, though, are less easily assigned 'wage' values such as cooking, the care of brethren's clothing, or tending of kitchen gardens, what Gillian Rose refers to as the "supposedly 'separate' spheres of private domesticity and public labour" (Rose 1993: 120–121). This intersection of production for the outside world and support for the community created unique communal spaces for Shaker women that combined domesticity, industry, and prayer.

Public and private dimensions

The assignment of gender in the landscape based on patterns of power and economics, as just discussed, only begins to describe the West Union community's cultural geography. As cultural geographer Gillian Rose has written, "Spaces constructed over many dimensions are necessary" (Rose 1993: 151). Similarly, geographer Alison Blunt believes the structuring of space has come to be viewed "less as a geography imposed by patriarchal structures, and more as a social process of symbolic encoding and decoding that produces 'a series of homologies between the spatial, symbolic and social orders'" (Blunt and Rose 1994: 3).[6] When considering the cultural geography of the West Union site as a "social process," the complex negotiations between public and private become important considerations. When Shaker women, in particular, are examined, their own unique spatial structure could be perceived as entirely private. Unlike many male members, they had no business dealings or contact with the outside world. When looking from within the sisters' part of the community, however, the women's space could be interpreted as entirely public. This is because the spiritual behavior of members, particularly women, was dictated through religious dogma in direct concert with their daily patterns of living and working, and individual expression was strictly prohibited.

A drawing reconstruction of the West Union site begins to illustrate some of the complex negotiations between gendered private and public space that occurred at a larger scale (see Figure 9.4). It shows the distinction between the open fields and industrialized sites where male brethren worked and the enclosed "family" areas of the village where women lived, worked, and prayed. A grand open public experience of the landscape experienced by male members, then, is contrasted with the confined private areas where female sisters lived. While the "family" framework could be perceived as a private "home" where small groups of women shared camaraderie and resources within a larger "public" Shaker community, it alternatively could be seen as a private form of "oppression" similar to that experienced by the many other women who lived in isolation on the western frontier. The isolation, fear, disease, and limited movement which women faced when they reached the western frontier was often in contrast to their hopes and dreams before their travel. Kolodny's research illustrates that the woman who moved to the western frontier before 1850 perceived her prairie landscape as a "ready-made park at the end of the trail" (Kolodny 1984: 97).[7] This is because novels of the time period perpetuated the idea of a cultivated garden. Rose further affirms the mythical views of frontier women:

> Always they dreamed . . . of locating a home and a familial human community within a cultivated garden, and this dream was part of their social location as white bourgeois women and the concomitant importance to them of the distinction between the public and the private.
>
> (Rose 1993: 111)

Similarly, Shaker women who established new communities in the West imagined they would parallel their former lives in the East – private "familial" groups within a larger "public" landscape. Other Shaker women hoped to increase their opportunities for leadership positions by relocating to the West because openings were so limited in the well-established Eastern Shaker communities that they left. Ruth Darrow was one such Shaker sister who in her twenties resettled in West Union, Indiana, from the East with hopes of bettering her position. She later did become an eldress but died early from malaria at only 27 years of age. Disease was just one of many real hardships that thwarted many women's idealized views of the western frontier.

Fragments of culture revealed

Often the desire to create a totally unified tourist experience means that only fragmented parts of a culture are revealed. Modern reconstructions of historic Shaker landscapes, for example, typically neglect women's histories by only restoring male barnyards and fields to the exclusion of women's outdoor workspaces. These include orchard and kitchen gardens near the dormitory kitchen, medicinal gardens adjacent to the herb shop where women sorted and bagged medicines, and outdoor drying racks for dyed woolens near the spin shop. By reconstructing, then, only select areas of the built environment visitors are left to assume that Shaker sisters engaged in only the household tasks that are visualized or "recreated."

The drawing reconstruction of the Center Family area at the West Union site shows how a site can be transformed by the inclusion of the vast garden landscapes worked by women. Because buildings are typically not hierarchical in the sense of having clearly defined front or back elevations, as one moves from garden to building, many points of entry allow for a contiguous experience of the landscape. Windows reinforce this idea further by being placed across from each other on opposite walls, giving a sense of transparency between building and landscape. Through the modern-day reconstruction of women's outdoor spaces of work, then, moving first into the "garden" and not the workhouse is the act of entering into a feminine social order.

The objects made by Shakers are probably the most familiar aspect of the culture, yet this focus at historic sites often limits the understanding of other aspects of the Society. Yaeger suggests that we are often misdirected into thinking that a constant flow of objects can recreate the experience of "place" (Yaeger 1996: 20) where "the illusion of a relation between things takes the place of a social relation" (Yaeger 1996: 22). Through her study of Susan Stewart's work on museum culture, Yaeger found that curators often "add motives or concepts to things" so that one object seems to attest to the social history of an entire culture "where an Ibo mask promises a covenant with all Ibo culture, while a ten-minute movie of China pretends to survey the vast differentiations of all Chinese history" (Yaeger 1996: 22).[8] At Shaker historic sites visitors have the advantage of being able to directly experience the historic landscape. Yet information imparted about the Society when visiting historic sites is largely centered on the beautiful yet functional objects men produced, from chairs to oval nested boxes to brooms. Often there is an underlying sentiment that men were the "artisans" while women were the "domestics" when witnessing the kinds of objects reproduced in workhouses or gift shops. The emphasis on objects as representative of the culture also has the effect of placing an overriding emphasis on interior spaces and not the shaping of the outdoor spaces. The Shakers' overriding emphasis on crafted and beautiful objects as a conduit to God is indeed something to be celebrated and imparted.

Because women's work, though, was centered on household duties and other tedious tasks, their direct economic value to the Society has largely been ignored.

Fragmented experiences of Shaker life also occur when there is little recognition of the unique enterprises or use of construction materials that arose from their special geographical location. At the West Union site, for example, agricultural activities were largely centered on growing fruit, the majority of which was for the commercial market. Proximity to the Wabash River contributed to their rich fertile farmland and the flat prairie landscape made large-scale cultivation much easier. In 1815 their orchard of 300 acres of apple and peach trees at the southern area of the village was considered "the best in the country" (Sheffler 1968: 68). A nursery held approximately 10,000 more fruit trees. Vernacular building traditions also influenced the physical structuring of the village. Influenced by regional customs in Vincennes, for example, several buildings were made of logs and plaster, as constructed by the military and French trappers. Later these buildings were often covered with clapboard. The West Union meeting-house was constructed in masonry in comparison to other Shaker villages in the east where wood clapboard structures were uniformly made. Overall, these examples begin to show that the physical landscape at each Shaker site was also an organically evolving place related to its unique geography and not just a place controlled by ideological beliefs.

Shaker identity through the moral landscape

Layered within the spiritual landscape of the Shakers is a tension between the "natural and cultivated" environment, most closely expressed through the creation of enclosures, rows of trees and other forms of boundary making. Because nature for the Shakers was one primary way in which "God protects and nourishes humans" (Ross 1991: 107), the wilderness territories that became their village sites first required the felling of trees – a necessary "purification" ritual so that the tree of life or *spiritual self* might dwell and thrive from within. Sweeping and other literal kinds of cleansing practices sustained the sanitization of the site as well as separation of the sexes in both living and working environments. Through these and other practices the purified state of both *site* and *soul* could be maintained. Though untamed nature could be "controlled," cultivated nature was more complex. Only useful, not ornamental flowers, vegetables, and herbs could be sown. For the sisters who were their primary keepers, flowers were a resplendent aid, allowing them more readily to recall their spiritual devotion to God. "Flowers," a Shaker sister writes:

> stand in similar relation to mankind. The worldly minded too often use but to abuse them, but they never were intended for such a purpose; they can be, and are, by some, used for a higher object. To me they

are the emblems of heavenly virtues, purity, meekness, innocence and love. These form and beautify the spiritual character.

(M. 1871: 24)

A drawing reconstruction of the West Union Center Family area illustrates how the Shakers' spiritually cleansed and well-ordered landscape was overlaid with not only complex gardens but also pre-industrial and household enterprises such as weave, spin, and herb shops, a cocoonery, and wash, fruit, and kitchen houses (see Figure 9.4). All the workshops and workhouses shown in the central region except the Doctor's Shop were organized and run by Shaker sisters. Their activities were located in close proximity to the meetinghouse – the communal place of worship – and the dormitory – where members and the central ministry slept and the community ate their meals. By locating the sisters' workshops near the meetinghouse, Shaker brethren, in control of building and planning activities, were symbolically locating them near the sacred village center. This was not because they lived on a higher spiritual plane but because both the brethren and sisters perceived women in need of protection from "the instincts of her own nature, and a slave to the baser passions of man" (Antoinette 1871: 20). Sisters were in need of more protection because evil lay in wait around them. Shaker sister Antoinette continues:

In this new order of things, woman must be purified and elevated: first, purified! . . . There is one great work to be accomplished by woman, on what we term the earthly (or generative) plane. She must work to roll back the flood of sensuality – the giant sin of our time, which finds its way into all ranks of society.

(Antoinette 1871: 20)

Various forms of other cloaking devices are used to protect sisters within the Center Family area. White picket fences and long lines of trees surround the buildings, and stone fence posts protect the primary north and south to east and west axis – all in orthogonal formation. The layering of apple trees in the Center Family area effectively veils the vision and inter-action between women and the few male workhouses which are located at the edge of the zone – the broom shop, smoke house, and cooper shop. Community worship was trickier. Inside the meetinghouse, specially designed for weekly prayer gatherings, separate groups of brothers and sisters were required to dance or "labor" in circular patterns with "hands moving up and down" ("The Shakers" 1857: 168) to the Heavens during their "visit to the Spirit Land" (Dewar 1975: 54). It was through the whirling bodily motion that each group stayed bound within their own uniquely gendered sacred sphere.

As spiritual experience occurred at all scales of the Shaker community, the metaphorical naming of Shaker villages as "The Garden of Eden,"

"New Jerusalem" or "Zion" continually reminded members that they were in a place quite separate from other outside earthly inhabitants "for the earth must, by true science, be subdued and redeemed unto God, as a new earth – New Jerusalem" (Evans 1869: 142). The connection of their communities with Heaven went far beyond a symbolic association:

> Our citizenship or community, (according to the original) is in the heavens. As individuals our names are enrolled with those of the city of God, the heavenly Jerusalem, and as a body, we are one with that community in the heavens . . . hence, through living upon the earth, we are not of the earth, but really of the heavens.
>
> ("Christian Distinction – No. 2" 1871: 21)

Because the Shakers saw themselves as a heavenly community, by default, then, the outside world and all of its participants, not just women as noted earlier, became immoral and debased.

Outsider views

The separatist views of Shakers, particularly during the earlier develop-ment of their society, meant that few outsiders had real encounters with members or saw their villages. This did not mean, though, that people in general were not aware of them and their basic religious practices. "There is no one," it was said in an 1823 *North American Review* article, "who has not heard of the mode, in which the Shakers perform their religious worship" ("Art. V" 1823: 91). National opinion, however, was "somewhat divided" about the "character of the Shakers, in the intercourse with the world" ("Art. V" 1823: 96). Women outsiders raised some of the most vehement misgivings about the Shakers, particularly the brethren. They expressed suspicions about the ironic or incongruous relationship between their professed spiritual beliefs and extremely profitable pre-industrial and agricultural enterprises. One such journalist referred to Shaker brethren who:

> have a perfect conviction that they have dived to the bottom of the well and found the pearl truth, while all the rest of the world look upon them as the bottom of a well indeed; but without the pearl, and with only so much light as may come in through the little aperture that communicates with the outside world.
>
> (Sedgwick 1849: 334)

In this instance brethren might have been segregated for more severe treat-ment because select numbers of them, unlike sisters, were allowed contact with outsiders. Shaker sisters were most likely spared attack because they were given little credit for their industrial enterprises and so they were not linked so directly with the profit motive. Also, most Shaker sisters adhered more closely to nineteenth-century society's generally prescribed role for women as caregiver and conveyor of religious beliefs and values. Another unidentified author in an 1883 issue of *New-England Magazine* raised doubts concerning the spiritual morality of male brethren. Sisters were conveyed as being pitiful and oppressed:

> The meager frames of the women, with the narrow shoulders and spare chests, the sallow and corpse-like complexions, please us, if possible, still less than the hypocritical visages of the men, overshadowed by the long sleek hair, and wearing an aspect of low cunning and assumed humility.
>
> ("Recent Travelers in America" 1833: 365)

Still other journalists were more accepting of the Shakers and not willing to probe their "thrifty, hardworking, inoffensive community," because they opened their doors to the friendless, orphans, and widows ("Art. V" 1823: 102). One last letter from sister Mary Settles, a Believer who resettled into the Pleasant Hill Shaker village in Kentucky after the demise of the West Union village, shows an insider's need to express the compassion and humanity of the Shakers. She wrote to the *Vincennes Commercial* in 1905 to refute claims about the Shakers she read in a *Sullivan Union* article by Mrs Eliza Speake, a female outsider. Sister Settles writes:

> I do not look upon the Shakers as the great part of the people do. The Shakers were not understood ... I have heard of many foolish and untrue statements made concerning the Shakers but I know the Shaker settlement has been a good home for many a widow and orphan, and I know that I enjoyed my life here and feel that I am competent to judge, as I lived in a city, was reared and educated there and I know that the Shakers are good people ... If there are any dark cellars in the house of worship they were used to keep fruit and vegetables in.
>
> (Settles 1905: 2)

While many outsiders were aware of the Shakers, most had not actually encountered members of the Society directly. Some, such as businessmen, journalists, and foreigners such as Harriet Martineau, made it a point to visit Shaker communities. Martineau, like many others, did not comprehend the sacred meaning of their landscape. She only could remark on the neat and tidy appearance of the Shakers' farms and buildings and could not perceive how the Shakers' very moral and religious identity was described through the "heavenly" landscape (Martineau 2000: 159). Geographer and traveler Timothy Flint, upon seeing the West Union Shaker village in 1823, remarked only that the place "exhibits the marks of order and neatness, that are so characteristic of those people everywhere"

(Flint 1970: 153). Few outsiders could have been expected to understand
this relationship because they did not have the religious indoctrination
or training so deeply integrated within the Society. One journalist in
1823, after pondering why so many people, particularly women and children, continued to live as Shakers, surmised that this "essential" training
had kept members together ("Art. V" 1823: 100–101). While the power of
religious indoctrination cannot be underestimated in shaping cultural views
of the landscape, modern historians have also explored other circumstances
which led women into the Society initially. Drawing from 85 testimonies
by Shaker women from the 1820s to 1840s, Campbell found that while
many women joined the Shaker community for religious reasons, more
complex circumstances were often involved in their decision making:
their economic opportunities were severely limited; their social or familial
structure became highly disrupted due to the death or desertion of a
husband; or as older children they had experienced abuse while appren-
ticed in farm positions outside their homes (Campbell 1978: 33–34).
So Shaker women did not leave the Society because in large part their
opportunities would have been very limited in the outside world.

Conclusion

Much is popularly known today about Shaker material culture and celibate rituals. Many assume that the equality represented in the sexual
division of dormitories and labor translated to daily life. Although Shakers
regarded themselves as closer to Heaven than non-Shaker communities, a
more complete view shows that they did not transcend the conventional
views of the time. Furthermore, the shape of their landscape can reveal
that the life of women, in particular, was not confined to their well-known
rituals. Industry, spirituality, and domesticity were combined both in their
interior spaces and their landscapes.

The understanding of how religious beliefs and practices structure the
landscape is a hidden history within the development of American culture.
The formation of gendered sacred landscapes, in particular, have remained
even more concealed from the public presence, limiting insight into
women's own unique spatial environments. The design and reconstruction
of outdoor historic museums, in particular, have an important role to play
in the process of uncovering the hidden histories of gendered sacred space.
Visitors must not only be more readily able to make connections with their
own individual histories or memories from the past but also question historical accounts and interpretations at reconstructed sites. In the larger
framework, it is important to understand how past religious beliefs are the
roots of our own contemporary culture and landscapes.

Collective experience invokes historic memories by revealing, defining,
and connecting Americans emotionally to their environments. Architects,
anthropologists, landscape historians, and cultural geographers in particular

have extended this proposition, by calling for more integration of women's neglected memories into the cultural landscape. The potential for modern reconstructions of sacred sites such as the West Union Shaker village, holds promise for a more complete understanding of the cultural landscape by integrating more complex histories such as those of Shaker women. This is not only because of the mythical appeal these sites already hold but also because of the complex meanings the sacred storied landscapes of these women represent.

Notes

1   See Page Putnam Miller's article "Landmarks of Women's History" in Miller (1992: 1–26). Miller's article gives a more detailed and sobering view of the few historic buildings in the United States committed to the interpretation of women's history. Less clear is the role that landscapes play in the preservation of women's history.
2   For a comprehensive discussion on how the Shakers utilized map-making to both symbolically and functionally document their villages, see Robert Emlen's book *Shaker Village Views Illustrated Maps and Landscape Drawings by Shaker Artists of the Nineteenth Century* (1987).
3   The term "family" was used to describe smaller groups of members within a Shaker community. They could be part of a different age group or at various stages of religious development.
4   With the death of Joseph Meacham in 1796, leadership of the Order went to Lucy Wright. Other than Ann Lee, Wright was the only other female named as leader of the Shakers. Wright's initiative and guidance led the expansion of the Society into the West, particularly in Kentucky, Ohio, Virginia, Tennessee, the Carolinas, Georgia, and Indiana. The Shaker village of West Union, Indiana, was founded under the support of "Mother Lucy." For a larger discussion of power relations in Shaker communities see Marjorie Proctor-Smith's book *Shakerism and Feminism: Reflections on Womens' Religion and the Early Shakers* (1991).
5   While the experiences of middle-class women in America were still confined to domestic life at the beginning of the nineteenth century, some younger women sought out new work opportunities. They became "students at female academies, teachers at common schools, workers in New England textile mills, and employees in workshops and factories." For further discussion of this issue see Nancy Woloch's book *Early American Women A Documentary History 1600–1900* (1992: 224–227).
6   Blunt makes reference to Shirley Ardener's work which also explores how certain activities relate to gendered places. Her 1981 book *Women and Space: Ground Rules and Social Maps* would be particularly useful.
7   For a selection of letters and diaries by frontier women see Annette Kolodny's book *The Land Before Her: Fantasy and Experience of the American Frontiers, 1630–1860* (1984).
8   See also pages 162–164 of Susan Stewart's book *On Longing: Narratives of the Miniature, the Gigantic, the Souvenir, the Collection* (1984).

Bibliography

Antoinette (1871) "God's Spiritual House, or the Perfected Latter Day Temple," *The Shaker (Albany, New York)* 1(3) (March): 20–21.

Ardener, Shirley (1981) *Women and Space: Ground Rules and Social Maps*, New York: St Martin's Press.

"Art. V" (1823) "Review of *The testimony of Christ's second appearing, containing a general statement of all things pertaining to the faith and practice of the Church of God, in this latter day*, by David Darrow, John Meacham and Benjamin S. Youngs," *North American Review* 16(38) January: 76–102.

Bhardwajand, S. and Rinschede, G. (1990) "Pilgrimage in America: An Anachronism on a Beginning," in G. Rinschede and S. Bhardwajand (eds) *Pilgrimage in the US*, Berlin: Dietrich Reimer Verlag: 9–14.

Blunt, A. and Rose, G. (1994) "Introduction: Women's Colonial and Postcolonial Geographies," in A. Blunt and G. Rose (eds) *Writing Women and Space Colonial and Postcolonial Geographies*, New York: Guilford Press: 1–25.

Campbell, D.A. (1978) "Women's Life in Utopia: The Shaker Experiment in Sexual Equality Reappraised – 1810 to 1860," *The New England Quarterly* 51(1) (March): 23–38.

"Christian Distinction – No. 2" (1871) *The Shaker (Albany, New York)* 1(3) (January): 21.

Conlin, M.L. (1963) "The Lost Land of Busro," *The Shaker Quarterly* 3 (Summer): 44–60.

"Co-operative Religious Organization of Shakers has Passed Away" (1905) *Sullivan Union* 8 March 1905: 9.

Dewar, K.P. (1975) *Sacred Space in Ritual Dance*, MS Thesis, Smith College.

Dowd, G.E. (1992) *A Spirited Resistance The North American Indian Struggle for Unity 1745–1815*, Baltimore MD: Johns Hopkins University Press.

Emlen, R. (1987) *Shaker Village Views Illustrated Maps and Landscape Drawings by Shaker Artists of the Nineteenth Century*, Hanover NH: University Press of New England.

Evans, F.W. (1869) *Autobiography of a Shaker, and Revelation of the Apocalypse*, Mt Lebanon NY: F.W. Evans.

Flint, T. (1970) *A Condensed Geography and History of the Western States or The Mississippi Valley*, 2 vols 1823, Gainesville FL: Scholars' Facsimiles & Reprints.

Hanson, S. and Pratt, G. (1995) *Gender, Work, and Space*, London: Routledge.

Hayden, D. (1995) *The Power of Place: Urban Landscapes as Public History*, Cambridge MA: MIT Press.

Janzen, D. (1992) *An Archaeological Survey of the West Union Shaker Village, Knox County, Indiana*, Indianapolis IN: Indiana Historical Society.

Kolodny, A. (1984) *The Land Before Her: Fantasy and Experience of the American Frontier, 1630–1860*, Chapel Hill NC: University of North Carolina Press.

A Look Inside a Shaker Kitchen (1996) leaflet, Pleasant Hill: Harrodsburg KY: Shaker Village of Pleasant Hill.

M. (1871) Dialogue "Flowers," *The Shaker* 1(3) (March): 23–24.

Martineau, H. (2000) *Retrospect of Western Travel, 1838*, Armonk, England: M.E. Sharpe.

Miller, P.P. (ed.) (1992) *Reclaiming the Past: Landmarks of Women's History*, Bloomington IN: Indiana University Press.

Nash, C. (1994) "Remapping the Body/Land: New Cartographies of Identity, Gender, and Landscape in Ireland," in Alison Blunt and Gillian Rose (eds) *Writing Women and Space Colonial and Postcolonial Geographies*, New York: Guilford Press: 227–250.

Procter-Smith, M. (1985) *Women in Shaker Community and Worship A Feminist Analysis of the Uses of Religious Symbolism*, Lewiston ME: Edwin Mellen Press.

—— (1991) *Shakerism and Feminism: Reflections on Women's Religion and the Early Shakers*, Old Chatham NY: Center for Research and Education Shaker Museum and Library.

"Recent Travelers in America" (1833) *New-England Magazine* 5(5) (November): 361–368.

Rose, G. (1993) *Feminism & Geography The Limits of Geographical Knowledge*, Minneapolis MN: University of Minnesota Press.

Ross, E. (1991) "Diversities of Divine Presence: Women's Geography in the Christian Tradition," in J. Scott and P. Simpson-Housley (eds) *Sacred Places and Profane Spaces: Essays in the Geographies of Judaism, Christianity and Islam*, New York: Greenwood Press: 93–114.

Sassen, S. (1996) "Identity in the Global City: Economic and Cultural Encasements," in P. Yaeger (ed.) *The Geography of Identity*, Ann Arbor MI: University of Michigan Press: 131–151.

Sebree, M.S. (1956) "Prosperity, Then Troubles: Lot of Busseron Township. Sect," *Vincennes Commercial* (April 1): 11.

Sedgwick, C.M. (1849) "Magnetism Among the Shakers," *The Living Age* 21(261) (May 19): 334.

Setta, S.M. (1979) "Woman of the Apocalypse: The Reincorporation of the Feminine Through the Second Coming of Christ in Ann Lee," PhD thesis, Philadelphia PA: Pennsylvania State University.

Settles, Sister Mary (1905) "Response to Mrs. Eliza Speake at Carlisle," *Vincennes Commercial* 26 (March): 2.

"The Shakers" (1857) *Harper's New Monthly Weekly Magazine* 15(86): 164–177.

Sheffler, D.M. (1968) "A Historical Study of the Shakers at the Busseron Settlement in Indiana," Masters thesis, Bloomington IN: Indiana State University.

Stewart, S. (1984) *On Longing: Narratives of the Miniature, the Gigantic, the Souvenir, the Collection*, Baltimore MD: Johns Hopkins University Press.

Warren, S. (1997) "This Heaven Gives Me Migraines: The Problems and Promise of Landscapes of Leisure," in J. Duncan and D. Ley (eds) *Place/Culture/Representation*, New York: Routledge: 173–186.

"Who are the Shakers?" (1871) *The Shaker* 1(3) (January): 1.

Woloch, N. (1992) *Early American Women A Documentary History 1600–1900*, Belmont CA: Wadsworth.

Yaeger, P. (1996) "Introduction: Narrating Space," in P. Yaeger (ed.) *The Geography of Identity*, Ann Arbor MI: University of Michigan Press, 1996.

10 Cleaning house

Or one nation, indivisible

Jessica Blaustein

Introduction

It was October 1892, Dedication Day for Chicago's World's Columbian Exposition. Four hundred years after Christopher Columbus "discovered" America, thousands of people gathered from all over to recite the recently drafted Pledge of Allegiance to a living representation of the nation, an impressive spectacle composed of hundreds of schoolgirls. Dressed in red, white, and blue, their small bodies were arranged to form the flag of the United States of America.[1]

The Fair officially opened to the public in May of 1893.[2] An "orgiastic assemblage of the rich and monumental," in the words of renowned critic of American urban planning, Jane Jacobs (Jacobs 1961: 24), Chicago's Columbian Exposition has invited a broad range of commentary and criticism concerning its participation in the discourses of gender, race, nation, and civilization. But the romance between two major characters, sex and architecture, in this bizarre national fairy tale, is a story left largely untold. The following pages attend closely to the relationship between the architectural production of space, the social organization of sexual difference, and the self-conscious construction of national identity. For what tends to be overlooked in even the most thorough analyses of the marginalization of women at the Exposition is precisely the *centrality* of a spatial discourse of femininity to a white heterosexual nationalist project.

Part one of this study explores the 1893 World Fair's architectural construction of a national identity. Through a close and critical reading of the architects' rhetoric as well as the built space itself, I understand the central grounds of the Exposition, the "White City" (as it was called then and as it is remembered now), as a model home, a national domestic space built by a self-declared family of man. Part two situates the proper "Woman" in this national household, linking the architects' concern for aesthetic coherence with the disciplined display of a certain kind of female body. In short, this paper critically investigates the processes by which the Columbian Exposition put on a public face of the nation: the evanescent white facades, symbolizing the heights of civilization, powerfully framed aesthetic coherence against the visual and sensual disorder associated with living, breathing, desiring bodies. My readings of the architecture of the World's Fair are at the same time analyses of normative standards of hygiene, of inhabitability, of gender identity, and of sexual activity that, in a myriad of ways, continue to organize social space today. With every stroke, this work insists upon a mutually constitutive relationship between material productions of space and structures of morality that assign value to practices of gender and sexuality. For such structures are not imposed from above; they are *erected*, propped up in this particular case by a family of man desperately trying to keep their house clean.

Part one

Setting

The Columbian Exposition staged a ground-breaking event for a city with a distinct history of dirt, disorder, and disaster. Mid-nineteenth-century Chicago's mud caused buildings to sink and rendered the few sidewalks that did exist non-traversable (Muccigrosso 1993: 18–19). In 1857, 600 workers employed by George Morton Pullman quite literally raised the foundations of the city's buildings with jackscrews four feet above the muck to enable the construction of new sidewalks.[3] The smell no doubt accentuated the muddy landscape: Packingtown, the Union Stockyards, opened in December 1865, and "regularly conveyed its odors as well as its meats northward to the city proper" (Muccigrosso 1993: 20). In 1871, the Great Fire of Chicago killed 300 people, left 100,000 more homeless, and destroyed 2,000 acres of the city's downtown including 17,000 buildings and one-third of the city's wealth. And yet, because a large majority of Chicago's manufacturing capacity as well as its livestock, lumber, and grain holdings were left unharmed, a good portion of the city bounced back relatively quickly (Muccigrosso 1993: 22).

If, by the 1890s, Chicago could boast a visibly modern landscape, it was a city-space clearly divided between the newly cleansed and ordered living and working arrangements for which it wanted to be known, and the multiple realities of overcrowding, disease, and corruption. Struggling to produce itself as a metropolitan city, Chicago proved to be the ideal site for the World's Fair. Henry Van Brunt, one of the architects involved in the design of the White City, emphasized the region's raw potential when he wrote:

> Chicago is the nucleus of a vast interior country, newly occupied by a prosperous people, who are without local traditions; who have been absorbed in the development of its virgin resources; and who are more abounding in the out-of-door energies of life, more occupied by the

practical problems of existence, more determined in their struggle for wealth and knowledge, than any people who ever lived.

(Van Brunt 1969: 306)

In its very early stages of construction, "engaged in the comparatively coarse work of laying the foundations and raising the solid walls of material prosperity" (Van Brunt 1969: 306), Chicago signified the nation's interiority – a site to be developed, refined, perfected, domesticated. Van Brunt continued:

> This nation within a nation is not unconscious of its distance from the long-established centers of the world's highest culture, but it is full of the sleepless enterprise and ambitions of youth . . . it knows its need of those nobler ideals and higher standards.
>
> (Van Brunt 1969: 306)

The higher stages of societal development required certain *finishing touches*: "The new nation," he wrote, "is now ready to adorn the great fabric, to complete and refine it, and to fit it for a larger life and a wider usefulness" (Van Brunt 1969: 306). Significantly, this important work was to be performed not by engineers, but by *decorators*: the nation he compared to "a machine which requires only those more delicate creative touches necessary to bring its complicated adjustments into perfect working condition, so that it may become as effective as a part of the civilizing energy of our time" (Van Brunt 1969: 306).

Family of man

Van Brunt's rhetoric radically departed from conventional languages of heroic masculinity. And, yet, the "delicate creative touches" were to be performed by white *male* bodies. Refiguring domesticity on a national scale, he identified himself as a member of a household of architects who adorned the country (from its very heart) with civilizing standards and ideals. The men were hand-picked by Chief of Construction Daniel Hudson Burnham and his consulting architect John Wellborn Root, along with landscape architects Frederick Law Olmstead and Henry Codman.[4] Van Brunt, one of the appointed few, affectionately recalled the gathering with these words:

> By a remarkable fortune, the architects to whom the five buildings on the great court were assigned constituted a family, by reason of long-established personal relations and of unusually close professional sympathies. Of this family Mr Hunt was the natural head; two of its members, Post and Van Brunt, were his professional children; Howe, Peabody, and Stearns, having been pupils and assistants of the latter,

may be considered the grandchildren of the household; while McKim, who had been brought up under the same academical influences, was, with his partners, of the same blood by right of adoption and practice. Collaboration under such circumstances, and under a species of parental discipline so inspiring, so vigorous, and so affectionate, should hardly fail to confer upon the work resulting from it some portion of the delightful harmony which prevailed in their councils.

(Van Brunt 1969: 236)

They were the perfect millennial family, "composed of equal parts of white supremacy and powerful manhood" (Bederman 1995: 31) (Figure 10.1). Together they composed a magnificent White City at the center of which stood the Court of Honor, a half-mile long formal basin framed by seven colossal buildings symbolizing the highest scientific, artistic, and technological achievements of civilization: Manufactures, Mines, Agriculture, Art, Administrations, Machinery, and Electricity. The fabricated landscape was self-contained and self-sufficient – with its own transportation, sewage, police, and even governmental systems.

Figure 10.1 Directors, artists, and sculptors of the 1893 Chicago World's Columbian Exposition

Courtesy of Chicago Historical Society (ICHi-02207)

Site

According to the *Official Guide to the World's Columbian Exposition*, "the site of the fair [had] been universally pronounced an ideal one" (*Official Guide* 1893: 17). "A sad wasteland of sand, sparse vegetation, and marshes" (Muccigrosso 1993: 51), "a flat uninteresting piece of sandy soil with some scrubby trees" (Bolotin and Laing 1992: 8), Chicago's Jackson Park between 56th and 57th Streets was far from picturesque. But to its future inhabitants, it offered a clean slate.[5] As architect Louis Sullivan put it, "the site was to be transformed and embellished by the magic of American prowess" (Sullivan 1956: 317). Literary giant William Dean Howells' Altrurian Traveler visited the Fair and described the complex preparation process as one of purification:

> There were marshes to be drained and dredged before its pure waters could be invited in. The trees which at different points offer the contrast of their foliage to the white of the edifices, remain from wilding growths which overspread the swamps and sand dunes, and which had to be destroyed in great part before these lovely groves could be evoked from them. The earth itself, which now of all the earth seems the spot best adapted to the site of such a city, had literally to be formed anew for the use it has been put to.
>
> (Howells 1961: 23)

Much of the World's Fair literature similarly portrays the alteration of the landscape. The *Official Guide*, for example, goes to great lengths to illustrate the drainage and "perfect sewerage" systems through which the water and refuse were "rendered entirely inoffensive" (*Official Guide* 1893: 17).[6] The enthusiasm for a space so completely untainted was not merely aesthetic, for in the eyes of its designers and many of its visitors, the site represented an idealized body free of dirt, disease, and messy complications. It was a paradise with "no shadow," as the fictionalized Altrurian Traveler put it, with "no hint of the gigantic difficulties of the undertaking" (Howells 1961: 23), manufactured to produce comfort and pleasure by masking its history, hiding the labor involved in its preparation, and containing its excesses in an orderly system of underground pipes.[7]

Order

In his papers, Van Brunt is very careful to assert the artfulness of the Fair's design and construction. Upon a blank canvas, the architects constructed a dream world, "a vast phantasm of architecture, glittering with domes, towers, and banners" (Van Brunt 1969: 233), seemingly untied to the whims of the earth. But he stresses that the White City was a *rule-bound* aesthetic composition, an "orderly result of forethought, ... a result of logical processes and not a mere matter of taste, a following of fashion, or an accident of invention" (Van Brunt 1969: 232). The aesthetics were resolutely classical. He points repeatedly to the Court of Honor's "perfect harmony," "uniform and ceremonious style," "historical authority," and "academic discipline" (Van Brunt 1969: 245, 233). And he dwells on the importance of "the principle of symmetry," "the balanced correspondence of parts on each side of a center line" (Van Brunt 1969: 245) that governs the buildings' facades. Montgomery Schuyler, newspaper journalist and architectural critic, certainly did not share Van Brunt's neoclassical tastes, and yet he affirmed that the White City's success was a result of the classical formulas governing its design. The uniform cornice line of sixty feet ensured a continuous and regular skyline, and the classical aesthetic guaranteed a sense of unity (Schuyler 1961: 560). It was "a triumph of *ensemble*," he wrote, wherein "the whole is better than any of its parts and greater than all its parts, and its effect is one and indivisible" (Schuyler 1961: 559).[8]

Pedagogy

The White City was a society by design, "the effect of a principle, and not the straggling and shapeless accretion of accident" (Howells 1961: 23), whose clean lines filtered the chaos and contingency of the everyday. Strategically set apart from the rest of the world, its buildings and landscapes performed at least two critical pedagogic functions. First, the classical aesthetic disciplined and ordered the "impure and unhealthy vernaculars" practiced by many architects of the day so that "true architecture" might replace "undisciplined invention" and "illiterate originality" (Van Brunt 1969: 233–4). Second, the White City was supposed to transform its inhabitants into a well-rounded citizenship. Van Brunt preached:

> The Exposition will furnish to our people an object lesson of a magnitude, scope, and significance such as has not been seen elsewhere. They will for the first time be made conscious of the duties, as yet unfulfilled, which they themselves owe to the civilization of the century. They will learn from the lessons of this wonderful pageant that they have not as yet taken their proper place in the world ... they will obtain in short a higher standard by which to measure their own shortcomings and deficiencies.
>
> (Van Brunt 1969: 307–8)

He described the Fair as a "university ... majestically housed ... open to all, where the courses of instruction cover all the arts and sciences, and are so ordered that to see is to learn": "To walk in these grounds will be in itself an education, as well as a pleasure of the most ennobling sort" (Van Brunt 1969: 311–12). The Fair's pedagogic purpose was made explicit from the very first page of the *Official Guide to the World's Columbian Exposition*.

It read:

> The first duty of the visitor who is desirous of obtaining the best possible results from a visit to the World's Columbian Exposition, be his time brief or unlimited, is carefully to study the accompanying map. This is an absolute necessity to one who would not travel aimlessly over the grounds and who has a purpose beyond that of a mere curiosity hunter. It is presumed at the outset that the great majority of visitors are those who seek to enlighten themselves regarding the progress which the world has made in the arts, sciences and industries. To him who enters upon an examination of the external and internal exhibit of this the greatest of World's Fairs a liberal education is assured. It is the aim of this volume to aid in such endeavor – to clear the way of obstacles – to make the pathway broad and pleasant.
>
> (Official Guide 1893: 5)[9]

Performance

If to see is to learn, the Exposition's pedagogy accordingly took the form of a spectacle. What the fairgoers acquired at this magnificent university was a powerful *image* of the nation.[10] Benedict Anderson's now classic analysis of the nation as "an imagined political community" makes powerfully clear that "all communities larger than the primordial villages of face-to-face contact (and perhaps even these) are imagined. Communities are to be distinguished, not by their falsity/genuineness, but by the style in which they are imagined" (Anderson 1983: 6). The White City was intended to shine. The architects of the Columbian Exposition produced a very special effect, in Schuyler's words, "an effect of the whole [that] does not depend upon the excellence of the parts" (Schuyler 1961: 567–8). Above all else, the grounds were designed to communicate unity, or at least the *possibility* of unity.[11] He explains:

> The great advantage of adopting a uniform treatment, even when the uniformity is so very general as is denoted by the term classic, and even when the term has been so loosely interpreted, is that the less successful designs do not hinder an appreciation of the more successful, nor disturb the *general sense of unity* in an extensive scheme, which is so much more valuable and impressive than the merits of the best of the designs taken singly.
>
> (Schuyler 1961: 567 (my emphasis))

The architects of the Columbian Exposition carefully and elaborately engineered an aesthetically coherent national identity. For their materials, they used iron, wood, glass, and, according to the *Official Guide*, 30,000 tons or 2,000 carloads of a substance called "staff" (*Official Guide* 1893: 21).

Invented in France and used in the Paris Exposition of 1878, the "durable, cheap and easily produced white admixture [of] plaster, cement and fiber" (Muccigrosso 1969: 74) could be "cast into molds in any desired shape" (*Official Guide* 1893: 22) when mixed with water. The neoclassical facades of the White City, in other words, were quite simply outfits dressing wood and iron structures. Flexible forms to be manipulated for appropriate effects, the classical orders could "be used without a suggestion of any real structure" (Schuyler 1961: 562). Because staff had a murky light tan color, in fact, the "whiteness" of the city needed to be applied – as a second layering, an outfit over an outfit – with a powered paint sprayer (*Official Guide* 1893: 22; Muccigrosso 1969: 74).

And so, it turns out, the White City was not timeless – especially in Chicago. According to the *Official Guide*, "no one seem[ed] to know how long [staff could] withstand the effects of the [city's] changeable climate" (*Official Guide* 1893: 187). Of course, this was part of the point; the ideal city was not designed to withstand ordinary conditions; its "holiday buildings" served only "festal and temporary" (Schuyler 1961: 571) functions. Despite their differences, Schuyler and Van Brunt both distinguished between the built environment of the work-a-day built environment and the Exposition's architectural spectacle.[12] If "real" architectural facades expressed an underlying structure, these classical outfits reflected a "monumental dignity" (Van Brunt 1969: 234–5) divorced from any foundations.[13] In fact, in order for these decorative envelopes to successfully play their designated roles, structures *had* to be completely masked. Otherwise, the illusion of congruity presented to the observers would shatter.

Transcendence

In order to express the nation, the producers of the 1893 spectacle felt the need to transcend "personal idiosyncrasies and accidents of mood or temperament" (Van Brunt 1969: 247) in terms of architectural style. The apparent coherence of national identity was a collective *aesthetic* production, but the thin layer of plaster, cement, and fiber masked more than the crude underlying structures of iron and wood. The Exposition's seamless white facades also abstracted the public face of the nation from the racialized, sexed, and gendered bodies that composed it. The *Official Guide* boasts of "a universal congress which is no respector of geographical boundary, race, color, party or sex" and of "the golden age of American enterprise, American industry and American development . . . where all is peace and prosperity from the Atlantic to the Pacific, from the Lakes to the Gulf of Mexico" (*Official Guide* 1893: 7, 9). The language of universalism in the architects' rhetoric and the Fair's tourist guides blatantly contrasts with the Exposition's rigid hierarchical display of races and civilizations.[14] The white neoclassical city insisted at once upon its neutrality *and* its superiority as it forcefully segregated social groups and identities within and outside of its boundaries.

But there is an interesting twist to this familiar story. Van Brunt's discussion of the Women's Building marks a strange shift in his aesthetic principles, from a disembodied, transcendental aesthetic to a more particular, identity-bound one. In the second part of this paper, I will try to understand the critical role that sex plays at the edges of a deceptively stable masculine household.

Part two

Sex

Van Brunt preaches a universal architecture everywhere else, but when he examines the work of Sophia G. Hayden, the architect of the Women's Building, he maintains that "an architectural composition, like any other work of art, is always more or less sensitive to the personal qualities of the designer," and he announces "an irresistible impulse to look for the distinctive characteristics in which the feminine instinct may have betrayed itself" (Van Brunt 1969: 272). For him, "it is eminently proper," that the Women's Building "reveal the sex of its author" (Van Brunt 1969: 273). He admires the building for the way it tempers its "evident technical knowledge" with "a certain delicacy and elegance of general treatment, a smaller limit of dimension, a finer scale of detail, and a certain quality of sentiment, which might be designated, in no derogatory sense, as graceful timidity or gentleness" (Van Brunt 1969: 273). These qualities "differentiate it from its colossal neighbors" whose "imposing proportions" (Van Brunt 1969: 273) the *Official Guide* reviews in significant detail (*Official Guide* 1893: 18). "The design is rather lyric than epic in character," Van Brunt says, "and it takes its proper place on the Exposition grounds with a certain modest grace of manner not inappropriate to its uses and to its authorship" (Van Brunt 1969: 275).

Without a doubt, the Women's Building tends to be over-sexed by its past and present commentators and critics. David F. Burg, in his informative study of the Exposition called it "one of the lightweights, a mere 199 by 388 feet" (Burg 1976: 143). The *New York Times* drew attention to its "sublimely soft and soothing atmosphere of womanliness" (Weimann 1992: 427).[15] Many of the women felt the same way. Bertha Honore Palmer, the President of the Board of Lady Managers, apparently referred to the building as the Ladies' "lovely child" (Weimann 1992: 155).[16] Unlike the imposing and dignified Administration Building ("the gem of all the architectural jewels of the Exposition" according to the *Official Guide* (1893: 19, 27), the Women's Building was described by another guide as "a white silhouette against a black background of old stately oaks ... a daintily designed building ... encompassed by luxuriant shrubs and beds of fragrant flowers" (*Picturesque Chicago* 1892: 291).[17]

These overwhelmingly warm receptions of the Women's Building *in terms of its essential femininity* firmly situate it within a separate sphere – a female world that complements but does not interfere with the world of men, a "Cult of True Womanhood" that, by the end of the nineteenth century, frequently designates a powerful, and often explicitly public, space for women based on some essential difference from the male-dominated political sphere and marketplace.[18] If the activities of the Board of Lady Managers and their Women's Building posed any threat to the structures and powers that be, these anxieties were most certainly contained by a rhetoric (authored by both women and men) that reiterated the differences between the sexes and designated their separate spheres accordingly. This is not to suggest that such spheres actually existed so definitively, for the Fair's gendered geography was constructed and maintained by a multiplicity of contradictory and unstable processes. Any material and discursive designations of differences between two distinct sexes tend to expose their own boundaries and limitations. As proper femininity takes its place in this spectacle therefore, it is rather precariously propped up amidst a myriad of other possible gender and sexual identities that differently intersect race, class, and nation.

And so why and how is the sex of the Women's Building so significant? What is the proper place of the woman in the Columbian Exposition? If this household featured a family of man adorning the nation with their aesthetic sensibilities, what role does *she* play in this curiously domestic scene? The *Official Guide* marvels at the presence and visibility of women – and by this the Commissioners mean white bourgeois women – in the Fair, bragging on their behalf that "they are capable, in almost every department of human activity, of competing with man" (*Official Guide* 1893: 107). And yet their activities were assigned to a spatially segregated zone. The Women's Building was not originally conceived as an exhibition space for women's work; it became so only as a result of women's exclusion from other spaces.[19] As historian Gail Bederman explains, "Despite [the Board of Lady Managers'] best efforts, the Women's Building was perceived as a place apart" (Bederman 1995: 33–4), and its contents were deemed qualitatively different from the contents of other buildings. In this sense, women's work was contained. This study, however, directs attention away from the modes by which the proper woman is *contained* toward the ways in which she is *containing*. Indeed, white bourgeois women inhabited a place distinct from men at the Columbian Exposition, but it was the discourse of femininity itself that functioned to establish this difference. If, at the turn of the century, the ideology of white womanhood reigned over the interior spaces of the home – and by extension, the nation – the Exposition's femininity performed a surprising amount of work on the outside and at the borders of a decidedly masculine-gendered domesticity.[20] The *Official Guide*'s lengthy description of the Statue of the "Republic" that faces the Administration Building in the Court of Honor is striking in this regard:

This figure is sixty-five feet tall, is perfect in symmetry, and was designed by Daniel C. French, of New York. The arms and hands are upraised toward the head . . . in harmony with the beautiful buildings which surround the Grand Basin, the Statue is of true classic build . . . From the chin to the top of the head is 15 feet, and the arms are 30 feet long. A line around the head and hair measures about 24 feet, and the nose is 30 inches in length. Four men could find sitting room on one of the hands, and it would take a wedding ring ten and a half inches in diameter to fit her finger. The length of her forefinger is 45 inches. There is a stairway through the inside of the figure, and the man who attends to the electric lights in the diadem clambers up a ladder through the neck and out through a doorway in the crown of the head.

(*Official Guide* 1893: 25)

A passage like this deserves a book in itself, but let it be said that the gigantic, abstract, and idealized female form frames a space to be explored, inhabited, and domesticated by men (Figure 10.2).

Figure 10.2 The White City's Court of Honor: Grand Basin looking westward from the Statue of the Republic

Courtesy of Chicago Historical Society (ICHi-02524)

Ornament

For every building in the White City, according to the *Official Guide*, the decoration is understood to express the purpose of the building on its surface; ornament, that is to say, has an illustrative function. As discussed in part one, the decorative facades do not in fact reflect an inner structure; rather, they are the vehicles for certain ideals and values. The Fair's official discourses articulate the place of proper womanhood on the Exposition's landscape in a strikingly similar manner. At its *surfaces*, on its face, so to speak, the "Woman" illustrates the purpose of the nation. Like the white facades, her identity is cast to accessorize a certain function.[21]

At the "very edge of the civilized White City, far from the manly Court of Honor" (Bederman 1995: 34), and "situated directly east of the Midway Plaissance" (*Official Guide* 1893: 107), the Women's Building, the last of the Main Exhibition Buildings before the foreign nations, functioned as the decorative envelope of the Exposition's classically composed center. Habitual alignments of ornament with the trivial or unnecessary mislead in this regard, for this is no minor role. Strategically positioned at the border between the civilized and the savage, the white woman policed a national space, the space of "proper" citizenship. In this sense she was the most important element of the White City, its finishing touch against which the exotic and the sexual were articulated. It was her duty to differentiate the inside from the outside, to delineate, through her symmetrical decorous feminine aesthetic, an interiority wherein resided the nation's citizens. "Clad confidently in snow-white Italianate plaster," the Women's Building terminated the Midway Plaissance and asserted cultural superiority . . . but "not . . . too strongly," another critic made clear: "Everything was done in moderation; no Moorish hip-swaying here, no gentleman drinking gin" (Miller 1990: 229) (Figure 10.3).

Hootchie Kootchie

Moorish hip-swaying happened *outside* the house, beyond the frame of the aesthetically coherent White City. The *Official Guide* described the Midway Plaissance as "a strip connected with but not necessarily part of the Exposition proper" where the visitor could find the "minor buildings, special exhibits, etc." (*Official Guide* 1893: 193, 15). "It is lined with picturesque structures, among which rise minarets, mosques, domes, towers, and castle battlements almost without number" (*Official Guide* 1893: 193). Because the admissions to shows and special exhibits along the Midway were separately charged, its attractions were figured as private in relation to the Exposition proper.[22] In this sense, the White City's order domesticated a public sphere, a theoretically democratic national space, while the Midway Plaissance exoticized a series of spaces open to the public but cloaked in a curious kind of privacy – the peep show variety, not interiority proper. As opposed to the aesthetically coherent White City, the Strip, its sexuality unbound,

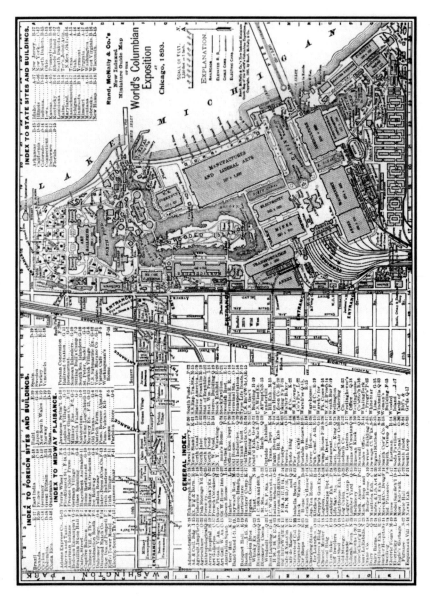

Figure 10.3 Souvenir map of the World's Columbian Exposition
Rand McNally & Co.

signified disorder, play, *tactility*. Among the spectacles of excess, hootchie kootchie, or belly dancing, considered by Julia Warde Howe (the author of "The Battle Hymn of the Republic") to be "simply horrid," indecent, "a most deforming movement of the whole abdominal and lumbar region" (Muccigrosso 1993: 167).

Chaos

Howe's voice emerges from an extremely interesting space: a visually becoming femininity (firmly?) situated between an appalling hootchie kootchie and the nation. Mediating the composed and the deformed, the Women's Building functioned, in other words, as a screen and spectacle, not of excess, but of decency.[23] Judy Sund argues that "the bawdy perform-ances of the Fair's outer edge conjured up a sort of womanliness that stood in dichotomous relation to the static and decorous feminine embodiments at the center" (Sund 1993: 457), but the virgin-whore complex in fact does not adequately capture the Fair's spatial morality. For it was not female promiscuity that rested at the very other end of the Midway Plaissance; it was, rather, a bewildering sexual ambiguity. There resided the savage Dahomens; the "least manly of all the Midway's denizens, [they] seemed to lack gender difference entirely" (Bederman 1995: 36). The *New York Times* gawked at:

> the Dahomey gentleman, (or perhaps it is a Dahomey lady, for the distinction is not obvious,) who may be seen at almost any hour . . . clad mainly in a brief grass skirt and capering nimbly to the lascivious pleasings of an unseen tom-tom pounded within . . . There are several dozen of them of assorted sexes, as one gradually makes them out'. Asserting that he could only "gradually" make out the difference between the sexes, the columnist suggests that savages' sexual difference was so indistinct that the Dahomens might have a larger "assortment" of sexes than the usual two.
>
> (Bederman 1995: 36)[24]

The border between the civilized White City and its outside functioned as the mark of sexual difference itself. The visual spectacle of femininity, there-fore, maintained national coherence through the production of the difference between sex one and sex two. The molded white facade of proper womanliness enveloped an aesthetically coherent domestic space with distinct gender boundaries.

Conclusion

The 1893 World's Columbian Exposition's spectacular production of a national identity was at every scale a performance of the proper relation

between two distinct sexes. If white-washed womanhood provided the decorative frame for the fair's national pedagogy, the red, white, and blue bodies
of schoolgirls seemed to be its basic building blocks. On Dedication Day,
little girls enacted the nation on a diminutive scale. And for six months in
one of the most charming exhibits of a White City engineered and maintained by a family of man, several other little girls, daughters of immigrants,
kept a very tiny house clean.

Right next door to the Women's Building, it was a live demonstration
of Emily Huntington's kitchen garden, an entertaining and "novel method
of developing in little girls a 'knowledge of household duties'" (Weimann
1992: 339).[25] If the rest of the White Cityscape masked its labors, its
excesses, and its unsightly waste, the Broom Brigade performed its curiously choreographed work before the public's watchful and adoring gaze
(Figure 10.4).

> *Look* (through the large glass windows):
>   The little girls did their work to piano music and sang nursery rhymes
> altered for the occasion ... each little girl wore a quaint costume of
> white muslin cap and apron, proudly displaying a badge of miniature

Figure 10.4 The Broom Brigade, Children's Building (photographer C.D. Arnold)
Courtesy of Chicago Historical Society (ICHi-25109)

knife, fork and spoon crossed and tied with ribbon ... In the center of their room were low tables with tiny beds on them. Elsewhere in the room were tiny brooms, dishes, washtubs, and scrubbing boards ... The little beds were all mussed up and, two little girls to a bed, they started to make them up. First the mattresses were turned and punched to a degree of softness. The sheets were spread with the hem turned the right way, the blanket put on, spread, and then came the tucking-in process. No danger of that tucking coming out ... After the beds were made ... half the people who watched wished they were tiny enough to crawl right into them ... While the beds were being made, other little girls were sweeping the room. They did not sweep around the middle, but went into corners in a way which, if it is carried into later life, will cause some men to call them blessed. Others got down on their knees and scrubbed, and some went to the washtubs and, with sleeves rolled up over pretty little arms, made dirty doll clothes look as white as snow.

(Weimann 1992: 341)

Notes

1 According to Robert W. Rydell, Francis J. Bellamy drafted the Pledge of Allegiance for the 1893 Exposition "to make the event truly national in character" (1984: 46). Distributed by the Federal Bureau of Education to teachers around the country, the Pledge served an explicitly pedagogical function (1984: 46). See also Miller 1990: 207; *Memorial Volume: Dedicatory and Opening Ceremonies of the World's Columbian Exposition*, ed. Joint Committee on Ceremonies, Chicago, 1893; *The Youth's Companion* 65 (8 September 1892): 446–7; and White (1893).
2 The Exposition stayed open through October of 1893, and apparently the buildings were destroyed by arsonists on July 5, 1894.
3 In the following year, of course, Pullman's employees were to make United States history again. The 1894 Pullman Strike was the nation's first national strike.
4 Together, the appointed group selected the following architects and firms: Richard Morris Hunt; McKim, Mead and White; Peabody and Stearns; George B. Post; Van Brunt & Howe; Adler and Sullivan; William Le Baron Jenney; Henry Ives Cobb; S.S. Beman; and Burling and Whitehouse.
5 In fact, the *Official Guide* boasted of the millions spent to wipe the slate clean: over $4,000,000 on "laying out the grounds and beautifying them with lawns, flower-beds, shrubbery, etc," and after the site was chosen for the fair, over $5,000,000 on "laying out and beautifying the Exposition grounds" (1893: 17).
6 See Muccigrosso (1993: 163) for further descriptions of the Exposition's sanitation technologies.
7 The Altrurian Traveler goes on to explain: "Every night the whole place is cleansed of the rubbish which the visitors leave behind them, as thoroughly as if it were a camp. It is merely the litter of lunch-boxes and waste paper which has to be looked after, for there is little of the filth resulting in other American cities from the use of the horse ... wheeled chairs pushed about by a corps of high school boys and college undergraduates form the means of transportation by land for those who do not wish to walk" (Howells 1961: 24).
8 Daniel Burnham, Director of Works, would go on to design *The Plan of Chicago* with Edward H. Bennett in 1909. In his introduction to the proposal, Burnham

wrote, "The origin of the plan of Chicago can be traced directly to the World's Fair Columbian Exposition ... the beginning, in our day and in this country, of the orderly arrangement of extensive public grounds and buildings" (1993: 4).

9 Interestingly, Henry Adams found the pathway anything *but* broad and pleasant. He looked to the fair for instruction, but met instead a "Babel of loose and illjoined, such vague and ill-defined and unrelated thoughts and half-thoughts and experimental outcries as the Exposition, had ever ruffled the surface of the Lakes" (1961: 339–40). "Education ran riot at Chicago," he claimed, "Men who knew nothing whatever – who had never run a steam-engine, the simplest of forces – who had never put their hands on a lever – had never touched an electric battery – never talked through a telephone, and had not the shadow of a notion what amount of force was meant by a *watt* or an *ampere* or an *erg*, or any other term of measurement introduced within a hundred years – had no choice but to sit down on the steps and brood as they had never brooded on the benches of Harvard College" (1961: 342). Adams's words speak to an acute crisis of masculinity that is specific to the Industrial Revolution and that he shares with the architects of the Exposition. Whether or not the self-made man ever existed, these "men who knew nothing whatever" certainly usher in his replacement. One might also read this crisis of the self-made man into Frederick Jackson Turner's famous paper, "The Significance of the Frontier in American History," presented at the Exposition in 1893 and first published in *The Proceedings of the State Historical Society of Wisconsin*, 14 December 1893. See Turner, *The Frontier in American History* (1921).

10 They actually acquired multiple images of the nation. At the Columbian Exposition, "America" was packaged and sold in a way that it had never been before. M. Christine Boyer explains that "the preservation of nostalgic forms of architecture or the reproduction of stereotypical urban scenes in illustrated views, album cards, stereoptical photographs, and picture postcards – all commercial exploits of the nineteenth century – offered the spectator a packageable and consumable manner of looking at cities" (1994: 301). The first official American picture postcards were published at the Chicago World's Fair, and they were sold from vending machines: "Architectural and city views along with patriotic places and heroic men soon became the stereotypical forms and illusionistic spaces into which traditions and memories were poured, as if into so many molds" (1994: 302). See also Bolotin and Laing (1992: 150) for an inventory of the fair's memorabilia.

11 Even Henry Adams praised the fair's attempts to express the nation as a unified whole: Of the classicist architecture, he wrote, "Chicago tried at least to give her taste a look of unity" (1961: 340). And "Chicago was the first expression of American thought as a unity; one must start there" (1961: 343).

12 Schuyler parts ways with Van Brunt in his passionate warnings against any applicability of the Fair's aesthetic to what he calls "true" American architecture. Insisting the fair was to be judged according to different standards and not to be imitated in real life, he warned of the threat posed by cheap imitations and spin-offs in the age of mechanical reproduction, and he asked, "Who would care to have the buildings reproduced without the atmosphere of illusion that enveloped them at Jackson Park and vulgarized by being brought into the light of common day?" (1961: 574). Schuyler's wishes were certainly not granted. For direct evidence of the impact of the Columbian Exposition on American urban planning, see Burnham and Bennett (1993). In the interest of beauty, Burnham and Bennett's 1909 plan for Chicago provided absolutely no housing for the poor. Their defense? Wide streets and sunlight can disinfect slums, and "the appeal of a grand visionary city center carried more prestige and brought in more money than cautious programs for municipal housing reform" (Wright

1980: 211). For a discussion of the impact throughout the country, see Wright (1980). Wright explains that by the end of the century, "civic grandeur was the architectural course toward progress in which most city officials chose to invest" (1980: 211).

13 Architect Louis Sullivan's critique of the Fair's illusive qualities deserves consideration in this regard. In *The Autobiography of an Idea*, he famously declares that, after the Columbian Exposition, "architecture died in the land of the free and the home of the brave" (1956: 324). He refers to the Fair as a sickness, a contagion, a denial of the real: "The damage wrought by the World's Fair will last for half a century from its date, if not longer. It has penetrated deep into the constitution of the American mind, effecting there lesions significant of dementia" (1956: 325). It was Sullivan who proclaimed that architectural form should follow function, an axiom from which followed modern architecture's aggressive assertions of "truth" in the face of eclecticism (assertions not without their own contradictions and attendant politics of gender, sexuality, race, and nation).

14 See for example, Ida B. Wells, *The Reason Why the Colored American Is Not in the World's Columbian Exposition* (1893), a pamphlet Wells self-produced and distributed to protest about the Exposition's racist exclusionary practices.

15 Weimann is quoting a 25 June 1893 *New York Times* article subtitled "Exhibits Which Prove that the Sex is Fast Overhauling Man": "The atmosphere of the entire building is not ... equal suffrage and ... woman's right to invade the domain of man, but the sublimely soft and soothing atmosphere of womanliness."

16 See also Weimann (1992: 61, 262).

17 The interior and the layout of the Women's Buildings were also thought to be distinctly feminine. The *Official Guide* announces that "the tone of the entire interior is beautifully softened" (1893: 111). Burg explains that the interior "was probably the most profusely ornamented of all the building interiors" (1976: 166). And from Rydell: "The visitor might pleasantly end the tour by resting and dining in the garden cafe of the Women's Building – a quiet oasis in green at which, reputedly, the best food at the fair was to be found" (1984: 46).

18 Here, I can only gesture toward a wide range of texts in history and literary criticism that have emerged over the last thirty years. For a thorough and highly influential analysis of the Cult of True Womanhood, see C. Smith-Rosenberg, *Disorderly Conduct* (1985). For an excellent overview of "separate spheres" criticism in the field of literary criticism, see C.N. Davidson and J. Hatcher, "Introduction," in *No More Separate Spheres* (2002).

19 The Board of Lady Managers, established in response to women's requests for involvement, wanted to mount women's work throughout the White City with the Women's Building serving as a museum base. They "planned to place placards throughout the White City informing fair-goers what proportion of each exhibit was produced by women's labor" (Bederman 1995: 33).

20 For an excellent analysis of the relationship between domesticity and liberal individualism's ideology of privacy, see Gillian Brown (1990). Brown's analysis is relevant here insofar as she discusses the various ways that domesticity enables and protects individual privacy as distinct from an increasingly chaotic masculine-gendered marketplace. Brown explains that even though women themselves are not granted the individual rights that men are in the latter half of the nineteenth century, their "feminine" virtues are absolutely necessary for the production and maintenance of a distinctly American individuality. The 1893 World's Fair, in this sense, can be understood as an attempt to produce a stable national *domestic* space in the face of a fluctuating market economy. And social norms of femininity, I want to argue, play an incredibly important role in the maintenance of this order. See also Amy Kaplan, "Manifest Domesticity" (2002)

for a reading of the complex relationships between domestic interiority and American imperialism. Analyzing nineteenth and twentieth century US concepts of the nation as technologies of domestic enclosure and imperial expansion, she argues that the internal constitution and stability of the national household wholly depended upon "the imperial . . . project of mapping the boundaries between the domestic and the foreign" in the border zones, and upon strongly differentiating an interiority from an outside at its (always conflicted) edges (2002: 4).

21 See Judy Sund (1993: 450) for a discussion of the absence of concrete, living women in the Fair's discourses of the "the generic woman," a "malleable feminine shell" with an "ever-flexible identity."

22 According to Bolotin and Laing, "fifty cents entitled the visitor to view all of the Great Buildings, the State and Foreign buildings, all of the grounds and even the Midway Plaissance. But many of the private shows and special events along the Midway charged an additional fee – from a dime to two dollars" (1992: 127).

23 See Mary P. Ryan's very interesting discussion of nineteenth-century ceremonial performances of gender in *Civic Wars* (1997). Ryan explains that "the appearance of gender symbolism after the Civil War was updated and inflated, such that projections of difference between male and female assumed an unprecedented centrality in civic culture" (1997: 246). Discussing the increasing visibility of icons of femininity into the nineteenth century, she briefly outlines two functions of the "whole repertoire of feminine imagery" at this time: First, she explains, femininity functioned as a transcendental symbol "transmit[ting] meaning beyond and above the field of social difference" (1997: 250). And second, she served to unify specific ethnic identities around "common national origins, biological descent, or class position" (1997: 250). Ryan sees the relationship between these two functions as paradoxical to say the least. Within the terms of Ryan's analysis, then, this paper attempts to grasp the paradox that Ryan only posits in the most general sense – to understand how a self-consciously American white bourgeois femininity can be both transcendental and (aggressively) sexed at the same time.

24 Bederman is quoting *The New York Times* June 19, 1893; from *The Best Things at the World's Fair* (1893): "'The natives of Dahomey, male and female, give exhibitions, consisting of war songs and dances, and showing their methods of fighting . . . They are a savage looking lot of females, masculine in appearance, and not particularly attractive . . . The men are small and rather effeminate in appearance" (quoted in Bolotin and Laing 1992: 130). See also Rydell 1984: 66.

25 In the words of journalist Marian Shaw: "A few steps across the way bring us to the Children's Building, where we linger long enough to see the provision which has been made for the care, instruction, and amusement of the little folks, while their parents are viewing the fair without anxiety as to their well-being. This building provides a well-protected and pleasant playground upon the roof, a gymnasium on the lower floor . . . a kitchen-garden where little girls are taught to perform simple household tasks, a kindergarten" (1992: 64). The kitchen garden was a play on the word kindergarten. It was devised by Emily Huntington at the Wilson Industrial School of New York: "Her pupils, the children of immigrants, knew nothing about 'homemaking.' Her desire was to teach them, in pleasant ways" (Weimann 1992: 339). Apparently, Huntington adapted educator Froebel's geometric forms "to little pots, brooms, and scrubbing brushes . . . Poor children, she said, could be taught to make their homes more comfortable; they could also be fitted for domestic service" (1992: 339). By 1885, Kitchen Garden Associations spread throughout the country. For the Children's Building exhibit, arrangements were made with the parents of twenty-five little girls to allow them to come to the Fairgrounds every day at three o'clock where they were drilled

by Miss Huntington. "They were not the same 25 every day and throughout the Fair: 175 girls were given kitchen garden instruction over the span of the Fair" (1992: 341).

References

Adams, H. (1961) *The Education of Henry Adams: An Autobiography (1918)*, Boston MA: Houghton Mifflin.

Anderson, B. (1983) *Imagined Communities: Reflections on the Origin and Spread of Nationalism*, New York: Verso.

Bederman, G. (1995) *Manliness and Civilization: A Cultural History of Gender and Race in the United States, 1880–1917*, Chicago: University of Chicago Press.

Bolotin, N. and Laing, C. (1992) *The Chicago World's Fair of 1893: The World's Columbian Exposition*, Washington DC: Preservation Press.

Boyer, M.C. (1994) *The City of Collective Memory: Its Historical Imaginary and Architectural Entertainments*, Cambridge MA: MIT Press.

Brown, G. (1990) *Domestic Individualism: Imagining Self in Nineteenth-Century America*, Berkeley CA: University of California Press.

Burg, D.F. (1976) *Chicago's White City of 1893*, Lexington KY: University Press of Kentucky.

Burnham, D. and Bennett, E.H. (1993) *Plan of Chicago (1909)*, C. Moore (ed.), New York: Princeton Architectural Press.

Davidson, C.N. and Hatcher, J. (2002) "Introduction," in *No More Separate Spheres*, Durham NC: Duke University Press, pp. 7–26.

Howells, W.D. (1961) *Letters of an Altrurian Traveler (1893–4)*, Gainsville FL: Scholars' Facsimiles & Reprints.

Jacobs, J. (1961) *The Death and Life of Great American Cities*, New York: Random House.

Kaplan, A. (2002) *The Anarchy of Empire in the Making of U.S. Culture*, Cambridge MA: Harvard University Press.

Miller, R. (1990) *American Apocalypse: The Great Fire and the Myth of Chicago*, Chicago: University of Chicago Press.

Muccigrosso, R. (1993) *Celebrating the New World: Chicago's Columbian Exposition of 1893*, Chicago: Ivan R. Dee.

Official Guide to the World's Columbian Exposition (1893), Chicago: Columbian Guide Company.

Picturesque Chicago and Guide to the World's Fair (1892), Baltimore: R.H. Woodward.

Ryan, M.P. (1997) *Civic Wars: Democracy and Public Life in the American City during the Nineteenth Century*, Berkeley CA: University of California Press.

Rydell, R.W. (1984) *All the World's a Fair: Visions of Empire at American International Expositions 1876–1916*, Chicago: University of Chicago Press.

Schuyler, M. (1961) *American Architecture and Other Writings*, vol. 2, W.H. Jordy and R. Coe (eds), Cambridge MA: Harvard University Press.

Shaw, M. (1992) *World's Fair Notes: A Woman Journalist Views Chicago's 1893 Columbian Exposition*, St Paul, MN: Pogo Press.

Smith-Rosenberg, C. (1985) *Disorderly Conduct*, New York: Oxford University Press.

Sullivan, L.H. (1956) *The Autobiography of an Idea (1924)*, New York: Dover Publications.

Sund, J. (1993) "Columbus and Columbia in Chicago, 1893: Man of Genius Meets Generic Woman," *The Art Bulletin*, LXXV.3: 443–66.

Turner, F.J. (1921) *The Frontier in American History*, New York: Henry Holt.

Van Brunt, H. (1969) *Architecture and Society: Selected Essays of Henry Van Brunt*, W.A. Coles (ed.), Cambridge MA: Harvard University Press.

Weimann, J.M. (1992) *The Fair Women*, Chicago: Academy Chicago.

Wells, I.B. (1893) *The Reason Why the Colored American Is Not in the World's Columbian Exposition*, Chicago: Ida B. Wells.

White, A. R. (1893) *The Story of Columbus and the World's Columbian Exposition, Embracing Every Historical Fact and Event Connected with the Life and Adventures of the Great Navigator, Together with a Detailed Account and History of "The Worlds Fair", Including the Dedicatory Exercises*, SI: Juvenile Book Company.

Wright, G. (1980) *Moralism and the Model Home: Domestic Architecture and Cultural Conflict in Chicago 1873–1913*, Chicago: University of Chicago Press.

11 "Virgin land," the settler-invader subject, and cultural nationalism

Gendered landscape in the cultural construction of Canadian national identity[1]

Paul Hjartarson

Canada consists of 3,500,523 square miles mostly landscape.
(One of the "Algomaxims" featured at the
Algoma Sketches and Pictures Exhibition held at
the Art Gallery of Toronto in 1919)[2]

I am not the wheatfield
nor the virgin forest
(Adrienne Rich)

In Canada, as in most settler-invader colonies,[3] cultural representations of the land are central to the national imaginary. Nowhere is this more apparent than in the continuing Canadian fascination with the nationalist school of landscape painters known as the Group of Seven. The Group of Seven was formed in 1920 just as the National Gallery of Canada, which had been incorporated by an Act of Parliament in 1913, was establishing itself under the leadership of its first Director, Eric Brown. Although Group members, all of them male, produced a wide range of images, including portraits, urban scenes, and some abstract paintings, the Group of Seven became increasingly identified in the 1920s with its representations of northern wilderness landscapes.[4] "The extraordinary continuing appeal of the Group of Seven to Canadians," W.H. New argues in his 1989 history of Canadian literature, "owes much to the nationalist climate of the 1920s and its governing equation between nation and landscape" (New 1989: 144). The 1920s' equation of nation with landscape can be traced not just to the nationalism of the period but specifically to the work of the National Gallery. That the new gallery actively promoted the nationalism that developed out of Canada's participation in the First World War is apparent from its annual reports. As Joyce Zemans notes, in its *Annual Report* for 1921–2 the Gallery lists as its "key responsibilities" "nation-building and the establishment of a common heritage for Canadians" (Zemans 1995: 11). While the Group's yearly exhibitions developed its position at the fore-front of Canadian art, Eric Brown and his staff at the National Gallery began early in the postwar decade to build a national tradition around the Group. By 1939, the year in which he died, Brown had established the pre-eminence of the National Gallery in the field of the visual arts in Canada and he had done so, in part, through his support and advance-ment of the Group's work. In "Art Museums and the Ritual of Citizenship" Carol Duncan points out that "museums can be powerful identity-defining machines": "To control a museum," she argues:

> means precisely to control the representation of a community and some of its highest, most authoritative truths. It also means the power to define and rank people, to declare some as having a greater share than others in the community's common heritage – in its very identity.
> (Duncan 1991: 101–2)

In this chapter I examine the National Gallery's championing of the Group of Seven, analyze the 1920s' equation of nation with wilderness landscape, and consider the importance of gender and race in that equation. At least four arguments underlie this paper: that all nations are "invented"; that culture is inextricably bound up in the construction of national identity; that all nationalisms are gendered; and that, as Joanne P. Sharp argues in "Gendering Nationhood" whereas "men are incorporated into the nation metonymically," so that "the nation is embodied in each man and each man comes to embody the nation," women "are scripted into the national imaginary" not as "equal to the nation" but as "symbolic of it" (Sharp 1996: 99).

   Although I have focused this chapter on the equation of nation with land-scape in the 1920s, that equation still shapes the cultural construction of Canadian national identity today. In 1988 Canada opened the new home of the National Gallery in the nation's capital, Ottawa. In Moshe Safdie's design for the gallery and in Cornelia Hahn Oberlander's landscape archi-tecture, the 1920s' equation of nation with landscape was reaffirmed. The site chosen for the new gallery is on Nepean Point, just downstream from, and within sight of, the nation's Parliament Buildings. The most prominent feature of Safdie's architectural design is the "Great Hall," a "pinnacled glass polygon that dominates the exterior of the building just as it provides the main focus for the interior"; that feature is repeated in the smaller entrance pavilion (see Figure 11.1). In *A Place for Art: The Architecture of the National Gallery of Canada*, Witold Rybczynski points out that "in general shape, although not in materials, the conical Great Hall gently mimics the Gothic Revival Parliamentary Library (1859–66) – the only one of the Parliament Buildings to survive the disastrous fire in 1916 –

Figure 11.1 Aerial view, National Gallery of Canada
Ottawa Canadian Aerial Photo Corporation

designed by Thomas Fuller and Chillian Jones" (Rybczynski 1993: 101).
Both the placement of the building on the site and its design reinforce that
connection. After entering the smaller pavilion on Sussex Drive, visitors
ascend a long ramped colonnade to the Great Hall. When they enter the
Great Hall, they are rewarded with an impressive view of the Parliament
Buildings; in the foreground is the Parliamentary Library. The linking of
art with nation is unmistakable.[5]

What, then, of landscape? I want to highlight just one of several things
that could be discussed under this heading. As visitors ascend the ramped
colonnade, a view of a taiga garden opens to the left. In designing that
terraced garden, which uses existing and excavated rocks and which
features, among other things, Siberian juniper, bog rosemary, red-leafed
roses, dogwood, Austrian pine, Arctic bluegrass, cotton grass, wild straw-
berries, Canadian dwarf cinquefoil, and Vancouver jade kinnikinick, the
landscape architect Cornelia Hahn Oberlander sought to recreate the
northern Canadian landscape painted by the Group of Seven and specifically the "wild and savage country" depicted by A.Y. Jackson in his *Terre
Sauvage* (see Figure 11.2). In a report on the National Gallery published in
the December 1988 issue of *Landscape Architecture* Nancy Baele writes:

Since the gallery opened in May, more than 700,000 visitors have climbed the long, ramped glass colonnade to the Great Hall with its 140-foot towering glass cupola echoing the architecture of the Parliamentary Library. The view they see as they make their ascent is Oberlander's taiga landscape. It is a simulation of the Canadian north found in A.Y. Jackson's famous painting *Terre Sauvage*, which belongs to the gallery's collection of the Group of Seven, the first Canadian nationalistic artistic movement. The Vancouver-based landscape architect's design has the effect of a familiar Canadian icon, representing the elemental, untamed aspect of the northern terrain that stretches from British Columbia to Newfoundland. The landscape has paths, flat rocks that can be used as a sitting area, and massed areas of colour – iris, bog cranberries and wild grasses. Its composition is made dramatic by the diagonal lines of stunted Austrian pines bent westerly in bonsai treatment, as though continually resisting wind.

(Baele 1988: 39)

Figure 11.2 Terre Sauvage, A.Y. Jackson, 1913, oil on canvas, 128.8 × 154.4 cm

National Gallery of Canada, Ottawa; courtesy of the estate of the late Dr Naomi Jackson Groves

To Baele, Oberlander describes the site as follows:

> It was a found landscape. I discovered when digging for soil depth that
> the area was covered with flat rocks, so I exposed them and reposi-
> tioned others – some weighing as much as ten tons – that had been
> excavated during the construction.
>
> (Baele 1988: 39)

All elements of the new gallery have been shaped – like Oberlander's
ten-ton rocks and bonsaied Austrian pines – to reiterate the cultural
construction of the nation as landscape. Exhibitions such as *The Group of
Seven: Art for a Nation*, which in 1995 marked the seventy-fifth anniversary
of the Group's formation in 1920, and a vigorous reproductions program
reinforce the equation of nation with landscape.

In "The National Longing for Form" Timothy Brennan argues –
following Benedict Anderson – that, as "imaginary constructs," nations
"depend for their existence on an apparatus of cultural fictions in which
imaginative literature plays a decisive role"(Brennan 1990: 49); moreover,
observing that the rise of the novel parallels the development of nation-
states, he asserts that the novel "was crucial in defining the nation as
an imagined community" (Brennan 1990: 48). If the rise of nation-states
in Europe coincides with the rise of the novel, it also parallels the development of modern European landscape painting. In the opening pages
of "Imperial Landscape" W.J.T. Mitchell poses the following question:
"Is it possible that landscape, understood as the historical 'invention' of
a new visual/pictorial medium, is integrally connected with imperialism?"
(W. Mitchell 1994: 9). "[T]he posing of a relationship between imperialism
and landscape," he remarks:

> Is not offered . . . as a deductive model that can settle the meaning of
> either term, but as a provocation to an inquiry. If Kenneth Clark is
> right to say that "landscape painting was the chief artistic creation of
> the nineteenth century," we need at least to explore the relation of this
> cultural fact to the other "chief creation" of the nineteenth century –
> the system of global domination known as European imperialism.
>
> (W. Mitchell 1994: 10)

In the closing pages of the article he argues that "a close reading of specific
colonial landscapes may help us to see, not just the successful domination
of a place by imperial representations, but the signs of resistance to empire
from both within and without" (W. Mitchell 1994: 21). In a discussion of
New Zealand landscapes and "the contested territories of Israel and Palestine" (W. Mitchell 1994: 20), Mitchell concludes that landscape in both
places "is central to the national imaginary, a part of the daily life that
imprints public, collective fantasies on places and scenes" (W. Mitchell
1994: 27–8).

That Mitchell chose New Zealand as one of the two places to test his hypothesis is revealing. New Zealand, like Canada, is a settler-invader colony, that is, a colony in which the invading Europeans killed, displaced, or marginalized the indigenous peoples to become a majority, non-indigenous population. For such colonies, national identity has been, and remains, a problem. In *The Post-Colonial Studies Reader* Bill Ashcroft, Gareth Griffiths and Helen Tiffin observe:

> Settler colony cultures have never been able to construct simple concepts of the nation, such as those based on linguistic community or racial or religious homogeneity. Faced with their "mosaic" reality, they have, in many ways, been clear examples of the *constructedness* of nations. In settler colony cultures the sense of place and placelessness have been crucial factors in welding together a communal identity from the widely disparate elements brought together by transportation, migration and settlement. At the heart of the settler colony culture is also an ambivalent attitude toward their own identity, poised as they are between the centre from which they seek to differentiate themselves and the indigenous people who serve to remind them of their own problematic occupation of the country.
>
> (Ashcroft *et al.* 1995: 151–5)

For settler-invader colonies, in short, the construction of national identity has proved an ongoing problem. The source of the settlers' difficulty can be traced not just to the diverse cultural backgrounds and languages of the immigrant population but also to their "problematic occupation of the country." (In this context, Mitchell's decision to yoke New Zealand with Israel is telling.) In a study of settler colonies, Anna Johnston and Alan Lawson argue that the settler-invader subject should be conceived as "uneasily occupying"

> a place caught between two First Worlds, two origins of authority and authenticity. One of these is the originating world of Europe, the Imperium – the source of its principal cultural authority. Its "other" First World is that of the First Nations whose authority they not only replaced and effaced but also desired.
>
> (Johnston and Lawson 2000: 370)

As Lawson points out elsewhere, "The address of the settler is toward both the absent(ee) cultural authority of the imperium and the effaced, recessive cultural authority of the Indigene" (Lawson 1995: 29). On the one hand, the settler-invader represents and very often mimics imperial authority and authenticity; on the other, he effaces indigenous authority and appropriates indigenous authenticity. "The national," Lawson argues, "is what replaces the indigenous and in doing so conceals its participation in colonization by nominating a new colonized subject – the colonizer or invader-settler" (Lawson 1995: 30).

Canadian history texts still commonly trace the nationalism of the 1920s to Canada's participation in the First World War and teach that the nation "came of age" on the battlefields of France. The story, as I see it, is more complicated. At the turn of the century Canada was a self-governing British colony known as a "dominion" but a colony nonetheless. In the early decades of this century it had undergone tremendous changes. As a result of immigration, the nation's population had grown at unprecedented rates. In the prairie provinces alone, the population grew from 400,000 in 1901 to 2 million in 1921. Most of that immigration occurred prior to 1914 and the majority of the new immigrants were from non-English-speaking, central European countries against which, in 1914, Canada went to war. At the same time, the colony was rapidly industrializing, a process accelerated even faster when war was declared. With rapid industrialization came rapid urbanization. At the turn of the century, most Canadians lived in the country; by 1931, the majority lived in the nation's cities. With rapid industrialization also came increasing union militancy. The success of the Russian Revolution, combined with concerns about the wartime loyalty of the colony's central European immigrant population, heightened concerns about labor unrest. The business elite's worst fears seemed to be realized in 1919 when workers in Winnipeg declared a general strike. To many men, the changes demanded by the increasingly organized and vocal women's movement, and by women's increasing presence in the workforce, were no less revolutionary. The success of the suffrage movement between 1916 and 1922 raised both expectations and anxieties. "The end of the Great War," historian Tom Mitchell argues:

> did not mark a simple and uncomplicated return for Canadians to a condition of normalcy . . . Canada's business elite and its allies among Canada's middle-class progressives, interpreted the post-war malaise as a crisis of citizenship which threatened the stability of the social order and the destiny of the emergent Canadian nation. They sought to address this condition by casting the post-war order in a particular idiom of nationalism informed by a common Canadianism rooted in Anglo-conformity and citizenship framed in notions of service, obedience, obligation and fidelity to the state. This approach to Canadian nationhood and citizenship furnished a rationale for the suppression of labor radicalism, ignored group claims to expanded citizenship, and consolidated class inequalities.
>
> (T. Mitchell 1996–7: 21)

Canada's business elite and its allies among Canada's middle-class progressives not only conceived "the post-war malaise" as a failure of nationalism and citizenship but traced that failure to the disloyalty of "immigrant aliens" (T. Mitchell 1996–7: 16). A conference on citizenship held in Winnipeg in 1919, in the aftermath of the Winnipeg General Strike, and attended by 1500 of Canada's Anglo-Canadian business and professional elite, recommended that the immediate crisis be addressed through a restriction on immigration, a restriction that would allow time to transform "aliens" into loyal citizens. As Mitchell argues, in the wake of the First World War middle-class English-Canadian nationalism dislodged social reform as the content of Canadian progressivism.

"All nations," Anne McClintock argues in *Imperial Leather: Race, Gender and Sexuality in the Colonial Contest,* "depend on powerful constructions of gender" (McClintock 1995).

> Despite many nationalists' ideological investment in the idea of popular *unity,* nations have historically amounted to the sanctioned institutionalization of gender *difference.* No nation in the world gives women and men the same access to the rights and resources of the nation-state.
> (McClintock 1995: 353)

Nationalism in the 1920s in Canada was informed by prevailing conceptions of both gender and race; it conceived the "emerging" nation as Anglo-Saxon, English-speaking, Protestant, and thus – of course – as morally and sexually pure. The same conception of the nation which saw "alien" immigrants, particularly female immigrants, as the problem, valorized Anglo-Saxon, English-speaking, Protestant women as the biological and cultural reproducers of the nation. As such, they were the focus of increased regulation. In a study of the moral reform movement in Canada from 1885 to 1920, Mariana Valverde argues that the purpose of the social purity movement, which worked largely with immigrants and particularly with women, "was not so much to suppress as to re-create and re-moralize not only deviants from its norms but, increasingly, the population of Canada as a whole" (Valverde 1991).

> The entities being regulated were in the first instance the characters of individuals, with particular emphasis on sexual and hygienic habits; but the nation was also seen as held together by a common subjectivity, whose constant re-creation at the individual level ensured the continued survival of the collectivity. The collectivity thus organized had very specific class, gender, and racial/ethnic characteristics, generally supporting the domination of Anglo-Saxon middle-class males over all others but allowing women of the right class and ethnicity a substantial role, as long as they participated in the construction of women in general as beings who, despite their heroic and largely unaided deeds in maternity, were dependent on male protection.
> (Valverde 1991: 33)

Postwar Canadian nationalism located moral and sexual purity, and thus "continued survival of the collectivity," in representations of virgin land.

One of the speakers at the 1919 Winnipeg conference on citizenship was Dr Henry Suzzallo; he was the president of the University of Washington and a leading figure in the suppression of the Seattle General Strike in March of that year. In his conference speech, Suzzallo declared:

> You can control a man [*sic*] in two ways – by putting a club on one side of his head, or putting an idea inside his head. One is external control and the other is internal control. One is coercive and the other is educational control.
>
> (quoted in T. Mitchell, 1996–7: 17)

According to Mitchell, "Suzzallo's formulation provided a concise and graphic formulation of ruling-class strategy. Hegemony would be achieved through a mixture of naked coercion and manufactured consent" (T. Mitchell, 1996–7: 17). While, in the short term, Canada's ruling class would rely on "external control," restricting immigration and deporting radicals, their long-term strategy – developed on the advice of Suzzallo and others – would focus on "internal control," that is, on the use of public schooling as an instrument of state formation. Taking up the analysis of political philosopher Claude Lefort, art historian Rosalyn Deutsche characterizes totalitarianism as "an attempt to reach solid social ground." Totalitarianism, she argues:

> invest[s] "the people" with an essential interest, a "oneness" with which the state identifies itself, thus closing down public space where state power is questioned and where our common humanity – the "basis of relations of *self* and *other*" – is settled and unsettled.
>
> (Deutsche 1996: 326)

In responding to the postwar crisis, Canada's business class moved in the direction of totalitarianism, investing "the people" with "an essential interest," closing down public space and, with the aid of their allies among Canada's middle-class progressives, mounting a wide-ranging and pervasive nationalist educational campaign to inculcate the moral character appropriate to citizenship in the dominion.

Middle-class English-Canadian nationalism found its icons of moral and sexual purity in the wilderness landscapes painted by members of the Group of Seven. Why this equation of nation with landscape? I want, by way of conclusion, to outline three contexts within which an answer might be framed. The first sees the equation of nation with landscape as a product of the emerging sciences and, following Benedict Anderson, of the related discourses of the geological survey, map, and museum. At the outset of *Imagined Communities* Anderson states as the first of three paradoxes regarding the nation, "The objective modernity of nations to the historian's eye vs their subjective antiquity in the eyes of nationalists" (Anderson 1991: 5). Nationalists need to create an ancient past for the nation. As Tom Nairn points out in "The Curse of Rurality," nationalists usually locate those origins in some representation of the rural and of "peasantry."[6] For settler-invader cultures, however, this presents a problem: how do you break with the authority of the imperial power and, at the same time, shape a past in a nation supposedly devoid of any? Canada found its ancient past – what, following Walter Benjamin, Ann McClintock terms "archaic time" (McClintock 1995: 40) – and its heritage, in the Precambrian rock formation known as the "Canadian Shield." As Suzanne Zeller points out:

> The basis of science as it was practiced in Canada during the Victorian age was inventory, systematic surveys of the land and its resources with the ultimate goal of assessing its material potential. ... Both inventory and nation-building were important organizational processes which Canadians believed would arm them to meet the challenges of industrialization and modernization.
>
> (Zeller 1987: 269)

The Geological Survey of Canada was founded in 1842 to map the colony's mineral wealth; by 1867, it was a vigorous organization engaged in systematic exploration, making maps, producing reports, and maintaining a public museum, which in 1927 became the National Museum of Canada. As the science of geology developed, the Canadian Shield, once thought of as wasteland, was identified as some of the oldest rock on earth and found to contain considerable mineral wealth, timber, and potential for hydroelectric development. By the 1920s all three were being exploited at an unprecedented rate. If the Geological Survey gave the colony an ancient past and a heritage, Group of Seven landscapes shaped that past into icons for the nation. While the "Canadian Shield" was depicted as virgin territory, it also held the promise of the nation's future development and its wealth. The implied narrative was progress through the exploitation of the nation's natural resources.

The second context addresses the question of ownership. If Canadian writers and painters equate nation with landscape in the 1920s, whose nation is it? As geographer Denis E. Cosgrove points out, landscape painting:

> represents a way in which certain classes of people have signified themselves and their world through their imagined relationship with nature, and through which they have underlined and communicated their own social role and that of others with respect to external nature.
>
> (Cosgrove 1984: 15)

In landscape painting, Cosgrove argues, "Subjectivity is rendered the property of the artist and the viewer – those who control the landscape – not those who belong to it" (Cosgrove 1984: 26). Feminist scholars have long engaged in a critique of the subjectivity derived from such visual pleasure. As Deutsche points out:

> For years, feminist theories have differentiated vision – pleasure in looking – from the notion of seeing as a process of perceiving in the real world. The image and the act of looking are now understood to

be relations highly mediated by fantasies that structure and are structured by sexual difference. Visual space is, *in the first instance*, a set of
social relations; it is never innocent.

(Deutsche 1996: 197)

One of the fantasies almost invariably at play in landscape painting involves
positionality; landscape artists typically position the viewer on high ground
and create the illusion of command and control. That mastery typically
extends from visual space to the field of knowledge as a whole. What
Deutsche states concerning the contemporary desire for a "commanding
position" in the postmodern battle over representation is relevant here:

> If representations are social relationships, rather than reproductions of
> preexisting meanings, then the high ground of total knowledge can
> only be gained by an oppressive encounter with difference – the rele-
> gation of other subjectivities to positions of subordination or invisibility.
>
> (Deutsche 1996: 198)

If, as Deutsche argues, "visual space is, in the first instance, a set of social
relations," then the wilderness landscapes of the Group of Seven are more
than simply "beautiful paintings." Art historians have frequently noted the
absence of human and animal life from Group landscapes. The absence
of human life suggests that the land is uninhabited and therefore open to
claim. Group of Seven landscapes sanction what Anne McClintock terms
"the myth of the empty lands."

> The myth of the virgin land is also the myth of the empty land,
> involving both a gender and a racial dispossession. Within patriarchal
> narratives, to be virgin is to be empty of desire and void of sexual
> agency, passively awaiting the thrusting, male insemination of history,
> language, and reason. Within colonial narratives, the eroticizing of
> "virgin" space also effects a territorial appropriation, for if the land is
> virgin, colonized peoples cannot claim aboriginal territorial rights, and
> white male patrimony is violently assured as the sexual and military
> insemination of an interior void.
>
> (McClintock 1995: 30)

Deutsche, in the essay "Agoraphobia" and throughout her writing, places
debates about aesthetic questions "within the context of broader struggles
over the meaning of democracy" (xxii). Focusing on the work of artists in
a 1982 exhibition titled *Public Vision*, she writes:

> [these artists] did not confine their analysis of the politics of the image
> in what appears inside the borders of a picture, within the visual field.
> Instead, they turned their attention to what is invisible there – the
> operations that generate the seemingly natural spaces of the image and
> the viewer.
>
> (Deutsche 1996: 296)

The work of these artists, she argues, interrogates the "modernist model of visual neutrality":

> Visual detachment and its corollary, the autonomous art object, emerged as a constructed, rather than given, relationship of external-ity, a relationship that produces – not is produced by – its terms: dis-crete objects, on the one hand, and complete subjects, on the other. These subjects are not harmless fictions: They are, rather, relationships of power – masculinist fantasies of completion achieved by repressing different subjectivities, transforming differences into otherness, or subordinating actual others to the authority of a universal viewpoint presupposed to be, like the traditional art viewer, uninflected by sex, race, an unconscious, or history.
>
> (Deutsche 1996: 296)

Group of Seven paintings are typically represented as "autonomous art objects." The analyses of McClintock and Deutsche serve as necessary reminders that the modernist assumption of visual neutrality both conceals a relationship of power and licenses a "masculinist fantasy of completion."

If art historians have noted the absence of human and animal life from Group landscapes, they have also remarked on the tendency of Group members to foreground in their wilderness paintings a solitary tree (see Figure 11.3). In "Sighting the Single Tree, Sighting the New Found Land," Matthew Teitelbaum remarks:

> [A]lthough Biblical and European antecedents for the imaging of the single tree has long been a cornerstone of nineteenth-century cultured learning, it was not until the 1920s, and the frequent public exhibition of their pictures, that the Group of Seven and Tom Thomson popu-larized the image of the lone pine tree for the Canadian public. In the popular imagination and the mythos of a new Dominion, the twisting, windswept single tree, with its art nouveau and German romantic inspi-rations, came to symbolize a pioneering spirit crystallizing at the edge of an unknown space.
>
> (Teitelbaum 1991: 71)

The image of the solitary tree has a long history in western culture. In Romantic art the tree can become "a sentient almost human presence" and according to Jonathan Bordo, its use in Group landscapes is, in effect, "two-sided, marked by palpable human absence and concealed human presence" (Bordo 1992–3: 115). The tree is thus, among other things, a signpost or marker announcing the presence of western *man* and claiming the land. Within the context of 1920s' nationalism and the transformation of the dominion into a nation-state, the paintings constitute a land claim which remains the subject of litigation in Canadian courts.

Figure 11.3 Stormy Weather, Georgian Bay, F.H. Varley, 1921, oil on canvas
    132.6 × 162.8 cm

National Gallery of Canada, Ottawa; estate of Kathleen G. McKay

The third context sets the equation of nation with wilderness land-scape within "antimodernism." The concept of antimodernism I have in mind was developed by historian T.J. Jackson Lears and taken up in the Canadian context by Ian McKay. According to Lears, antimodernism can be defined, in its dominant form, as "the recoil from an 'overcivil-ized' modern existence to more intense forms of physical and spiritual experience"(Lears 1981: xiii). In *The Quest of the Folk* McKay writes:

> From the late nineteenth century on, and across the western world, scepticism about "progress" and fear that unprecedented social and economic changes were destroying the possibility of "authentic" experience (and even undermining the bases of selfhood itself) shaped social thought and cultural expression across a wide ideological spectrum.

(McKay 1994: 31)

Antimodernism is an expression of that skepticism about progress and
that fear of change. Lears argues, however, that antimodernism is more
than simply "a 'reaction' against modernizing tendencies." "By exulting
'authentic' experience as an end in itself," he argues:

> Antimodern impulses reinforced the shift from a Protestant ethos of
> salvation through self-denial to a therapeutic idea of self-fulfilment in
> *this* world through exuberant health and intense experience. Anti-
> modernists were far more than escapists: their quests for authenticity
> eased their own and others' adjustments to a streamlined culture of
> consumption.
>
> (Lears 1981: xiv)

The 1920s' equation of nation with landscape conceives the Canadian
north as, in effect, a vast "therapeutic" space in which modern city-dwellers
could find refuge from the materialism and godlessness of inauthentic
modern life. In this view, the unpeopled northern landscape is natural,
unspoiled, virginal: it is the source of spiritual value and, indeed, of existence itself, a place where life is simple, where men are pioneers and women
are the land to be claimed, civilized, and made productive. This is a view
of the nation produced by and for city-dwellers: all members of the Group
of Seven were city-dwellers; most of them were, moreover, born in cities.
The vision of the nation as virginal wilderness is inextricably bound up
with the rise of the tourist economy in Canada and was commodified by
both state and private tour operators, most notably the Canadian Pacific
Railway. Above all, however, in turning away from industrialization, urbanization, and change of all kinds while, at the same time, facilitating "a
streamlined culture of consumption," antimodernism foreclosed the public
space in which those changes – including changes championed by the
women's movement – might be debated.

   In all three contexts, the northern wilderness becomes what McClintock
terms "anachronistic space:"

> The colonial journey into the virgin interior reveals a contradiction,
> for the journey is figured as proceeding forward in geographical space
> but backward in historical time, to what is figured as a prehistoric zone
> of racial and gender difference. One witnesses here a recurrent feature
> of colonial discourse. Since indigenous peoples are not supposed to be
> spatially there – for the lands are "empty" – they are symbolically
> displaced onto what I call *anachronistic space*, a trope that gathered . . .
> full administrative authority as a technology of surveillance in the late
> Victorian era. According to this trope, colonized people – like women
> and the working class in the metropolis – do not inhabit history proper
> but exist in a permanently anterior time within the geographical space
> of the modern empire as anachronistic humans, atavistic, irrational,

bereft of human agency – the living embodiment of the archaic "primitive."

(McClintock 1995: 30)

In colonial discourse, women, like indigenous peoples, exist in space rather than in time. As McClintock argues, "Women are represented as the atavistic and authentic body of national tradition (inert, backward-looking and natural), embodying nationalism's conservative principle of continuity." Men, for their part, "represent the progressive agent of national modernity (forward-thrusting, potent, and historic), embodying nationalism's progressive, or evolutionary principle of discontinuity" (McClintock 1995: 359). In this regard, it is worth noting that the Group's equation of nation with virginal wilderness landscape involved not only a land claim but, in keeping with that claim, a celebration of pioneering. "These are pioneering days for artists," the catalogue for the second Group exhibition declares, "and after the fashion of pioneers we believe whole-heartedly in the land" (Lord 1974: 136). Stories of the Group painters braving the elements to sketch the northern wilderness had its literary counterpart in narratives of settlers risking all to create a new life for themselves and others on the land.

The celebration of pioneering proved an effective means of countering the growing strength of the women's movement: the equation of nation with virgin land and the celebration of pioneering served not to exclude women but to keep them in their place. McClintock writes:

> For women, the myth of the virgin land presents specific dilemmas, with important differences for colonial or colonized women. . . . Women are the earth that is to be discovered, entered, named, inseminated and, above all, owned. Symbolically reduced, in male eyes, to the space on which male contests are waged, women experience particular difficulties laying claim to alternative genealogies and alternative narratives of origin and meaning. Linked symbolically to the land, women are relegated to a realm beyond history and thus bear a particularly vexed relation to narratives of historical and political effect. Even more importantly, women are figured as property belonging to men and hence as lying, by definition, outside the male contests over land, money, and political power.

(McClintock 1995: 31)

While the Group of Seven represented its members as cultural pioneers, socially and politically they did little to support the women's movement: their Toronto home, the Arts and Letters Club, restricted its membership to men, as did other cultural institutions, such as the University of Toronto's Hart House, which collected and exhibited paintings by Group members and where several of the painters collaborated in the creation of set designs

for theatrical productions. Above all, the Group of Seven was instrumental
in forging the link between nation and virgin wilderness, an equation that
gendered the landscape and effectively kept women in their place.

   In forging the equation of nation with wilderness landscape the English-
Canadian elite both redefined the field of cultural production in national
terms and asserted patriarchal ownership over the means of production
and consumption. As Joyce Zemans (1995) and others have demonstrated,
the National Gallery of Canada played a central role in that work. In
representing Canada as unpeopled, virgin territory, the national culture
not only laid claim to the land but relegated First Nations people to the
"anachronistic space" of non-European prehistory and equated women
with the land to be colonized. In the final analysis, the question that needs
to be asked is less how the nation came to be equated with wilderness landscape than why that construction of national identity continues to have
such force. Peter Hulme defines "colonial discourse" as "an ensemble of
linguistically based practices unified by their common deployment in the
management of colonial relationships" (Hulme 1986: 2). Certainly the trope
of virgin land is one of those practices. Taking up Hulme's formulation,
Lawson argues that "postcolonialism is driven by its engagement with the
power that remains in the discourse" (Lawson 1995: 24). Feminism itself
has a considerable stake in the critique of colonial discourse. In "Gendering
Nationhood" Joanne P. Sharp writes:

> Out of this ambivalent political process there emerges a feminist
> analysis which is not solely concerned with constructing an identity of
> "women *as* women" . . . but with mapping the complex societal rela-
> tionships which construct dominance and subjugation: dominance not
> as a monolith but as overdetermined by a number of subjugations,
> one of which is centered around the construction of gender norms
> and differences. Feminism, as I understand it then, is involved in the
> project of disentangling power and dominance, in denaturalizing and
> opposing the apparently "natural" gender relations supporting of [*sic*]
> and supported by other forms of subjugation.
>
> (Sharp 1996: 107)

To analyze the cultural representations of national identity is to engage in
a similar mapping of complex societal relationships and to recognize the
central role that gender and race play in the construction of difference.

Notes

1   This paper is part of a larger project on the cultural construction of Canadian
    national identity in the 1920s. Development of the project has been facilitated
    by a three-year research grant from the Social Sciences and Humanities Research
    Council of Canada and by the continuing support of my own university, includ-
    ing the granting of a McCalla Research Professorship. I should also like to

acknowledge the invaluable assistance of two staff members at the National Gallery of Canada, archivist Cindy Campbell and curator Charles C. Hill, and of my friend and colleague Daphne Read.

2   This exhibition, which featured the work of J.EH. MacDonald, Lawren Harris, and Frank Johnston, was a precursor to the founding exhibition of the Group held at the Art Gallery of Toronto in May of 1920.

3   For a discussion of the category of settler-invader colony, see Johnston and Lawson (2000). Although Johnston and Lawson themselves point out both that "settler-invader" is "the more historically accurate term" to describe these colonies and that the displacement of First Nations peoples is cultural and symbolic as well as physical, they choose early in their essay to drop the term "invader." For their explanation, see p. 362.

4   The founding members of the Group were: Franklin Carmichael, Lawren Harris, A.Y. Jackson, Frank Johnston, Arthur Lismer, J.E.H. MacDonald, and F.H. Varley. Tom Thomson, who died in 1917, was, in many ways, the absent center of the Group. Membership in the Group changed over the years; for example, Johnston resigned from the Group after the first exhibition and in 1926 A.J. Casson was appointed a member. The group disbanded in 1933.

5   In an interview with Safdie published in 1984, John Strasman remarks: "I can see the entrance pavilion, the colonnade and the great hall being rented out by the Prime Minister's office for receptions." Safdie replies: "There's already been a decision about that and we've met with protocol and the great hall will be used for state events like dinners for visiting heads of state. We've already met with the security people. We have VIP entrances, and the whole sequence of how you have a state dinner is worked out for the great hall" (Strasman 1984: 30).

6   Nairn argues that Ernest Gellner's "'modernization theory' in his original formulation could not help over-stressing the elements of literary modernity themselves – machine industry, the transformation of vernaculars into literacy, the inventive 'rediscovery' of the countryside by the new 'clerks' of national movements and so on. Greater attention is invariably given to these motives and instruments of change than to the 'raw material' itself – that is, the peasants masses who underwent the change and ... got themselves not only made over into Frenchmen and Czechs but idealised into the very source of nationhood." According to Nairn, "Whatever happens to them in the urban vaudeville, or the TV soap opera, 'traditions' are also a real matrix borne forwards from the past time by individuals and families. The kind of remaking which features in modern nationalism is not creation *ex nihilo*, but a reformulation constrained by determinate parameters from the past. And the past which has mainly counted here – and given its 'bite' and sentimental incontrovertibility to all ethno-nationalist belief structure – is that of peasant existence" (Nairn 1997: 104).

References

Anderson, B. (1991) *Imagined Communities: Reflections on the Origin and Spread of Nationalism*, revised edition, London and New York: Verso.

Ashcroft, B., G. Griffiths and H. Tiffin (eds) (1995) *The Post-Colonial Studies Reader*, London and New York: Routledge.

Baele, N. (1988) "Northern Terrain," *Landscape Architecture*, 78: 38–40.

Bordo, J. (1992–3) "Jack Pine – Wilderness Sublime or the erasure of the aboriginal presence from the landscape," *Journal of Canadian Studies*, 27(4): 98–128.

Brennan, T. (1990) "The National Longing for Form," in Homi K. Bhabha (ed.) *Nation and Narration*, London and New York: Routledge, 44–70.

220   *Paul Hjartarson*

Cosgrove, D.E. (1984) *Social Formation and Symbolic Landscape*, London and Sydney: Croom Helm.

Deutsche, R. (1996) *Evictions: Art and Spatial Politics*, Cambridge MA and London: MIT Press.

Duncan, C. (1991) "Art Museums and the Ritual of Citizenship," in Ivan Karp and Steven D. Lavine (eds) *Exhibiting Cultures: The Poetics and Politics of Museum Display*, Washington DC: Smithsonian Institution Press, 88–103.

Hulme, P. (1986) *Colonial Encounters: Europe and the Native Caribbean, 1492–1797*, London: Methuen.

Johnston, A. and A. Lawson (2000) "Settler Colonies," in Henry Schwarz and Sangeeta Ray (eds) *A Companion to Postcolonial Studies*, Oxford: Blackwell, 360–76.

Lawson, A. (1995) "Postcolonial Theory and the 'Settler' Subject," in Diana Brydon (ed.) *Testing the Limits: Postcolonial Theories and the Canadian Literatures*, special issue of *Essays on Canadian Writing*, 56: 20–36.

Lears, T.J.J. (1981) *No Place of Grace: Antimodernism and Transformation of American Culture, 1880–1920*, New York: Pantheon.

Lord, B. (1974) *The History of Painting in Canada: Towards a People's Art*, Toronto: NC Press.

McClintock, A. (1995) *Imperial Leather: Race, Gender and Sexuality in the Colonial Contest*, London and New York: Routledge.

McKay, I. (1994) *The Quest of the Folk: Antimodernism and Cultural Selection in Twentieth-Century Nova Scotia*, Montreal and Kingston: McGill-Queen's University Press.

Mitchell, T. (1996–7) "'The Manufacture of Souls of Good Quality': Winnipeg's 1919 National Conference on Canadian Citizenship, English Canadian Nationalism, and the New Order After the Great War," *Journal of Canadian Studies*, 31(4): 5–28.

Mitchell, W.J.T. (1994) "Imperial Landscape" in W.J.T. Mitchell (ed.) *Landscape and Power*, Chicago: University of Chicago Press, 5–34.

Nairn, T. (1997) "The Curse of Rurality: Limits of Modernisation Theory," in *Faces of Nationalism: Janus Revisited*, London: Verso, 90–112.

New, W.H. (1989) *A History of Canadian Literature*, London: Macmillan Education Ltd.

Oberlander, C.H. (1985) "National Gallery of Canada: Landscape Concept," *Landscape Architectural Review*, Nov./Dec.: 5–7.

Rybczynski, W. (1993) *A Place for Art: The Architecture of the National Gallery of Canada/ Un Lieu Pour L'Art: L'Architecture du Musée des beau-arts du Canada*, Ottawa: National Gallery of Canada.

Sharp, J.P. (1996) "Gendering Nationhood: A Feminist Engagement with National Identity," in Nancy Duncan (ed.) *Body Space: Destabilizing Geographies of Gender and Sexuality*, London and New York: Routledge, 97–108.

Strasman, J. (1984) "An Interview with Moshe Safdie," *The Canadian Architect*, 29 (February): 22–31.

Teitelbaum, M. (1991) "Sighting the Single Tree, Sighting the New Found Land," in Daina Augaitis and Helga Pakasaar (eds) *Eye of Nature*, Banff: Walter Philips Gallery, 71–88.

Valverde, M. (1991) *The Age of Light, Soap, and Water: Moral Reform in English Canada, 1885–1925*, Toronto: McClelland and Stewart.

Zeller, S. (1987) *Inventing Canada: Early Victorian Science and the Idea of a Transcontinental Nation*, Toronto: University of Toronto Press.

Zemans, J. (1995) "Establishing the Canon: Nationhood, Identity and the National Gallery's First Reproduction Programme of Canadian Art," *Journal of Canadian Art History*, XVI.2: 7–35.

Part IV
Writing home

12 The manly map

The English construction of gender in early modern cartography

Dalia Varanka

Questions of gender in cartography most often focus on the sex of people involved in the cartographic process. These areas of research include the history of women cartographers (Tyner 1997: 46; Ritzlin 1989: 5; Hudson 1989: 29), the cartography of issues centered on women (Seager and Olson 1986; Seager *et al.* 1997; Rocheleau *et al.* 1995: 62), and women in the cartographic labor force (McHaffie 1996). Such studies examine the experiences of men and women in map-making, but not the map itself. When studies have addressed gender and contemporary map design qualities, the focus has been on cognitive aspects of and potential differences between the sexes in the cartographic process of making, analyzing, or reading maps (Golledge and Gilmartin 1986; Gilmartin and Patton 1984). Such studies tend to polarize the experience of gender in cartography as either male or female, when in fact people display properties of both genders. This paper examines facets of the historical basis of contemporary cartographic design principles, especially where these principles are gendered. By examining gender as social constructions of femininity and masculinity and how they have been incorporated into mapping, instead of as the interaction of individuals with maps (although sexual stereotypes act predominantly on individuals on the basis of their sex), we more closely approach the complex array of skills and relationships within which people live, while gaining insight into the stereotypical experiences of individuals. The premise is that early modern map design was not gender neutral in terms of broader social forces shaping gender identity. Historical evidence suggests that modern cartographic design principles originated along the lines of ideals of masculinity.

The rise of modern English prose style in the seventeenth century is described by language historians as a movement toward simplicity, plainness, directness, and utility of communication (Adolph 1968; Wanning 1936; Shapin 1996; Clark 1986). This style, called "Plain Style" by Robert Adolph, was contemporary with and largely influenced by, the Royal Society's implementation of the new scientific philosophy of Francis Bacon. The Royal Society favored the style based on principles of general representation for its "manliness," in contrast to philosophies of the classics and of pre-empirical science. These masculine design principles, which also are called "Plain Style" in this paper, are described and compared to characteristics of a style that was considered "feminine." Semiotic analysis of the way these principles function in cartography is developed by coupling descriptions of maps and texts with historical details about seventeenth-century English maps and texts to explore the spread of an ideological affect that today seems neutral and objective. The conflicted historical setting of the rise of English Empiricism drove the process of assigning different cartographic styles to public and private spheres, though this process served to exclude cartographic features from mainstream practice. This paper also attempts to evaluate the legacy of these conventions in terms of ranges of representational design options. Some conclusions will be drawn about the possible implications of the "masculine" cartographic style with the intention of offering a more flexible and diversified approach to these issues.

Plain Style principles

The cultural transitions historically associated with the rise of English Empiricism, or the Scientific Revolution, have been described as the change from hermeneutic, or symbolic, to positivist knowledge (Foucault 1970; Easlea 1980; Merchant 1995: 75). Ideas of objectivity, limits, and perception radically changed the organization and even the nature of knowledge. In contrast to the modern positivist idea that things are outside of oneself and, therefore, can be represented very generally, such as trees or houses, the symbolic sense of objectivity was to perceive through the senses something about objects much more immediate, sensually substantive, and essentially connected to the mind – this tree, this house. This earlier thinking encouraged a worldview of things as individual and unique. In a basic way, the realistic nature of knowledge remained the same, but the hermeneutic approach focused on the limits of an object through substantive and specific description, rather than atomistic or mechanical analysis.

Although it seems contradictory, substantive and specific description enabled the capture of elusive or abstract qualities beyond the immediate, definite limits. As Medcalf explains, the substantive quality of the object, with its clearly defined limits, made more real the implication of what was beyond the limits (Glanvill 1970: xxii). This approach was one way to handle abstractions that are often elusive or difficult to express. Metaphors, a powerful device of the symbolic style, seemed so real because they made abstract qualities substantive by offering limits to the specific and unique nature of those qualities. Both styles of representation presented objects of the material world within their limits. Both the scientific and the hermeneutic viewpoints conveyed the same facts, but when specifically conceived (as in the case of the hermeneutic), the limits could be transcended. When positively conceived, the image is made more general and, therefore, cannot be transcended. The transcendent is imagined but contained in the generalization.

In their work towards the establishment of the new scientific method in the late seventeenth and early eighteenth centuries, the members of the Royal Society of London believed that the laws of the universe made the material world factual and objective. Older forms of scientific philosophy were criticized for attaching perceptions or meanings to material objects, causing the knowledge of these objects to become distorted. The distortions lead to extensive discussions and arguments about impractical abstractions. The attempt to avoid verbosity led the Society to advocate a certain prose style that they considered to be as important as the new empirical method derived from Francis Bacon. A contemporary description of the experimental prose style could be read from Sprat's *History of the Royal Society*:

> to reject all the amplifications, digressions, and swellings of style: to return back to the primitive purity, and shortness, when men deliver'd so many things, almost in an equal number of words. They have exacted from all their members, a close, naked, natural way of speaking; positive expressions; clear senses; a native easiness: bringing all things as near the Mathematical plainness, as they can: and preferring the language of Artisans, Countrymen, and Merchants, before that, of Wits, or Scholars.
>
> (Sprat 1667: 113)

New methods of scientific analysis were to discover the consistent, unchanging, and ultimately understandable truth about the material world. To aid this, representation of an object should be simple, plain, as close to material reality as possible, and without bias or influence from the viewer. It must be general so that it can be duplicated and reproduced. If it were too specific, it would become too impractical to communicate. John Locke, whose philosophy dominated this period in England, wrote:

> But yet if we would speak of things as they are, we must allow that all the art of rhetorick, besides order and clearness, all the artificial and figurative application of words eloquence hath invented, are for nothing but to insinuate wrong ideas, move the passions and thereby mislead the judgment, and so indeed are perfect cheats: and therefore however laudable or allowable oratory may render them in harangues and popular addresses, they are certainly, in all discourses that pretend to inform or instruct, wholly to be avoided: and where truth and knowledge are concerned, cannot but be thought a great fault either of the language or person that makes use of them.
>
> (Locke 1894: 146)

The rhetorical and figurative representation to which Locke refers means primarily the use of metaphor but also involves subjective, emotive qualities. An open revelation of the author's internal perceptions and thinking process was rejected because the nature of these processes is symbolic, nonlinear, and more difficult to follow.

Francis Bacon addressed this new scientific era in an early essay called "The Masculine Birth of Time, or Three Books on the Interpretation of Nature" (Farrington 1964: 61). Representing the masculine philosophy (Schiebinger 1989: 137), Plain Style came to be considered the "manly" form of speech and representation (Adolph 1968: 91). For example, Glanvill wrote in an address to the Royal Society:

> For I must confess that way of writing to be less agreeable to my present relish and Genius; which is more gratified with manly sense, flowing in a natural and unaffected Eloquence, then in the musick and curiosity of fine Metaphors and dancing periods.
>
> (Glanvill 1665: 27)

The main attack of the Plain Style was on oratory, metaphors, emotive statements, and on symbolic or rhetorical style. Manliness was contrasted to ornament and finery, which were considered feminine and juvenile, and morally wrong:

> They make the Fancy disgust the best things, if they come found, and unadorn'd: they are in open defiance against Reason; professing, not to hold much correspondence with that; but with its Slaves, the Passions: they give the mind a motion too changeable, and bewitching, to consist with night practice. Who can behold, without indignation, how many mists and uncertainties, these specious Tropes and Figures have brought on our Knowledge? How many rewards, which are due to more profitable, and difficult Arts, have been still snatch'd away by the easie vanity of fine speaking? For now I am warn'd with this just Anger, I cannot with-hold my self, from betraying the shallowness of all these seeming Mysteries; upon which, we Writers, and Speakers, look so bigg. And, in few words, I dare say; that of all the Studies of men, nothing may be sooner obtain'd, than this vicious abundance of Phrase, this trick of Metaphores, this volubility of Tongue, which makes so great a noise in the World.
>
> (Sprat 1667: 112)

Numerous passages from Sprat's *History of the Royal Society* (1667) and other contemporary texts make a convincing case that "manly," "feminine," and "juvenile" were familiar terms within the discourse of the Royal Society about their philosophy and were applied to style and ideas, not to humans (Schiebinger 1989: 138).

Shapin and Schaffer identify three types of technology that the Royal Society used to develop and further science (Shapin and Schaffer 1985: 25). In addition to a material technology that consisted of the actual hardware to perform scientific experiments, the Royal Society produced a literary technology for the dissemination of knowledge, and a social technology to fulfill the theoretical needs of considering and validating experimental science. The development of Plain Style was an aspect of the Royal Society's literary technology, which not only spread information but also was a tool of social affirmation of claims scientists made about knowledge.

The Plain Style map

Historians of cartography describe a "revolution" in cartography occurring in the late seventeenth century (Robinson 1982: 12). The Plain Style of the Scientific Revolution manifested itself in the contemporary early modern cartography. The map shown in Figure 12.1 was published by John Speed in about 1627. The map in Figure 12.2 was authored and

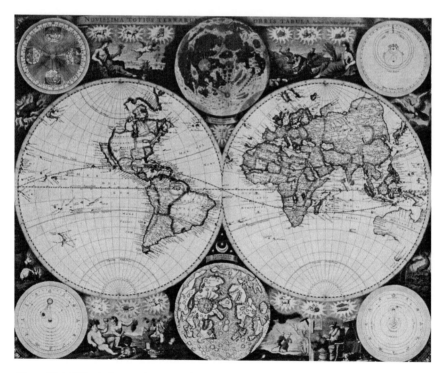

Figure 12.1 "Novissima totius terrarium orbis tabula," John Seller, in *Atlas Terrestris, c.*1676

Photograph by and permission to reprint requested from The John Carter Brown Library at Brown University

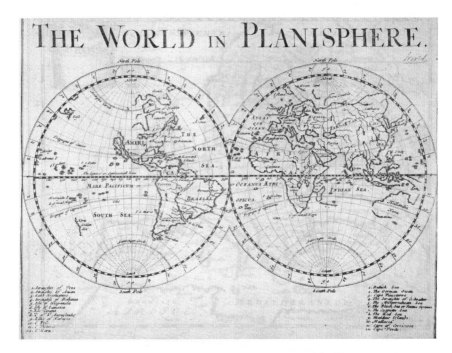

Figure 12.2 "The World in Planisphere" [the "cosmography"], Herman Moll, *c.*1700

Reproduced by the British Library and permission to reprint requested from the British Library, London, England

engraved by Robert Morden in various states after 1680. Morden's apprentice, Herman Moll, became the most prolific English mapmaker of this transitional period, and many of his maps carry acknowledgments of the work of Royal Society scientists. A glance at the two contemporary maps of the same place but of different traditions will confirm that the same trends are evident in maps as in the prose of that period, as analyzed above, but other detailed historical evidence supports the connection as well.

Historians of cartography describe the stylistic changes associated with the transition from symbolic to empirical science as a switch from art to science (Brown 1977: 174), and indeed, some of the qualities of cartographic style are related to art. The visual qualities on the maps and metaphors in the prose of atlases that were considered feminine and purged from the Plain Style, were most evidently iconology (Marchant 1986: 1) and classical rhetoric (Harvey 1993: 25), but also simply considered to be decoration used to market maps through fashion (Smith 1986: 106). Iconological imagery of the symbolic period before the Scientific Revolution served as metaphors by helping the reader to grasp abstractions and achieve a state of mind through attributing ambiguity to objects on the map (Harley 1983).

Imagery on the pages of atlases borrowed directly from iconological manuals, particularly in the design of cartouches (Welu 1987: 148). Although many cartouches, particularly of the Mannerist period, can be attributed to decoration pattern books, their more subtle implications were about intimacy with a place in the sense of a close, personal depiction (Boelhower 1988: 475; Rubasa 1985: 1).

But this does not mean that symbolic science abandoned realism (Woodward 1987: 2). The use of cartographic scale in England was already established by the end of the sixteenth century, and depictions on maps before the rise of the Royal Society were very exact in their rendering of specific places (Harvey 1993: 21; Stone 1978: 17). The argument of this paper is that by accepting subjectivity in mapping, a fuller realism was possible by the assertion of a complex context combining realism and ideology. This idea was also richly in the simultaneous depiction of alternative viewpoints through combinations of different scales of representation of places in "atlas factice" (unique compilations of maps, called "books of maps" or "sets of maps" in seventeenth-century England, assembled on the basis of individual or unique criteria and composed of maps in a wide variety of scales, styles, themes, and origins). Images and metaphors gave expression to the map reader's abstract experience and connected it to places that are the settings of lives. These elements are evident on most maps appearing before 1660 and in many maps drafted during a 60-year transition period after that.

The rise of Plain Style also was occurring in the aftermath of repeated civil wars and political conflicts over religion. Bacon's writings at the beginning of the seventeenth century united English with Biblical, rather than church, interests. From an imitation or coexistence of nature, Bacon's philosophy advocated that man is superior to nature, and the knowledge of nature was to worship God's creation. A history of evil was committed in the name of religious intellectualism. Greeks and scholars – the whole tradition of philosophy – was condemned not for intellectual but for moral reasons (Farrington 1964). Utilitarianism is moral, more so than quarrelling over opinions. Instead of an ascetic withdrawal from the world inherited from the church's intellectualism, an affirmation of experience began to regain ground. Alchemists and astrologers justified these practices as vehicles of God's revelation over man's reason, as were scripture and faith (Noble 1992: 178). Plain Style specifically intended to avoid the expression of personal and religious beliefs, for fear of their inflammatory nature. This would suggest why Plain Style mapping has difficulty communicating spirituality, or explain why it does not communicate spiritually at all.

Scientific method can be traced to many writers – Galileo, Descartes, Roger Bacon, Newton – but what made English Empiricism new was the context: that knowledge and power be shared with the populous, not just a privileged class of the enlightened (Farrington 1964: 22). Bacon's philosophy was a method of intellectual and manual work; England was to become wealthy through better resource exploitation. Seventeenth-century England was a supportive setting for the new manly science. The change to Baconian philosophy coincided with the transition from subsistence to capitalist economy and, in turn, conflicts between the landed gentry and the rising urban middle class. Although the popularization of the new philosophy was encouraged on democratic principles, in the sense that knowledge would disseminate to all social groups of any educated level, the Royal Society had the male professions of statesmen, merchants, and tradesmen in mind. Plain Style was structured by ideologies of utilitarian, commercial, and technological practices, and these were the social circles within which the dominant mapmaking forms were being nurtured (Skelton 1975: 14). The new philosophy turns to the domination of nature, not the reflection of it, and old constraints on the exploitation of the earth were eased.

Plain Style cartography became the mediating tool between the exploitation of the resources of the world. Soon after 1660, various applications of the Royal Society's empirical method of geography were practiced directly on Plain Style cartography. Geography was in places defined as the mathematical calculation of latitude and longitude, and the Royal Society sponsored empirical experiments that supported this. Mapping was the experimental method for the study of natural history. Examples of specific map projects the Society was involved in or sponsored directly include *The English Atlas* by Moses Pitt, under the direction of Robert Hooke (Taylor 1937: 529), and *A New General Atlas* by John Senex FRS, where the field collection of information from foreign places was reported to the Royal Society and incorporated into the published maps at the appropriate coordinates (Stearns 1936: 345). The accuracy of these coordinates were boasted as "truthful" in advertisements, lending greater moral authority to Plain Style mapping, assuring the safety of explorers on their journeys.

In addition to the materiality of new technology and commercial expansion the deeper theoretical underpinnings of cartography also would suggest Restoration period forces acting upon the establishment of Plain Style mapping. Modern semiotics, a foundation of theoretical cartography (MacEachren 1995; Wood and Fels 1986: 54), is rooted in philosophies of dualism developed by John Locke (Deely 1990; Locke 1894: 3). Locke, to whom an early map of Carolina is attributed (Black 1978: 109), included words in the more general category of all signs, including the cartographic. Thus strategies such as those implemented in this study have been described in the history of cartography and in cartographic theory as the "map as text" movement (Harley 1989a; Harley 1989b; Wood and Fels 1986: 54), which establishes the semiotic link between maps and language.

The view of dualism, the view that the universe is composed of separate ideas and that mental activity is the arrangement of those ideas in the mind, encouraged an atomistic depiction of the world as made up of clearly separable things: one word for each thing. In contrast to this, signs in the earlier period were believed to reflect the essence of things. In the animistic sense, objects were represented as coupled with other things with which they were associated, instead of as one positive entity. This new movement toward definitions or explications of names gained in the last part of the seventeenth century and brought on the more widespread use of legends for symbols on the map (Campbell 1962).

Without dualism, symbolic mapping was mapping from the perspectives of people. The elimination of symbolism and metaphysics brought the predominant focus on matter. The formation of generalized images required a dissociation of the inner self from the surrounding environment to form something separate seemingly unconnected to the reader's direct experience. Perception came to mean the presentation of an object to the senses rather than a full identification in the sense of mentally drawing a new object in. The contrast between the scientific and symbolic viewpoints was that things are known either superficially or essentially. The legacy that was rejected in Plain Style cartography was the exploration or expression of a connected state of being to the world and to others, though this continued on in the contemporary cartographic studies acceptable for women.

Feminine style

Gendered ideas of manliness and femininity were associated with the politics of the civil wars, religious dissent, and alternative science. Just as the new science intended to bring learning to the middle classes of England, the hermetic movement, the science of magic and resemblances, empowered lower classes of society and others excluded from intellectual circles with an alternative mode of learning, the sources of which lay in oral history (Thomas 1973: 270). Although the sex of individuals in scientific society should not be confused with the meaning of gender as social relationships, the number of women practicing alternative forms of science and writing books rose especially during the Interregnum as radical religious sects allowed a greater role for women (Noble 1992: 187; Jones 1988: 90). These movements outside the mainstream of learning became identified by association with women. Although many women were judged as witches and put to death, others continued with their work, especially as midwives and healers. To disassociate themselves from these popular movements, the Royal Society dismissed those practitioners, regardless of their sex, as "feminine."

Margaret Cavendish, Duchess of Newcastle, may have been the first author of a book of modern science for women (Meyer 1955: 2). Cavendish encouraged the great majority of women to participate in the new science, though many had little or no education at all. Margaret Cavendish wrote her books for women with a romantic style, explaining that because of their lack of previous education, women would otherwise find Plain Style boring.

Cavendish acquired some recognition for her manly style from contemporary intellectuals, though some, such as the poet John Dryden, framed her achievements within the terms of her marriage:

> Dryden was one of those who approved Margaret's endeavours, though he assignes the credit to Newcastle. He dedicated his play *The Mock Astrologers* to the Duke, referring to the Duchess as "a Lady whom our Age may justly equal with Sappho of the Greeks or the Sulpitia of the Romans, who by being taken into your bosom, seems to be inspired with your genius. And by writing the history of your life in so masculine a style, has already placed you in the Number of the Heroes."
>
> (Jones 1988: 159)

The Duchess pursued her enthusiasm for the new philosophy, though arguing that the effect of masculine science was to undermine the position of women (Cavendish 2001: 2).

Women were excluded from the Royal Society and apparently from any influence on the practice of map and atlas publication, with the exception of widows who continued their husbands' shops (Senex n.d.; Tyner 1997: 46). This exclusion did not prevent women from acquiring scientific abilities. Contributions to the new science originated not only in established centers of learning but in less formal places of social interaction as well (Schiebinger 1989: 17). A popular women's magazine of the early eighteenth century featuring a regular column posing puzzlers in algebra and calculus offers a reasonable indication that women did have the skills that would have been needed to do mathematical geography (Leybourn 1817). Many seventeenth-century atlases and books of astronomy and geography studied by this author had a woman's signature indicating "her book" (a common phrase included after the name). Reasons why the new science was sometimes regarded as feminine were that the lack of an education in the classics freed women from ideas from the past regarded as mistaken, and that scientific activity fit well into women's roles because the new science was at first an amateur activity done at home and was not directly within the established political or economic offices or spheres of influence (Wiesner 1993: 169).

Discourses regarding astronomy and geography were acceptable, even desirable, for women to practice at home in the company of men, probably for purposes of conversation. Samuel Pepys tells in his diary of his enjoyment in teaching his wife the use of globes (Wallis 1978: 6). These events took place between the two of them in their home and during leisure hours, indicating that these tasks were meant for domestic or social situations. A similar prose style used in books in this period, instructional dialogue between two fictional individuals, was not necessarily a gendered form of discourse but one that implied power relations. The dialogue form between a woman and a man was well enough established that Harris planned to use it for future scientific books for women.

> And as I think it practicable to explain and teach any Science in this Facetious way (*Facete enim & commode dicere quid vetat?*) so perhaps I may hereafter, if God grant me Health, Ease and Leisure, make some other Attempts of this kind. For the Lady may well be supposed, tho' the Sight of the Globes first struck her Fancy and turned her Desires this way, to have made Excursions into other Parts of Mathematicks, and to have discoursed with her Friend on those Subjects.
>
> (Harris 1719)

Harris believed a different written approach towards both presentation and content was necessary for women. The text of his geography book for women made heavy use of flourishes, manners, and poetry. Harris explains in his introduction:

> And that the Digressions, Reflexions, Poetry and Turns of Wit, are introduced to render Those [scientific] Notions pleasing and agreeable [to women], which perhaps without such a kind of Dress, would appear too crabbed and abstracted. However, I don't perplex my Fair Astronomer with any thing but the true System of the World: I mislead her by no Notions of Chrystalline Heavens, or Solid Orbs: I embarrass her with no clumsey Epicycles, or imaginary and indeed impossible Vortices: But I shew her at first the Coelestial World just as it is.
>
> (Harris 1719)

The last phrase suggests that although the prose style of the book is what the Royal Society would have rejected as feminine, the philosophy of the scientific concepts was within the terms of experimental science. In addition to the use of prose style, the language of the body of the work indicates that it is intended to be feminine by its heavy nuances of sexual stereotype. The pleasure of astrology is attributed to the lady's gender despite the fact that astrology was widely popular and considered conventional science at that time. The woman's voice in the dialogue gives planets personifications – Venus must have attendants (moons), which would be "an affront on our sex not to." Though curious, eager to learn and to go on reading, she says she is regretful and complains that it is barbarous that knowledge be kept between men alone.

Perhaps as a reaction against rhetoric or the voice of the individual, the manly map of modern science split its association with text, a form of dialogue between the author and the reader, in this period. Atlases and maps originally often had extensive sections of text, but a clear pattern of word and image dissociation occurs just roughly during 50 years (between 1676 and 1729) immediately after the establishment of the Royal Society in the 1660s (Varanka 1994). Even as late as in twentieth-century carto-graphic theory, cartography is treated separately, pointedly excluding text. In 1908 Max Eckert wrote that cartography has been raised to a "higher position" than other types of illustrative material because as products of the scientific process they are complete in themselves and no longer depen-dent on explanatory text (Eckert 1908: 344). This claim that cartography has no need to articulate its position verbally was followed by decades of research to improve the communicative efficacy of maps, though it was rarely studied in conjunction with prose.

A map of Virginia attributed to Virginia Farrar, illustrated in Figure 12.3, differs from other contemporary maps of the English colonies in its depiction of the close proximity of "The Sea of China and the Indies," just on the far side of the Appalachian mountains. The map was originally drafted in 1650, before the stylistic changes in map design, and reissued in 1651, 1652, and 1667 (Verner 1950: 281). By the time that this map was inserted in William Blathwayt's collection of maps, called *The Blathwayt Atlas*, the English government knew that extensive lands lay beyond

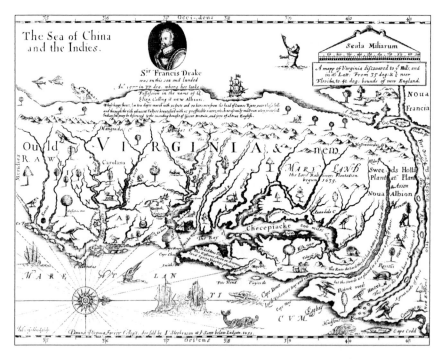

Figure 12.3 "A mapp of Virginia discovered to ye Hills, and in its Latt. From
    35.deg: & 1/2 neer Florida, to 41.deg: bounds of New England,"
    Virginia Farrar, from Black, *The Blathwayt Atlas*, 1975

Photograph by The Newberry Library. Permission to reprint requested from The John Carter Brown Library at Brown University

Virginia. The western orientation of the map, together with the shortened depiction of the continent, seems to betray English aspirations about the colony. The map can be interpreted variously: as an expression of the English desire to "contain" the colonies; as a "map" of life there; or as a symbol of English claims to the full extent of the land between the shores. But even when the map is interpreted as a pragmatic image, as in potential interpretations about its use to sway colonial settlement (Brod 2000: 172), it must be admitted that the map is an ideological artifact over an empirical product defined within the narrow terms of Plain Style cartography.

Discussion

The influence of Plain Style is evident throughout the modern period and well into the twentieth century. English maps and atlases of the early modern period, particularly those of Herman Moll and John Senex, continued to be printed into the early nineteenth century, serving directly as prototypes or starting points for newer mapping projects, but the social and cultural context of cartography indicates consistent trends with Plain Style principles, as well. Brian Harley identified the major themes of the history of cartography to have been cartographic "nationalism," the analysis of mathematical accuracy, and an exploratory record of the world (1989a: 1). These movements reflect that mapping structures were tailored to those realms of a society dominated by men.

Certain theoretical consequences of the Plain Style, though rooted in historical circumstances, are still present in our contemporary maps, though in a subtle way that seems quite natural. What does it mean that Plain Style mapping was propagated by men and became the dominant form of cartographic representation? Certainly, maps continue to be used for participating in the masculine-dominated spheres of influence. How do these gendered principles interact with the perception and intellectual processing of spatial symbolism by map users? Other studies, particularly about cognitive cartography, could shed light on this question, but they rarely actually address differences in human experience as critical criteria. At the real heart of these constructions of gender lie political and intellectual struggles.

Modern practitioners remained aware of the persistence of traits that were actively denounced or argued over in scientific mapmaking. The artificial yet persistent split between Plain Style cartography and humanistic elements in cartography surfaced in the "art vs. science" discussions of the twentieth century, in which "art" refers to subjective, emotive qualities of cartography (Eckert 1908: 344; Imhof 1963: 13; Morrison 1976: 84; Wright 1944: 527; Keates 1982: 127; MacEachren 1995: v). Postmodern analysis of cartography is restoring what was repressed in the modern era, as studies of maps as texts, including maps within the scope of combined word and image studies (Gilmartin 1982: 145). The view of the map as a text incorporates personal situation and historical context with scientific cartography and in this way attempts to resolve the subjective worlds with Plain Style mapping, though often the exploration of truth in mapping is still only discussed within the context of science (MacEachren 1994). The modern, or two-part, semiotic approach to cartography, based on the work of Roland Barthes (Wood and Fels 1986: 62), contrasts with the American semiotics of Charles Peirce, which separated and broke the implicit "naturalization" of meaning from the sign. By disassociating meaning from the sign itself, Peirce's tri-part semiotics, in contrast to Locke's dualism, made representations less and recognized alternative contexts to meaning than that represented by the sign.

The attributes of feminine-style cartography, particularly the importance of personal narratives and the influence of interpersonal relationships within the community are qualities that are consistent with tenets of gender or feminist research (Staeheli and Lawson 1995: 321). These differences in values become clear in Boyle's "Physico-Mechanical Touching the Spring of the Air" in which he relates that the Royal Society conducted experiments late at night with the purposeful exclusion of women to avoid the difficulty of feelings, after an occasion when women objected to the suffocation of a bird for science (Haraway 1997: 31; Merchant 1995: 63). The feminine style had a distinctive way of writing, of being open to qualitative analysis, of abstractions of the sense of self, and of being practiced in the time and space of everyday life. It was identified with people who were mostly excluded from elite institutions. The social context that is missing in Plain Style cartography can be viewed more broadly than the people and institutions that produce scientific knowledge. It speaks directly to the character of the researcher. Donna Haraway and others discuss the theoretical implications of the "modest witness" of experimental science, in which the presence of the investigator is silenced, and thus has no part in the findings (Haraway 1997: 23). It is a way of objectifying scientific findings and making them impersonal.

From the strictly literal standpoint, Plain Style stripped something away socially from modernity. Apparently the gains made from positivism, which are generality and verifiability, require a loss by excluding other assets and qualities regarded as feminine. Instead of using resemblance, we practice a map style that is general and not overly specific yet gives the impression of great detail, built primarily by its technical context. It invites the stereotyping of places because people must fill in the detail mentally themselves, a common device used and noted in both literature and art (Stafford 1984). The lack of text to explain logical connections results in the repression of thoughts or ideas by which associations are made and makes the reader vulnerable to power relations implied by hierarchy, a central principle in modern cartographic design analysis. Without the revelation of the individual self or the subjective viewpoint of the author, not only can those thoughts not be considered in discussions but the outward materialism of the map provides little with which the reader can personally connect. Generalization hides imperfections, in contrast to the individual, which brings out those specific qualities. Atomistic viewpoints of the world make the world look discrete and unique, but individual subjectivity is what makes the world truly unique. The result is a cartography that helps us to communicate ideas with others but not to connect with nor to explore the world within ourselves.

Bibliography

Adolph, R. (1968) *The Rise of Modern Prose Style*, Cambridge MA: MIT Press.

Black, J. (1978) "The Mapping of the English Colonies in North America: The Beginnings," in N. Thrower (ed.) *The Compleat Plattmaker, Essays on Chart and Mapmaking in England in the Seventeenth and Eighteenth Centuries*, Berkeley CA: University of California Press.

Boelhower, W. (1988) "Inventing America: A Model of Cartographic Semiosis," *Word and Image*, 4: 475–497.

Brod, R. (2000) "Maps as Weapons in the Conquest of the Old American Northwest, 1609 to 1829," unpublished thesis, Chicago: University of Illinois.

Brown, L. (1979) *The Story of Maps*, New York: Dover Publications.

Campbell, E. (1962) "II: The Beginnings of the Characteristic Sheet to English Maps," *The Geographical Journal*, 128: 411–415.

Cavendish, M. (2001) *Observations Upon Experimental Philosophy*, E. O'Neill (ed.), Oxford: Cambridge University Press.

Clark, J. (1986) *Revolution and Rebellion: State and Society in England in the Seventeenth and Eighteenth Centuries*, Cambridge: Cambridge University Press.

Deely, J. (1990) *Basics of Semiotics*, Bloomington and Indianapolis IN: Indiana University Press.

Easlea, B. (1980) *Witch-hunting, Magic & the New Philosophy: An Introduction to Debates of the Scientific Revolution*, Highlands NJ: Humanities Press.

Eckert, M. (1908) "On the Nature of Maps and Map Logic," *Bulletin of the American Geographical Society*, 40: 344–351.

Farrington, B. (1964) *The Philosophy of Francis Brown: An Essay on its Development from 1603–1609 with New Translations of Fundamental Texts*, Chicago: University of Chicago Press.

Foucault, M. (1970) *The Order of Things: An Archaeology of the Human Sciences*, New York: Pantheon.

Gilmartin, P. (1982) "The Instructional Efficacy of Map in Geographic Text," *Journal of Geography*, 81: 145–150.

Gilmartin, P. and Patton, J. (1984) "Comparing Sexes on Spatial Ability: Map Use Skills," *Annals of the Association of American Geographers*, 74: 605–619.

Glanvill, J. (1665) "To the Royal Society," *Scepsis Scientifica*, London.

Glanvill, J. (1970) *The Vanity of Dogmatizing: The Three Versions*, with a critical introduction by Stephen Medcalf, Brighton: The Harvester Press.

Golledge, R. and Gilmartin, P. (1986) "Maps, Mental Imagery and Gender in the Recall of Spatial Information," *American Cartographer*, 13: 335–344.

Haraway, D. (1997) *Modest_Witness@Second_Millennium. FemaleMan©_Meets_Onco Mouse™, Feminism and Technoscience*, London: Routledge.

Harley, J. (1983) "Meaning and Ambiguity in Tudor Cartography," in S. Tyacke (ed.) *English Map-making 1500–1650, Historical Essays*, London: The British Library Board.

Harley, J. (1989a) "Deconstructing the Map," *Cartographica*, 26: 1–20.

Harley, J. (1989b) "The Myth of the Divide: Art, Science, and Text in the History of Cartography," paper presented at the 13th International Conference on the History of Cartography, Amsterdam.

Harris, J. (1719) *Astronomical Dialogues Between a Gentleman and a Lady: Wherein The Doctrin of the Sphere, Uses of the Globes, And the Elements of Astronomy and Geography are Explain'd, In a Pleasant, Easy and Familiar Way*, London.

Harvey, P. (1993) *Maps in Tudor England*, Chicago: University of Chicago Press.

Hudson, A. (1989) "Pre-Twentieth Century Women Mapmakers," *Meridian*, 1: 29–32.

Imhof, E. (1963) "Task and Methods of Theoretical Cartography," *International Yearbook for Cartography*, 3: 13–25.

Jones, K. (1988) *A Glorious Fame: The Life of Margaret Cavendish, Duchess of Newcastle, 1623–1673*, London: Bloomsbury.

Keates, J. (1982) *Understanding Maps*, New York: Wiley.

Leybourn, T. (1817) *The Mathematical Questions Proposed in the Ladies' Diary, and Their Original Answers, Together with Some New Solutions, From its Commencement in the Year 1704 to 1816*, London: J. Mawman.

Locke, J. (1894) *An essay concerning human understanding*, collated and annotated, with prolegomena, biographical, critical and historical, by Alexander Campbell Fraser, 2 vols, Oxford: Clarendon Press.

MacEachren, A. (1994) *SOME Truth with Maps: A Primer on Symbolization and Design*, Washington DC: Association of American Geographers.

MacEachren, A. (1995) *How Maps Work: Representation, Visualization and Design*, New York: The Guilford Press.

McHaffie, P. (1996) "Situating Public Cartographic Labor: (Wo)manufacturing Meaning," paper presented at the Association of American Geographers annual meeting, Charlotte NC.

Marchant, H. (1986) "A Momento Mori or Vanitas Emblem on an Estate Map of 1612," *Mapline*, 44: 1–4.

Merchant, C. (1995) *Earthcare: Women and the Environment*, New York: Routledge.

Meyer, G. (1955) *The Scientific Lady in England 1650–1760, An Account of Her Rise, with Emphasis on the Major Roles of the Telescope and Miscroscope*, Berkeley CA: University of California Press.

Morrison, J. (1976) "The Science of Cartography and Its Essential Processes," *International Yearbook for Cartography*, 16: 84–97.

Noble, D. (1992) *A World Without Women: The Christian Clerical Culture of Western Science*, New York: Oxford University Press.

Ritzlin, M. (1989) "Women's Contributions to North American Cartography: Four Profiles," *Meridian*, 2: 5–15.

Robinson, A. (1982) *Early Thematic Mapping in the History of Cartography*, Chicago: University of Chicago Press.

Rocheleau, D., Thomas-Slayter, B. and Edmunds, D. (1995) "Gendered Resource Mapping: Focusing on Women's Spaces in the Landscape," *Cultural Survival Quarterly*, 18: 62–68.

Rubasa, J. (1985) "Allegories of the Atlas," in Francis Barker *et al.* (eds) *Europe and its Others: Proceedings of the Essex Conference on the Sociology of Literature, July 1984*, Colchester: University of Essex.

Schiebinger, L. (1989) *The Mind has no Sex?: Women in the Origins of Modern Science*, Cambridge MA: Harvard University Press.

Seager, J. and Olson, A. (1986) *Women in the World: An International Atlas*, New York: Simon & Schuster.

Seager, J., Wilson, A. and Jarrett, J. (1997) *The State of Women in the World Atlas*, 2nd edition, New York: Penguin.

Senex, M. (no date) "A Catalogue of Globes, Maps &c. Made by the late John Senex, F.R.S. and continue to be sold by his Widow Mary Senex, at the Globe, over-against St. Dunstan's Church in Fleet-Street", London.

Shapin, S. (1996) *The Scientific Revolution*, Chicago: Chicago University Press.

Shapin, S. and Schaffer, S. (1985) *Leviathan and the Air-pump : Hobbes, Boyle, and the Experimental Life : Including a Translation of Thomas Hobbes, Dialogus physicus de natura aeris by Simon Schaffer*, Princeton NJ: Princeton University Press.

Skelton, R. (1975) *Maps: A Historical Survey of Their Study and Collecting*, Chicago: University of Chicago Press.

Smith, D. (1986) "Jansson versus Bleau," *The Cartographic Journal*, 32: 106–114.

Sprat, T. (1667) *History of the Royal Society*, London.

Staeheli, L. and Lawson, V. (1995) "Feminism, Praxis, and Human Geography," *Geographical Analysis*, 27: 321–338.

Stafford, B. (1984) *Voyage into Substance: Art, Science, Nature and the Illustrated Travel Account 1760–1840*, Cambridge MA: MIT Press.

Stearns, R. (1936) "Joseph Kellogg's Observations of Senex's Map of North America (1710)," *The Mississippi Valley Historical Review*, 23: 345–354.

Stone, J. (1978) "Origins and Sources of the Blaeu Atlas of Scotland with Particular Reference to 'Extima Scotia'," *Imago Mundi*, 26.

Taylor, E. (1937) "Robert Hooke and the Cartographic Projects of the 17th Century (1666–1696)," *Geographic Journal*, 90: 529–540.

Thomas, K. (1973) *Religion and the Decline of Magic*, London: Penguin.

Tyner, J. (1997) "The Hidden Cartographers," *Mercator's World*, 2: 46–51.

Varanka, D. (1994) "Editorial and Design Principles in the Rise of English World Atlases," unpublished thesis, The University of Wisconsin-Milwaukee.

Verner, C. (1950) "The Several States of the Farrer Map of Virginia," *Studies in Bibliography (University of Virginia Bibliographical Society)*, 3: 281–284.

Wallis, H. (1978) "Geographie is Better than Divinitie," in N. Thrower (ed.) *The Compleat Plattmaker, Essays on chart and mapmaking in England in the Seventeenth and Eighteenth Centuries*, Berkeley CA: University of California Press.

Wanning, A. (1936) "Some Changes in the Prose Style of the Seventeenth Century," unpublished thesis, Cambridge University.

Welu, J. (1987) "The Sources and Development of Cartographic Ornamentation in the Netherlands," in D. Woodward (ed.) *Art and Cartography*, Chicago: The University of Chicago Press.

Wiesner, M. (1993) *Women and Gender in Early Modern Europe*, Cambridge and New York: Cambridge University Press.

Wood, D. and Fels, J. (1986) "Designs on Signs/Myth and Meaning in Maps," *Cartographica*, 23: 54–103.

Woodward, D. (1987) *Art and Cartography*, Chicago: The University of Chicago Press.

Wright, J. (1944) "Map Makers are Human," *Geographical Review*, 32: 527–544.

13 The importance of being provincial
Nineteenth-century Russian women writers and the country

Hilde Hoogenboom

Around the mid-nineteenth century, a number of Russian women writers made the Russian landscape figure prominently in their works and placed heroines in the country. In digressions on the countryside, in plots, and especially in choosing provincial heroines to be writers, poets, and narrators, they represented themselves as provincial women writers. This phenomenon, which has been dubbed "the provincial tale" (Kelly 1994: 60), cannot and should not simply be explained as a question of provenance. Although most began their lives in provincial towns and cities, at least half later lived and wrote in either Moscow or St Petersburg, often moving back and forth between city and country. This subtle dissonance between these writers' actual as opposed to literary lives, aside from once again reminding us not to read writers' (especially women's) lives directly into their works, reveals two things: the constructed nature of their self-representations and, indirectly, the pressing problem for Russian women writers of how in fact to formulate their literary identity.

Women's literary history has approached the problem of a literary identity for women writers in several fruitful ways. Miller proposes a dual strategy of overreading with gender in mind and reading women's fiction and non-fiction together, or intratextually, to counter what she calls "the problems of imagining public female identities" (Miller 1988: 60). In nineteenth-century English literature, Gilbert and Gubar argue that women writers formed their literary identities by looking to earlier women writers as models, a process that they term "anxiety of authorship" and contrast with Harold Bloom's idea of the Freudian anxiety of influence for men writers (Gilbert and Gubar 1979: 49). More recently, in both nineteenth-century English and French literatures, feminist literary historians have turned to questions of literary production, social and cultural capital, and constructions of gender (Ezell 1993; Cohen 1999; Cohen and Dever 2002).

In contrast to these European scenarios, Russian literature developed a good deal later and therefore by the late eighteenth and early nineteenth centuries, men and women writers simultaneously confronted the anxiety of authorship and of establishing public identities in Russian literary culture

(Lotman 1992a). Thus unlike Bloom's and Gilbert's and Gubar's theories, the problem was an absence of identifiable models for writers. Moreover, these writers were almost all from a small noble elite and thus shared a good deal of social and cultural capital. Perhaps most important, noblewomen also owned a substantial share of real capital, for unlike their European counterparts, married women had long exercised complete control over their property and by 1800, women controlled "as much as a third of the estates in private hands, as well as serfs and urban real estate" (Marrese 2002: 2). At the same time, the importation of the new European cult of domesticity – by way of Paul I's wife Maria Fëdorovna (1759–1828) (formerly Princess Sophia-Dorothea of Württemberg, who was in charge of noblewomen's education) as well as the writings of Madame de Genlis (and other conduct literature at the beginning of the nineteenth century), along with George Sand's novels in the 1830s and the idea of women's emancipation – feminized the image of Russian noblewomen. These discourses disguised them as "women," minus the power of their continued historical, political, cultural, and literary responsibilities and thus real influence as noblewomen. This layering of new ideas with old traditions occurred at the same time as noblewomen first began to write and publish in larger numbers, and resulted in a sense of *dédoublement* that permeated their self-representations, writings, and aesthetics.

How did writers attempt to model themselves in barely the second generation of Russian writers? Iuri Lotman posits the rise of the importance of the writer's personality as a justification for choosing a field of endeavor besides God and country (Lotman 1992a). In his 1923 essay "Literature and Biography," the Russian formalist critic Tomashevslyi argues against the tendency to view literature as an extension of the writer's biography when he posits that writers generate their own literary biographies at certain periods in literary history. Thus, the literary biography as distinct from the actual facts of a person's life can become a significant literary fact and part of the writer's oeuvre. While both Lotman and Tomashevslyi limit their discussions to men, Russian women writers at mid-century were engaged in a similar literary process – the invention of a recognizable literary identity, which found sustained expression through evocations of the Russian countryside.

Like Poland and France, Russia organizes its cultural life in spatial terms, specifically in the distance between the center and the periphery (Shchukin 1998). The competition between two centers – St Petersburg and Moscow – actually heightens the sense that there should be only one center. In a rich analysis of Russian provincialism, Irina Savkina (1996) examines the conception of distance from the capital in Maria Zhukova's work as an essential, though unexamined, opposition between the provinces and the capital in Russian culture. When Russian women wrote about the countryside in the nineteenth century, St Petersburg and Moscow hovered in the background. Savkina argues that in works by women, this opposition between city and country underscores the marginality of provincial heroines as both provincial and female (Savkina 1998: 56).

This opposition in women's writings also emphasizes the marginality of the woman writer to a Russian literary culture overwhelmingly centered around St Petersburg and Moscow. Nevertheless, I think that by constructing themselves at the margins as provincial, such women writers as Zhukova (1804–55), Elena Gan (1814–42), Nadezhda Sokhanskaia (1823–84), Nadezhda (1822–89) and Sof'ia Khvoshchinskaia (1824–65), and Marko Vovchok (1833–1907) in fact could find their literary voices on aesthetic issues.[1] In her study, *The Provincial Women of Russian Literature (Women's Writing of the 1830s and 1840s)*, Savkina posits that the provincial theme carried fewer taboos for a patriarchal cultural tradition than the more obviously suitable theme of women's emancipation and that women writers identified with, and worked out their identity through, their beleaguered provincial heroines (Savkina 1998: 63). But was this only the path of least resistance? Savkina leaves open the question of the importance of the countryside for women writers' aesthetic principles and positions within the ubiquitous Russian literary battles.

Aesthetic debates in Russian literary culture about the countryside were also about the country, or Russia. In his original study significantly linking the development of nineteenth-century Russian national identity with realist writings about the countryside by Pushkin, Gogol, Goncharov, Aksakov, Turgenev, and Tolstoy, Christopher Ely argues that Russians imported European thinking about nature and defined themselves and Russia by contrast, as different from, and either better or worse than, Europe (Ely 2002). Comparing Russia to the dramatic, picturesque, European landscape, Russians made a virtue out of the deficiencies of their landscape: "The vast openness of the level, unforested steppes and plains, the uncultivated, unmanicured nature of dense forests only lightly brushed by human habitation, the seasonal extremes, and even the distinctive emptiness and impoverishment of the stubborn agricultural lands" (Ely 2002: 7). In *Dead Souls*, for example, Nikolai Gogol seems to acknowledge such a new tradition:

> Hardly had the town retreated when (quite the usual thing among us) there unrolled on both sides of the road vistas of a wild preposterousness: hummocks; fir groves; small, squat, sparse undergrowths of young pines; charred stumps of old ones; wild heather; and suchlike nonsensical rubbish.
>
> (Gogol 1996: 16)

Despite the emphasis on nature, the city is everywhere present because, as Ely concludes, educated Russians brought this new, visual vocabulary for the Russian landscape to bear on a central issue for the literate, urban elite: "How was an educated, Europeanized urban population to envision a unified society that included the mass of Russian peasants with whom it had almost nothing in common?" (Ely 2002: 88). Two very important European sources for Russian literature about nature, Jean-Jacques Rousseau (Savkina 1998: 62–5) and George Sand (Kelly 1994: 60), brought together issues of class, gender relations, and politics within the country-side that resonated deeply with such major Russian writers and thinkers as Herzen, Dostoevsky, Turgenev, and Tolstoy. At the very least, by writing about the Russian countryside, women writers implicitly stated their wish to be taken seriously as Russian writers alongside men. Clearly, in dis-cussions of these writers, the emphasis should fall equally on constructions of "Russian" and of "women."

Using male and female pseudonyms, Zhukova and Gan, in the 1830s and 1840s, and the next generation of women writers, in the 1850s and 1860s, established their literary identities through their aesthetic concerns as part of a literary world run by journals, critics, writers, and the censor. More specifically, in positioning themselves at the margins, in the country-side, women writers created a literary niche that was in fact central in aesthetic debates over Russian realism. Male critics who dominated the Russian literary scene claimed that women wrote simply or auto-biographically and therefore could not be good realist writers (Hoogenboom 1996: 10–16). According to the enormously influential critic Vissarion Belinsky, while men wrote about the whole of Russian society, women were limited by their experience to writing about women and domestic interiors: "a woman is locked in her very self, in her womanly and feminine sphere, and if she steps outside it, then she becomes some kind of ambiguous being" (Belinskii 1840). Not coincidentally, Belinsky and other critics rejected women's entry into realist writing just when realism would become the main path for writers' professional aspirations in creating a specifically Russian national literature on par with European literatures.

As Russia and its literary life revived after the death of Nicholas I in 1855, the number of women writers increased, as did the competition and criticism over their ability to correctly depict rural life. Ivan Turgenev's *Huntsman's Sketches* (1852) had established him as a touchstone on writing about provincial Russia. Turgenev must have felt the heat of competition, for he rushed to translate an exciting new collection of tales by Marko Vovchok (1859) from Ukrainian into Russian, and became her literary protector. Somewhat like Turgenev, a nobleman from Orël province who spent much of his time abroad in France, Vovchok had capitalized on her provincial origins. Vovchok, a masculine pseudonym for Mariia Markovich, grew up in the Russian provincial city of Orël, moved to the Ukrainian capital Kiev when she married in 1851, and did ethnographic work with her husband before moving to St Petersburg in 1859, when she entered Russian literary life; she then lived in Paris from 1860 to 1867, before returning to Russia and becoming a children's writer. In 1870 in a review titled "Female Heartlessness," the critic Nikolai Shelgunov, writing about

Vovchok's very successful collections of short stories about peasant life from a decade earlier, argued that the characters were aristocrats dressed up as peasants. Furthermore, Shelgunov argued that Turgenev did this better. However, Sokhanskaia and Khvoshchinskaia criticized Turgenev's depictions of peasants and their speech as urban and inauthentic, a criticism that indicates that they were competing with him over getting provincial life right. Tolstoy worried about Sokhanskaia's talent because she was too interested in Slavophile idealizations of country life, while Khvoshchinskaia criticized Sokhanskaia's conservative politics as manifested in her depictions of peasant women. In all these competitive debates about rural life, "realism" was invoked and it simply meant "good" writing.

As late as the 1830s, critics and writers did not recognize Russian women writers as actually having a specifically professional existence or identity. One can judge the extent to which women created a public identity *ex nihilo* from the first extended review of Russian women writers in 1834. Ivan Kireevsky does not use women's names, only hinting at whom he has in mind by noting their societal accomplishments, as if it would be indelicate to name names (Kireevskii 1984). In fact, most writers were nobles and, according to the cultural dictates of their class, were not supposed to have professions or earn money (especially women, unless they had to support their family) (Rosslyn 1996). A decade later, in the second important comprehensive review of Russian women writers, Belinsky maintains some mystery when he concludes that one would like to know more about Elena Gan, whom he considers the best (Belinskii 1843: 678). According to Tomashevsky, readers' need to know more, their desire for the illusion of a writer's life, is central to the creation of literary biography (Tomashevslyi 1923: 49). Paradoxically, for a woman writer, the atmosphere of secrecy and lack of information actually provides a not unusual literary biography – a somewhat blank one.

Although Belinsky was the most influential nineteenth-century critic on all issues, including women writers, an 1837 story by a minor writer reverberated through the 1860s among women writers. In "The Woman Writer," Verevkin declared, "A woman should not write," and his dictum was quoted by Gan, Zhukova, Sokhanskaia, and Nadezhda Khvoshchinskaia (Rakhmannyi 1837). Usually viewed as an attack on women writers, it is in fact aimed at the noblewoman writer. After a foolish education in the provinces reading novels, Vareta marries for money and moves to Moscow. Carried away by fame and ambition, she writes poems, novels, and plays, as she abandons her infant son to her husband's care. Their child dies while she presides at the opening night of her new play and ignores notes to come home; her husband leaves her and dies seven years later in a stage fire. Verevkin contrasts her heartlessness in life with her writing, where she hints at womanly sorrows, especially disappointed love. Women cannot simultaneously fulfill their duties as wife and mother, and have professional careers, unless they must support their families, according to Verevkin. Of course, in real life, neither noblewomen nor men directly cared for their children, who were left with nannies, governesses, and tutors, whom noblewomen would manage along with the other servants, serfs, and affairs of their homes and estates in town and in the country. New ideas about a woman's roles had supplanted noblewomen's actual duties in representations of women writers as emotionally cold and amoral society women.

How did women writers privately express their situation, keeping in mind that letters are conditioned by an addressee and by genre conventions (letters to one's editor, a lover, parents, or children, for example), and should not be read uncritically (Savkina 1998: 138–9)? They devised a liminal, doubled being, as echoed in the title of Karolina Pavlova's unusual novella, *A Double Life* (1844–7), where each chapter begins in prose and ends in poetry. The heroine's social debut in Moscow unconsciously triggers a poetic voice and romantic poetry that evoke nature to accompany the drawing-room machinations of the prose of this society tale. In a letter in 1839 to an unidentified friend, Elena Gan underscores her awareness of establishing and maintaining her problematic literary identity, and she solves the problem by creating a doubled identity:

> If you but knew how disgusting my title of *femme-auteur* is to me, how many curses and how much unpleasantness it has brought on me. No one wants to understand that *I* with a pen in my hands and *I* in my domestic daily life – am a completely double being.
>
> (G. 1911: 65)

Gan's use of French emphasizes her education and noble background, obstacles for her professional identity, which she masked as Zeneida R-va. A decade later, in 1848, at the beginning of her career, Nadezhda Sokhanskaia could only formulate rather than solve the problem of her identity in a letter to publisher and critic Petr Pletnev, on the impossible task of writing about herself in her *Autobiography*: "And the most unpleasant thing is that you say the truth, it seems, and it all comes out not quite right: you appear to be some kind of phoenix, which in reality could not exist" (Sokhanskaia 1896: 20). In Sokhanskaia's unhappy words, we sense the echo of Belinsky's phrase, "ambiguous being." Like Gan, she adopted a female pseudonym, Nadezhda Kokhanovskaia.

The contradictions between Gan's literary biography and actual biography center on her relationship to the provinces, and expose both the artifice and functions of her construction of the provincial heroine. Gan wrote two tales set in the countryside that formed the first literary biography of a Russian woman writer and set the tone for other writers. In accordance with his idea that women wrote autobiographically, Belinsky simply accepted this as Gan's real life and the general situation of women writers. Most critics and readers since then have done the same with few exceptions (Kelly 1994; Harussi 1981). In the opening frame of "Society's Judgment" (1840), under her usual pseudonym Zeneida R-va, Gan went public and wrote a famous, blistering attack on the difficult position of the Russian woman writer in a provincial town, a position that apparently mirrored hers. Like Gan, the female narrator is a writer married to a military officer stationed in the provinces and constantly moving from place to place. Provincial women look down on her as a freak of womanhood, while a visiting writer condescends to give her advice. She exposes the petty prejudices of provincial society, which ridicule the woman writer at every social occasion.

The provincial setting serves to underscore the woman writer's general social and professional isolation. In 1842, while she was dying of tuberculosis, Gan wrote a second tale, "The Futile Gift," about a lone woman writer, dying of tuberculosis, also set in the country. Again, the plot bore enough similarity with Gan's life that it was read as her swan song. Aided by a tutor and then a benefactress, a poet develops her talent in the haven of a good library in the countryside, but dies just as she is published in the city. Very probably, however, the model for the poet is in fact the Russian woman poet Elizaveta Kul'man, who had died at the age of 17 in 1825 from tuberculosis before ever publishing (Hoogenboom 2001). Here Gan works with what Tomashevsky identified as the more ready (male) model of the Romantic poet who dies young and unappreciated.

Suddenly in 1886 and 1887, over forty years after her death, conflicting biographical accounts contested Gan's tragic, provincial literary image in a way that revealed the tensions underlying this literary fabrication. At the center of the dispute was Starchevsky's account of her time in St Petersburg in 1836 and her supposedly amorous relationship with her new editor at *The Library for Reading*, Osip Senkovsky (1886). Gan's younger sister (Fadeeva 1886) and Gan's daughter (Zhelikhovskaia 1887), both writers, rejected Starchevsky's picture of a naive provincial woman seduced by literary life in the big city. These biographies, plus another fuller biography (Nekrasova 1886), together with Gan's letters (G. 1911) depict a life that was considerably more varied geographically and socially, with numerous long stays in St Petersburg, Odessa, and other health resorts in southern Russia frequented by the nobility. But the most contentious issues were questions over Gan's emotions: were they appropriate or inappropriate and could she control her feelings sufficiently? In *The Proper Lady and the Woman Writer* (1984), Poovey argues that English bourgeois society's fear of female desire and sexuality conditioned women's propriety and channeled emotions into meekness, self-effacement, and "indirection," both in life and on the page. In Russian society, however, the ability to control one's feelings reflects a woman's noble breeding. Gan's sister and daughter both argue that it was Senkovsky who was extravagant with his feelings, while Gan understood him perfectly and was not impressed by him or by Russia's greatest writer Pushkin, a known womanizer. Gan's clear-eyed views on society are evident in one letter, where she wrote about the advantages of being perceived as eccentric because she is a writer when she goes to half a dozen balls in the same black dress and falls asleep at them (G. 1911). In "Society's Judgment," provincial society rejects the woman writer, while in her letters, Gan disdains provincial society. For example, Gan keeps up appearances before her friend when she writes:

> wherever we are, I will positively not become acquainted with anyone; landowners of the lower ranks are not on my level, while those higher up look down on me, and I am not one to vie with them because I am too proud.
>
> (G. 1911: 66)

On the other hand, Russian society women were accused of heartlessness, and so Gan's sister and daughter both insist that she was a caring mother whose writing only took her away from them to make money for their education. Not surprisingly, Gan and other women writers at this time used the conventions of Sentimentalism to convey the depth and quality of their characters' feelings, and critics read those feelings back into women authors. In this way, in her biography, Nekrasova defended Gan using her works as if Gan herself were a sentimental heroine. Thus, the heart-breaking provincial isolation of Gan's heroines served to deflect criticism from Gan herself, who would not have been caught dead wearing her own heart on her sleeve.

Although sentimental novels are never just about feelings (Cohen 1999), Russian critics have tended to misread women's writings as mainly about emotions, which might be enhanced by the setting in nature. In his review of Gan, Belinsky on the one hand held up Gan as the model of oppressed Russian women writers but also trivialized Gan's concerted attempt to establish herself as a real author. He praised her highly as a writer primarily of love, but then criticized her for provincialism in her depiction of love and in her taste in Russian literature (Belinskii 1843: 675). Nevertheless, he was forced to concede that (unlike him) she read the great authors in Italian, German, English, and French – which Gan made sure her readers knew by liberally using epigraphs from European works in the original languages, an indication of her excellent noble education at home by her similarly well-educated mother.

In the same vein, Belinsky wrote slightingly of Zhukova that "the author has much soul, much feeling, and their burdensome fullness searches for the means to express itself in something external" (Belinskii 1840: 111). In her story "The Provincial Girl," Zhukova herself lightly mocks the tendency of provincial women to overdo their feelings as "over-indulgence and pomposity," thus exposing not only the womanly business of feeling, but also stereotypes of the provincial heroine, as clichés (1837–8: 222). Much more important are Zhukova's aesthetic digressions, which Belinsky and subsequent critics either did not recognize or ignored because of their assumptions that women wrote about love and not about aesthetics (Beletskii 1919; Hoogenboom 1998). Zhukova posits a crucial aesthetic connection between landscape, free time for feeling, and gender. As central to her realist aesthetics, Zhukova chooses a prosaic Russian landscape that is calculated to be boring. Zhukova's choice places her squarely in a developing tradition of descriptions of the Russian landscape as a vast, flat, dull open space, as opposed to the exotic European landscape that she had illustrated for her travel book about southern France (which Belinsky considered her best work) (1844). Moreover, rather than objectify the provinces as viewed from the center by urban passers-by, Zhukova subjectively identifies herself with provincial women as heroines and narrators (Savkina 1998: 63). For example, by comparison with Gogol, Zhukova's description of landscape is positive and immensely varied, a variety that Savkina finds typical for women writers. Zhukova distinguishes the different types of provincial towns – capitals of the larger "guberniia" as opposed to the smaller "uezd" (administrative units comparable to states and counties) – from villages and surrounding countryside, which is distinct from nature more generally (Savkina 1998: 63–4). As the following description by Zhukova indicates, she is emphatically *in* the landscape, not just passing through, unlike Gogol's hero Chichikov, who drives through the countryside in a carriage that Gogol introduces on the first page as essential to Chichikov's identity because it defines his outsider status as part of the center, and thus better than the periphery:

> Time flies in the capital in merriment, and in monotony in a provincial town; and in the latter, it seems, even faster. The years pass as nothing; you look back – many days! – but they do not distinguish themselves one from the other and the imagination slithers, unable to find a sign where one might stop. The life of a provincial person is like a path that winds among the fields in the flat plains of the Penza or Saratov regions. Exhausted by the monotony, the traveler walks and his eyes search in vain for something new in the distance: everything is the same. He descends a gradual slope and then goes up a hill: there he imagines will be revealed to him a valley, with copses, groves, hamlets, or a bright lake with inspiring skies, or blue mountains, forests; he speeds up his step; clambers up the hill: again the boundless, flat surface like the sea, of the plain covered in grasses that with the least puff of wind flow like waves into blue and green ebb-tides, joining up in the distance with the blue heavens. But in this monotonous picture is life, is delight; the ripening ear of grain and the blue cornflower between the slender trunks speak of the secret activity of nature; a whole world of insects buzzes between the flowers; scents, the patrons of the fields, play in the east, shining in the evenings as frisky summer lightning, and the peasant is glad, foretelling a fortunate blossoming.

And so it is in the life of a provincial person, if his days are poor in events, they are nevertheless filled with feelings; there feelings are deeper, more spiritual, there where they are more concentrated . . . I speak about provincial women, because in provincial towns, men are busy with service, the latest directive, the arrival of the governor, depositions and responses to questions and so on and so on . . . they have no time to occupy themselves with feelings.

(1838: 221–2)

After reading Zhukova's evocative homage to nature and landscape, it is hard to imagine that Belinsky called Zhukova primarily a writer of society tales and of the heart. Zhukova was born in this kind of area, and later lived in St Petersburg and then Italy, but as my title suggests (via Oscar Wilde), it was important for a woman writer to try to be seen as provincial.

I think these and other women found their place aesthetically by emphasizing precisely the liminal nature of their cultural, literary dislocation, and crossing boundaries (or breaking taboos) – in the first instance, by moving *between* the city and the country. In real life women writers moved back and forth, while as writers, they moved their heroines to the provinces, and in fiction, these heroines then moved back to town, in what Kelly calls the essential feature of the provincial tale, "the escape plot" (Kelly 1994: 63). Women's symbolic move to the country and back to the city represented a negotiation of tensions between accommodation and staking out literary territory.

In "The Provincial Girl," Zhukova's liminal narrator moves with the ease of her heroine Katia from the ballroom of one of St Petersburg's old families, to a provincial home, and to the fields. Movement between city and country abounds in *Evenings on the Karpov River* (1838–9), a collection of stories that includes "The Provincial Girl" and is set in a frame narrative: a group of people are telling stories to a female narrator while visiting the summer home of a benefactress on one of the islands at the periphery of St Petersburg – in other words, another highly marked boundary area. Essentially, the narrator is always walking, Zhukova's metaphor for her realist aesthetics, and negotiating various boundaries – spatial, social, cultural, literary, even formal (via the frame narrative).

Two decades later, in a tale with the telling title, "City Folk and Country Folk," Sofia Khvoshchinskaia, similarly underscores movement between the city and country (Vesen'ev 1863). Like Zhukova, she is critical of the aesthetic of feelings as women's (writers') work and considers the countryside a litmus test of being a good writer. In the tale, she mocks the turn-of-the-century writer Nikolai Karamzin, who in his 1792 tale "Poor Liza," created the archetypal Russian sentimental country heroine, who commits suicide after being jilted by the urban sophisticate, the nobleman Erast. Khvoshchinskaia brings a city writer named Erast to the countryside, where he is forced to live in his neighbor's bathhouse because he has mismanaged his estate, which is in ruins. In contrast, his neighbor, a widowed noblewoman, manages her estate well, but he only notices that she and her daughter are women. Naturally, he tries to educate them, following the liberal dictates of the time. His efforts end when the daughter Olga boxes his ears, an action representing, I think, a firm rejection of the aesthetic of feelings as innately womanly, and of patronizing male writers, and of their citified ideas about women.

Unlike Zhukova's more urbane female narrator, the Khvoshchinskaia sisters fashioned their narrative personas using the tone of a bemused male provincial in their fiction and essays. Both wrote extensively about provincial life and Nadezhda also wrote a series of critical essays, "Provincial Letters," in the style of Belinsky for which she adopted the role of the curious observer. In "City Folk and Country Folk," two sisters make a guest appearance as writing sisters visiting their old home in the provinces from the city. Unlike their male urban counterpart Erast, they are equally at home in both spheres. In real life, the Khvoshchinskaia sisters traveled each year to Moscow and despite their evenhanded treatment of provincial life in their fiction and criticism, in their letters they complained desperately of boredom and oppressiveness, and Nadezhda eventually moved to St Petersburg, first briefly in 1865, and then for the last five years of her life, where she died (Rosenholm and Hoogenboom 2001). After Nadezhda's death in 1889, her younger sister Praskovia, also a writer, wrote in her biography that the sisters preferred the anonymity of male pseudonyms because of the realities of provincial life for women writers, where people would ostracize them, in fear they would end up in the sisters' fiction (Khvoshchinskaia 1892: 1:i-xviii). Thus like Gan, Sof'ia and Nadezhda self-consciously constructed their reputations as provincial writers. They took a sympathetic yet objective tone as male narrators that served to make their vantage point from the periphery of Russian literary culture more compelling.

What did critics and writers find threatening about the periphery looking in at the center? Why was all this real and literary movement back and forth, between the city and the country, Europe and Russia, less acceptable, and therefore more taboo, for women? Women set themselves in opposition to critical and literary representations of women writers as feckless society women who might leave home, abandoning children and husbands in search of fame and fortune to satisfy their ambition. In literature, the virtuous nineteenth-century Russian heroine was depicted as most at ease in the countryside, while her less virtuous sister inhabited the domestic interiors of society life. Alexander Pushkin established this literary paradigm in *Eugene Onegin* (1830), with Tatiyana and her sister Ol'ga: "A wild creature, sad and pensive / Shy as a doe and apprehensive, / Tatiyana seemed among her kin / A stranger who had wandered in" (Pushkin 1830: 2:25). In contrast, Ol'ga is like any heroine in any novel, a kind of stateless cosmopolitan, which by comparison makes Tatiyana more natural and as it turns out, more Russian. Even when Tatiyana moves to Moscow and conquers high society by marrying well and comporting herself accordingly, she does not lose her true essence, one based in Russian nature. Moreover, Tatiyana is arguably a woman writer, producing the famous letter to Onegin in Chapter Three (Burgin: 1991).

Thus, by representing themselves as provincial, these women laid a claim to another, equally important kind of literary seriousness, not among men, but among women, writers. They entered a literary competition featuring bad versus good girls, or Europeanized versus Russian girls, or society versus provincial girls. By emphasizing nature and their place in the Russian landscape, these writers thus positioned themselves as moral Russian women – in the 1830s and 1840s as privately moral (i.e. not like George Sand or her heroines), and later, in the 1850s and 1860s, as civically moral. At the same time, movement between the city and country was a gesture of protest that showed that women could rightfully claim the larger social perspective that would qualify them as serious Russian writers.

Those serious Russian writers were by and large noblemen. In "The Poetics of Daily Behavior in Eighteenth-Century Russian Culture," Lotman determines that the nobleman's ability to *choose* his behavior formed the very basis of his daily life (Lotman 1992b). Here the undifferentiated "other" is the peasant, who has no choice. To the peasant, he adds the noblewoman, whose choices were not what he terms individual, but dependent on her age, by which he probably means before or after marriage. By the 1860s and 1870s, women resorted to fictitious marriages to gain personal and professional mobility, and escape the legal control of male relatives. The "escape plot" in fiction had its real-life counterpart, though these women often escaped to study medicine in Europe. Add to this the end to recent limitations on mobility under the reign of Nicholas I (1825–55), when fear of the spread of revolution in France closed down travel abroad except for health reasons. In other words, when women writers and their heroines traveled between the city and country, they flaunted their ability to choose, which reflected their noble estate as much as their gender identities, simply by virtue of making choices.

Note

1 Note on transliteration: I have used the Library of Congress system, without diacriticals. Names of well-known Russians appear as is customary in English (e.g. Dostoevsky, Tolstoy), though the list of works cited will use the transliterated name (Dostoevskii, Tolstoi).

Bibliography

Beletskii, A.I. (1919) "Epizod iz istorii russkogo romantizma: Russkie pisatel'nitsy 1830–60 gg.," Kharkov, IRLI, R.1, op. 2, No44a, 254.

Belinskii, V.G. [1840] (1953–9) "Povesti Mari'i Zhukovoi," *Polnoe sobranie sochinenii*, 13 vol., Moscow, vol. 4.

—— [1843] (1953–9) "Sochineniia Zeneidy R-voi," *Polnoe sobranie sochinenii*, 7: 648–78.

Burgin, D. (1991) "Tatiana Larina's *Letter to Onegin*, or *La Plume Criminelle*," *Essays in Poetics*, 16.2: 12–23.

Cohen, M. (1999) *The Sentimental Education of the Novel*, Princeton NJ: Princeton University Press.

Cohen, M. and Dever, C. (2002) *The Literary Channel: The International Invention of the Novel*, Princeton NJ: Princeton University Press.

Ely, C.D. (2002) *This Meager Nature: Landscape and National Identity in Imperial Russia*, DeKalb IL: Northern Illinois University Press.

Ezell, M. (1993) *Writing Women's Literary History*, Baltimore MD: Johns Hopkins University Press.

Fadeeva, N.A. (1886) "Zametki i propravki: po povodu stat'i 'Roman odnoi zabytoi romanistki'," *Istoricheskii vestnik*, 11: 456–64.

G., M. [M.O. Gershenzon] (1911) "Russkaia zhenshchina 30-kh godov," *Russkaia mysl'*, 12: 54–73.

Gilbert, S. and Gubar, S. (1979) *The Madwoman in the Attic: The Woman Writer and the Nineteenth-Century Literary Imagination*, New Haven CT: Yale University Press.

Gogol, N. (1996) *Dead Souls*, trans. by Bernard Guerney, New Haven CT: Yale University Press.

Harussi, Yael. (1981) "Hinweis auf Elena Gan (1814–1842)," *Zeitschrift für slavische Philologie*, 42.2: 242–60.

Hoogenboom, H. (1996) "A Two-Part Invention: The Russian Woman Writer and her Heroines from 1860 to 1917," Columbia University NY, dissertation.

—— (1998) "The Society Tale as Pastiche: Mariia Zhukova's Heroines Move to the Country," in Neil Cornwell (ed.) *The Society Tale in Russian Literature*, Studies in Slavic Literature and Poetics, vol. 31, Amsterdam: Rodopi Press.

—— (2001) "Biographies of Elizaveta Kul'man and Representations of Female Poetic Genius," in Marianne Liljeström, Arja Rosenholm, and Irina Savkina (eds) *Models of Self*, Helsinki: Kikimora.

Kelly, C. (1994) *A History of Russian Women's Writing 1820 – 1992*, Oxford: Oxford University Press.

Khvoshchinskaia, P.D. (1892) "Biografiia N. D. Zaionchkovskoi-V. Krestovskii psevdonim," *Sobranie sochinenii V. Krestovskogo (psevdonim)*, 5 vols, Moscow, 1: i–xviii.

Kireevskii, I.V. (1984) "O russkikh pisatel'nitsakh," *Izbrannye stat'i*, Moscow.

Lotman, Iu. M. (1992a) "Literaturnaia biografiia v itoriko-kul'turnom kontekste (K tipologicheskomu sootnosheniiu teksta i lichnosti avtora)," in *Izbrannye stat'i*, 3 vols, Tallinn: Aleksandra, I: 365–76.

—— (1992b) "Poetika bytovogo povedeniia v russkoi kul'ture XVIII veka," *Izbrannye stat'i*, 3 vols, Tallinn: Aleksandra, I: 253.

Marrese, M. (2002) *A Woman's Kingdom: Noblewomen and the Control of Property in Russia, 1700–1861*, Ithaca NY: Cornell University Press.

Miller, N. (1988) *Subject to Change: Reading Feminist Writing*, New York: Columbia University Press.

Nekrasova, E.S. (1886) "Elena Andreevna Gan (Seneida R–va) 1814–1842. Biograficheskii ocherk," *Russkaia starina*, 8: 335–54, 9: 553–74.

Pavlova, Karolina [1844–7] (1978) *A Double Life*, trans. by Barbara Heldt, Oakland CA: Barbary Coast Books.

Poovey, M. (1984) *The Proper Lady and the Woman Writer: Ideology as Style in the Works of Mary Wollstonecraft, Mary Shelley, and Jane Austen*, Chicago: University of Chicago Press.

Pushkin, A.S. [1830] (1995) *Eugene Onegin: A Novel in Verse*, trans. by James E. Falen, New York: Oxford University Press.

Rakhmannyi [Verevkin, N.V.] (1837) "Zhenshchina-pisatel'nitsa," *Bibliografiia dlia chteniia*, 23: 15–134.

Rosenholm, A. and Hoogenboom, H. (eds) (2001) *"Ia zhivu ot pochty do pochty . . ." Iz perepiski Nadezhdy Dmitrievny Khvoshchinskoi*, FrauenLiteraturGeschichte, vol. 14, Fichtenwalde: Verlag F.K. Göpfert.

Rosslyn, W. (1996) "Anna Bunina's 'Unchaste Relationship with the Muses': Patronage, the Market and the Woman Writer in Early Nineteenth-Century Russia," *Slavic and East European Review*, 74(2): 223–42.

R-va, Zeneida [E.A. Gan] [1840] (1987) "Sud sveta," in V. Uchenova (ed.) *Dacha na Petergofskoi doroge: Proza russkikh pisatel'nits pervoi poloviny XIX veka*, Moscow: Sovremennik.

—— [1842] (1991) "Naprasnyi dar," *Serdtsa chutkogo prozren'em: Povesti i rasskazy russkikh pisatel'nits XIX v*, Moscow: Sovetskaia Rossiia.

Savkina, Irina (1996) "Kategoriia *provintsial'nosti* v proze Marii Zhukovoi (povest' "Naden'ka)," *Slavica Tamperensia: Aspekteja*, Tampere, 5: 289–94.

—— (1998) *Provintsialki russkoi literatury (zhenskaia proza 30–40-kh godov XIX veka)*, Wilhelmshorst: Verlag F. K. Göpfert.

Shchukin, Vasilii (1998) "Krizis stolits ili kompleks provintsii?," *Novoe literaturnoe obozrenie*, 34: 350–4.

Shelgunov, N.V. (1870) "Sovremennoe obozrenie: Zhenskoe bezdushie (Po povodu sochinenii V. Krestovskogo-psevdonim)," *Delo*, 9: 1–34.

Sokhanskaia, N. (1896) *Avtobiografiia*, Moscow.

Starchevskii, A.V. (1886) "Roman odnoi romanistki," *Istoricheskii vestnik*, 8: 203–34, 9: 509–31.

Tomashevskii, Boris [1923] (1971) "Literature and Biography," in L. Matejka and K. Pomorska (eds) *Readings in Russian Poetics*, Cambridge MA: MIT Press.

Turgenev, I.S. [1852] (1979) *Zapiski okhotnika, Polnoe sobranie sochinenii i pisem*, vol. 3, Moscow: Nauka.

Vesen'ev, I. [S.D. Khvoshchinskaia] [1863] (1987) "Gorodskie i derevenskie," in V. Uchenova (ed.) *Svidanie: Proza russkikh pisatel'nits 60–80-kh godov XIX veka*, Moscow.

Vovchok, Marko [M.A. Markovich] [1859] (1957) *Sobranie sochinenii*, Moscow.

Zhelikhovskaia, V.P. (1887) "Elena Andreevna Gan: Pisatel'nitsa-romanistka v 1835–1842 gg.," *Russkaia starina*, 3: 733–66.

Zhukova, M.S. (1837–8) "Provintsialka," *Vechera na Karpovke*, vol. 2, St Petersburg.

—— (1844) *Ocherki Iuzhnoi Frantsii i Nitstsi*, 2 vols, St Petersburg.

14 "My garden, my sister, my bride"

The garden of "The Song of Songs"

Kenneth I. Helphand

A garden locked,
Is my own, my bride,
A fountain locked,
A sealed-up spring
     ("The Song of Songs"
       1982, 4: 12–13)[1]

גַּן | נָעוּל אֲחֹתִי כַלָּה
גַּל נָעוּל מַעְיָן חָתוּם:

The association of woman and the garden is nowhere more profound than in "The Song of Songs" (also known as *Shir ha Shirim* in Hebrew, "The Song of Solomon," ascribing authorship to King Solomon, and "The Canticle of Canticles," Latin for the *Song*.) The gendered landscapes of the poem represent women and the garden through body-landscape metaphors, and a garden tale of fulfillment of Eden's paradisiacal promise.

The biblical foundations of garden thought are critical to historical understanding; they may also provide alternative models, myths and theory for contemporary design. With the exception of the Garden of Eden of *Genesis*, until the Renaissance the garden of "The Song of Songs" was the most important western garden text. Numerous interpretive versions and editions have been penned and illustrated as authors and artists sought to give the poem both distinct voices and imagery. These illustrations of the text are a clue to its evolving meaning, and they also form a case study in visual representation.

As a landscape architect and landscape historian my concern focuses on the imagery of the garden, the garden/body relationship, and how the tale has been interpreted. A key question is the creative process of "translating" (i.e. illustrating) word to image.[2] I have examined modern illustrative versions, with attention to the intersection of word and image to explore how the body and the garden are visualized. This is a historical tradition and contemporary practice. There are dozens of illustrated versions, which range from popular romantic collections of love poetry to an edition illustrated by Matisse (Gregory 1962).

"The Song of Songs" has been approached from a variety of perspectives. There are scores of translations and the literature of religious exegesis and secular interpretation is extensive. In addition, the poem has been subject to diverse adaptations and presentations as poem, tale, play, and photographic essay.

There are many interpretations, with subtle variants. Jewish readings see "The Song" as an allegory of love between God and the people Israel, although God is never mentioned in "The Song." In this interpretation God is the lover and the people Israel the bride. This is a common Judaic image; for example, the Sabbath is also often personified as the bride who is welcomed each Friday evening. Marriage here is used as the symbol of the most sacred and fundamental union. However, the Judaic interpretation is not uniform. Some are based on images of the Shekinah, the divine presence, more particularly a female alter-ego of a male patriarchal God. In this interpretation the male and female spirits are the lovers to be reconciled and "The Song" represents such a balance of forces. In Jewish tradition "The Song" is also included in liturgy. In the Ashkenazi, or European tradition, it is chanted only at the end of Passover, but in the Sephardic tradition it is sung every Sabbath, reinforcing the bride interpretations (Alter 1985; Pope 1977).

Christian interpretation reads "The Song/Canticles" as an allegory of love, where the lovers are Christ and the Church, often personified and spoken of in bodily terms. Or it may also represent a union between the individual and Christ. In either interpretation images of marriage and love are fundamental. The most renowned commentator was St Bernard of Clairvaux in the twelfth century who wrote 86 sermons on the song. Modern scholars also interpret Mary as a figure who embodied the more ancient figure of an earth goddess. Historically archaic beliefs were appropriated and absorbed into official Church doctrine. For example in AD 431 bishops at Ephesus chose Artemis' festival day, August 15th, as the day to celebrate Mary's assumption into heaven.

In what is known as the Marian variation the lover of the song is the Virgin Mary. Here the connections to garden and landscape history become critical, for this imagery is fundamental to a basic medieval garden type, the *hortus conclusus* (Aben and de Wit 1999; Ricci 1996). The garden type itself is named directly for the Latin translation of the Hebrew *gan na'ul* of the locked garden in "The Song" is *hortus conclusus*. The garden is locked, meaning that access needs to be granted from those within, or an actual or symbolic "key" to the place, and its meaning, needs to be found. Similarly, the protective walls of the garden, which have their origins in the practicalities of garden creation, both contain the garden, its life and activities, and simultaneously act as protective barriers from the outside world.

The idea and imagery of the *hortus conclusus* exerted tremendous power on medieval and subsequent garden design. In the garden of "The Song"

there are significant echoes of *Genesis* and the paradisiacal imagery of the
Garden of Eden. The paradise garden and its meanings traverse the spectrum of time. On the one hand paradise is understood as the site of origins
and beginning, the primeval human home. On the other hand paradise
also represents the hoped for future, as the perpetual abode of an after-
life. In reality the design and experience of actual gardens can provide a
return to the idealized perfection of the past, or a foretaste, a worldly anticipation of what is to come. In all instances what is sought is a place that
embodies a state of being.

The Biblical paradise is actually modeled on the landscape of the desert
oasis, a place where life-sustaining water creates places that are in dramatic
contrast to their surroundings. These sites of abundant vegetation, flowing
waters, sensory delight, and gathering places of people were surely the
prototypes for the Edenic tale. In idealized form these were gardens that
were extant when "The Song of Songs" was written. Oases are places
of multi-sensory richness accentuated by their stark surroundings. They
offered physical pleasure, but also serve as a physical and emotional respite
from their surroundings. All of these qualities are then transposed to the
beloved female of "The Song." She, and her body, is a site of pleasure
and comfort.

In the Marian interpretation the Virgin Mary becomes the beloved of
the poem, the bride of Christ and Heaven's Queen (Daley 1986; Stokstad
and Stannard 1983). Metaphorically Mary is referred to as the *hortus
conclusus*, as the Madonna Mary personifies and embodies the beauty of
the garden and its natural imagery. Identified with her place of residence,
she becomes the object of veneration and her setting becomes a place of
desire, in a powerful combination of images that are simultaneously sacred
and erotic. Ultimately the Virgin Mary as the perfect and ideal woman
becomes associated with the love of woman in general, woman who is
idealized as the object of a chivalrous perfect love. The garden in turn
becomes a garden of love in an idealized spiritual state *and* in its pleasurable, even carnal pleasures. Thus, the *hortus conclusus* oscillates between the
virginal, pure, white pristine love of the Mary Garden and even the lusti-
ness of the Garden of Love. The *hortus conclusus* of the Mary Garden is
layered with symbolic iconography (Figure 14.1). In this image of the Mary
litany, "Mary with symbols of garden and other attributes," she is
surrounded by a host of emblematic associations. The *(h)ortus conclusus*, the
lily among the thorns, Tower of David, Fountain of the Gardens, and Well
of Living Waters are direct quotes from "The Song of Songs." To these
are added the City of God, a Fountain of Living Waters, Tree of Jesse,
Mystical Rose, Gate of Heaven, Star of the Sea, Unblemished Mirror,
Fountain of Life, Cedar of Lebanon, Sun and Moon.

In *The Paradoxes of Paradise* Francis Landy offers an intriguing thesis. She
interprets "The Song" as a version of Paradise, a kind of Eden II. Like
Adam and Eve in Eden, lovers in "The Song" can be viewed as archetypal

Figure 14.1 Mary with garden symbols and other attributes, 1578

figures, representing basic male and female aspects of all persons. She views the poem as a glimpse of the fulfillment of Eden's paradisiacal promise. "The Song of Songs" is *Genesis* inverted. The expulsion of *Genesis* sends us outside of the garden, "The Song" is a return to the Edenic garden and importantly it is through the female that man and woman reenter paradise (Landy 1979, 1983). The association of "The Song" and Eden are a reminder that the garden is the site of the most fundamental of human questions, the relationships between individuals, the limits of knowledge, the mystery of mortality.

Traditional religious interpretations typically retain the rich imagery of the poem but tend to divest it of its worldly meaning. The material realities of individuals, bodies, garden, and landscape are transcended in some fashion. The dramatic erotic aspects of the poem are often interpreted symbolically. There may be conscious repression or sublimation, but the sexual content is undeniable. Some see an urban worldly King Solomon and a female country lover, Shulamite, accompanied by a chorus of the daughters of Jerusalem. In the relationship between the lovers perhaps there is also recognition of the inseparable interconnection between city and country and even a reconciliation between their often antagonistic cultures and mores.

Some theorize that the song has its origins in songs composed for wedding feasts. There are historical affinities to the liturgy of ancient Canaanite or Babylonian fertility cults, where the lover is a goddess, such as Ishtar the moon goddess, and the male the Canaanite sun-god Tammuz, or Isis the Egyptian goddess (Pope 1977: 145). It is also clearly a love poem and has parallels with ancient wedding songs and Egyptian love poetry. There are records of many ancient Egyptian poems dating from *c.*1300 BC. One poem goes:

> (Girl) I am yours like the field
> > planted with flowers
> > > and with all sorts of fragrant plants
> > Pleasant is the canal with it
> > > which your hand scooped out,
> > > > while we cooled ourselves in the north wind:
> > > a lovely place for strolling about,
> > > > with your hand in mine
>
> > > > > > (Fox 1985: 26)

The garden is the poem's most significant image. In "The Song" the garden is anthropomorphized into both female and male bodies and the body is "gardenized" – we do not have a single word for this metaphor. The garden is a locale and setting and it is a metaphor for the female beloved, "The garden, my sister, my bride." The woman is the garden and also the source, the "fountain of gardens," the wellspring of all of the emotions of the poem. Significantly, the garden stands not only for the female person and body, but it also symbolizes the relationship that grows between the lovers – they are both in the garden, and they create the garden:

> Where has you lover gone
> O beautiful one?
> Say where he is and we will seek him with you
> My love has gone down to
> his garden, to the bed of spices,
> to graze and to gather lilies.
> My beloved is mine and I am his
> He feasts
> in a field of lilies.
>
> > > (Bloch and Bloch 1995: 91)

The lovers are within the garden and their growing love creates the garden. What happens when you are in love? There is the emotional interaction of lovers characterized by expectation, heightened awareness, focused attention, spiritual connection, carnal pleasure, erotic tension, excitation, fulfillment, and the delight of another's company. All of these emotional and physical responses apply to the characters in "The Song" *and* the experience of the garden. The places of love become associated with lovers and their personal stories and this also applies to the garden as a central image of the poem. Many commentators view the love of the poem as a force, one with the ultimate power to confront death and non-being. By extension, perhaps as locale and symbol, the garden too stands for these most powerful and positive forces:

> The germinal paradox of the Song is the union of two people through love. The lovers search for each other through the world and through language that separates them and enfolds them. The body is the medium for this search and is the boundary between the world and self. Thus the body comes to represent the self to the world, and the world to the self. It becomes the focus of metaphor, the conjunction of differentiated terms.
>
> (Landy 1987: 305)

The poem is replete with landscape references that are crucial as settings and central to the ambiance created by the poem. The landscape does have a literal aspect: places are named, real locales, actual mountains, plants, and animals. The mountains and streams of Lebanon, the sources for the waters of Northern Israel, are particularly prominent. Each of the primary settings described in the poem has an association with the lovers. There is a mutuality of person and place. The full range of a landscape and spatial spectrum is described. At one extreme there is a wilder and more remote natural landscape and its elements. These landscapes do not support the lovers' intimacy. However, over and over the poetic imagery of the poem suggests that "the lover is entirely assimilated into the natural world at the same time that the natural world is felt to be profoundly in consonance with the lovers" (Alter 1985: 194). The poem's central image is of a "middle landscape," a cultivated, domesticated countryside of vineyards, fields, pastoral in its aspects and emotional connections. For example, in the woodcuts by American artist Wharton Esherick (1887–1970) he concludes his illustrated edition with an image of such a middle landscape, "Solomon had a Vineyard" (Figure 14.2). Esherick's image is a classic scene of habitation (the King's palace?) set within a modestly rolling domesticated landscape of gardens, vineyards, groves, pastures, and fields viewed from upland meadows. This landscape is a mirror of the lovers' relationship. At the poem's conclusion all is in equilibrium: the lovers reconciled, and a balance of natural and human forces, a landscape of resolution.

Figure 14.2 "Solomon had a Vineyard," Wharton Esherick, 1927

Poet, translator, and commentator Marcia Falk says the landscape presented here is always relatively benign, that it is inviting, playful, even happy. Supportive of the lovers, it also symbolizes them in direct fashion. The lovers meet in gardens or natural "garden" spots, echoing garden historian Christopher Thacker's observation that "No doubt about it the first gardens were discovered" (Thacker 1979: 9). The illustrations by Ze'ev Raban (1890–1970) from a 1923 edition employ the idealized "Biblical" landscape of Palestine as a motif (Figure 14.3). In this figure the male and female lovers are speaking over a garden wall. The male lover stands atop a rock in an attempt to peer into the garden and encounter his love. Raban's illustrations are the rare version that uses the actual landscape that inspired the poem. He was a teacher at the Bezalel, the arts and crafts school in Jerusalem named for the first artist in the Bible. His is a romantic vision, but it has distinct recognizable elements. He shows a landscape of vineyards, date palms, cypresses, oaks, cedars, stone watchtowers, terraces,

Figure 14.3 Ze-ev Raban, 1923

wildflowers, grazing sheep, and goats. He is particularly faithful to the contrasting character of the city and a pastoral countryside and at the more intimate scale between the qualities of interior and exterior spaces. An astute observer can even identify specific places in contemporary Israel in his images, such as the city of Tiberias on the shore of the Sea of Galilee and the valleys surrounding Jerusalem.

> Follow the tracks that the sheep have made
> And feed your own little goats and lambs
> In the field where the shepherd's lie.
>
> (Falk 1993: 3)

> Come, love, let us go out to the open fields
> And spend our night lying where the henna blooms,
> Rising early to leave for the near vineyards
> Where the vines flower, opening tender buds,
> And the pomegranate boughs unfold their blossoms.
>
> (Falk 1993: 24)

The city and its streets are also included in the song. These places are the places least sympathetic to the lovers. A most interesting series of illustrations is in an edition by the British artist Arthur Wragg (1903–1976) that contemporizes the song, while presenting images that link it to an artistic tradition. He includes images of a female figure as a giant inflated parade float drawn through the city streets while city dwellers cavort below. Lovers look out from a bridge over a polluted stream surrounded by a skyline of smokestacks. Linking "The Song" to classical mythology and its Renaissance interpretation, he shows a figure of the Shulamite as Botticelli's *Birth of Venus* arising from the tenements where lovers eye each other and kiss across window to window from buildings that lean towards each other. Wragg's lover is portrayed as a Madonna where the *hortus conclusus* has been transposed to an urban rooftop with a garden of potted plants and an arbor of drying laundry (Figure 14.4).

Figure 14.4 Arthur Wragg, *c.*1930

I must rise and go about the city
the narrow street and squares, till I find
my only love.
I sought them everywhere
but I could not find him

(Bloch and Bloch 1995: 67)

At the center of this landscape spectrum is imagery of the garden. The garden is the world of nature, but much more in reality and its associations. The garden is perfected nature, where culture has refined by the hand and mind the world of plants, soil, water, and atmosphere. It is a private, intimate realm, and gardens are invariably beautiful, or they aspire to that condition. These same attributes are metaphorically extended to the female lover. The garden is the poem's most significant extended metaphor. "The Song" itself can be seen as the garden to be unlocked. That garden, "Enclosed and hidden, you are a garden," sits at the poem's center and one can imagine the first half of the poem as the path to it, the second as the journey through. The garden is the setting for the lovers, female and the male; its elements are described as body and the body as landscape. This anthropomorphic metaphor is not just descriptive of the lovers' bodies, but also of their relationship – their emotions as lovers. It is erotic, sexual, and physical. The rich descriptions of touch, taste, sight, smell, sound all recall Eden as a place of multisensory delight. It resonates in texts such as Alan of Lille in his *Anticlaudianus* of AD 1182 where he writes of "a place apart, far distant from our region, a place that smiles at the turmoils of our lands. That place by itself has the power of all other places combined." His description of this Edenic site notes that "That place of places has and holds everything that feasts the eyes, intoxicates the ear, beguiles the taste, catches the nose with its aroma and soothes the touch" (Alan of Lille 1973: 46–49). If the garden of "The Song" is resonant of Eden it may share its attributes. Eden is a microcosm of all of creation, yet it is without dimension, it is timeless and eternal, and it holds the secrets and mysteries of life. It is the first and ideal home of humankind.

The woman of "The Song" is garden and spring, she is place and source – the "fountain of gardens" – source of sources, a wellspring. The garden is a locale and a symbol for the female lover. The beloved, the woman, comes from outside the garden, but she simultaneously is the garden *and* the spring that waters it. In addition, in this rich layering of image and symbol the garden nourishes the lovers and is also nourished by the individuals and their encounter. The male lover enters the garden/female, and is engulfed by its/her qualities. In her "Paradise can be re-experienced . . . but she also represents something more ancient: the natural world from which man grew, and on which he feeds" (Landy 1983: 205). These gardens, people and place, are fertile and sensual, and the setting for and the symbol of satisfaction and fulfillment. In the poem there is a remarkable and surprising egalitarianism between the lovers. In a document over three millennia old the female lover is easily equal to a powerful royal male lover. Perhaps by extension the garden too is portrayed as a center of power and not just a place of respite. Landy writes:

> The Beloved is associated with the earth . . . In the Bible the earth is the feminine complement of God: the two combined to form man . . . The elaborate combination of parts of the body and geographic features, like those between the lovers' bodies, assert the indissolubility of man and the earth, as part of nature . . . the woman as earth is the trope.
>
> (Landy 1983: 314)

In many illustrated versions the artist addresses this question. In American Barry Moser's (b. 1940) illustrations of Marcia Falk's poetic translation the lovers appear to emanate from the earth as if they were a layer of soil or plants emerging from the fecundity of the ground.

There is a deep identification in "The Song" with the structures of the natural and the human worlds. Garden and landscape motifs abound in the poem. Flora and fauna are so dense that some describe "The Song" as nature poetry and there is a literature that enumerates each plant, flower, and species (Feliks 1983). There is clear delight in the specific beauty of nature, as well as metaphors of architecture and craft. For example, vines and the vineyard are described as setting and symbol. The vine, its tendrils and grape are common sexual symbols. Half of the references are specific to grapevines in the early stages of budding, at the early restrained, condensed time of spring, an obvious allusion to the anticipation and build up of sexual and erotic passion. The Marvin Pope translation is explicit in describing the body/landscape imagery:

> How fair, how pleasant you are!
> O love, daughter of delights,
> Your stature resembles the palm,
> Your breasts the clusters.
> Methinks I'll climb the palm,
> I'll grasp its branches,
> Let your breasts be like grape clusters
> The scent of your vulva like apples,
> Your palette like the best wine,
> Flowing (for my love) smoothly
> Stirring sleeper's lips.
>
> (Pope 1977: 11)

One of the most interesting aspects of "The Song" is the use of the Wasf, an Arabic word meaning description, but it has a more specific meaning as a literary device, a poem or a part of one which describes the body through a series of images (Falk 1982: 80). A common device in Arabic poetry, in Ancient Hebrew poetry it is found only in "The Song," although of course there may have been others. The Wasf is poetic description of the female or male body, typically going from head to toe or the reverse. One can understand the Wasf as both body and garden imagery, in its catalogue of metaphors. The images in "The Song" use all the senses and are known for their unlikely metaphors, as in the following examples:

> Your hair is like a flock of goats
> bounding down Mt Gilead
> Your teeth white ewes
> all alike
> that come fresh from the pond.
> The curve of your cheek
> a pomegranate
> in your thicket of hair
> > (Bloch and Bloch 1995: 93)

> Your navel is the moon's
> bright drinking cup.
> May it brim with wine!
> Your belly is a mound of wheat
> edged with lilies.
> Your breasts are two fawns
> twins of a gazelle.
> > (Bloch and Bloch 1995: 99)

H. Granville Fell's 1897 remarkable illustration shows conflated garden images at multiple scales (Figure 14.5). Its arts and crafts imagery of "A garden locked/A sealed up spring" is extraordinarily rich and a layered and suggestive iconography. The garden is simultaneously an island, an elevated mount, raised platform, as well as a gated fortress and walled garden, surrounded by a thorned hedge. The imagery recalls and links "The Song" to the interpretations of paradise that place the biblical Eden in remote locations such as an inaccessible island or a remote mountaintop. Dante places his paradise in such a locale. The paradise garden is always walled, often in dramatic contrast with its surroundings. Inside the wall is Fell's garden. It is richly vegetated with flowers, vines, fruit trees, and a lush profusion of roses. The rose had been sacred to the goddess Venus but had become associated with Mary, and was sometimes known as "Rosa Mystica" (Stokstad and Stannard 1983: 116). At the center of the garden the composition is a nude female figure with flowing hair. She has a bird in her hand: "unique is my dove, my perfect one." In contrast to Eden where Eve is portrayed with a serpent this female/Eve is accompanied by

Figure 14.5 Granville Fell, 1897

a dove, a sign of innocence. The female figure appears not only to be in the garden but to be presented as a great blossom growing out of it: she is both the inhabitant and the personification of the place. The entire image is elevated; one ascends steps up to the garden. The entwined tendril above her head suggests clouds and a sky hovering over the garden, which appears to be emerging as if it has been thrust up from the earth.

> Your limbs are an orchard of pomegranates
> And of all luscious fruits,
> Of henna and of nard –
> Nard and saffron,
> Fragrant reed and cinnamon,
> With all aromatic woods,
> Myrrh and aloes –
> All the choice perfumes

You are a garden spring.
A well of fresh water,
A rill of Lebanon

Awake, O north wind,
Come, O south wind!
Blow upon my garden,
That its perfume may spread,
Let my beloved come to his garden
And enjoy its luscious fruits!
I have come to my garden,
My own, my bride;
I have plucked my myrrh and spice,
eaten my honey and honeycomb,
Drunk my wine and milk.
            ("The Song of Songs" 1982: 341–342)

The poet Robert Bly speaks of the essence of a metaphor as being a psychic leap between object and the image (Falk 1982: 82). To be effective, acts of visualization need to take place. The gardens of the body and the body of the garden are disassembled and reconstructed in "The Song of Songs" (Landy 1983: 73). The garden becomes a body – the body a garden.

Notes

1 There are many translations of "The Song of Songs." I have used several in the paper to emphasize the diversity of interpretations.
2 In the context of courses on the History of Landscape Architecture and a Humanities course on the garden I have had students confront this same issue. They are asked to prepare illustrations of some portion of the text with a primary focus on the garden imagery of the poem. The choice of media is left open; thus people draw, paint, sculpt, collage, construct, photograph, and digitize. They are urged to aim for the spirit of the poem rather than a literal rendering, although some are very precise in their interpretation. Students come from all backgrounds and include those with no experience to those with advanced training in art and design. All have found the exercise to be both challenging and fulfilling, and a window into understanding the issues confronting artists as they have confronted the identical exercise. The project also makes them participants in this ongoing tradition.

Bibliography

Aben, R. and de Wit, S. (1999) *The Enclosed Garden*, Rotterdam: 010 Publishers.
Alan of Lille (1973) *Anticlaudianus*, Toronto: Pontifical Institute of Medieval Studies.
Alter, R. (1985) *The Art of Biblical Poetry*, New York: Basic Books.
Bloch, A. and Bloch, C. (1995) *The Song of Songs*, New York: Random House.
Crisp, F. (1924) *Medieval Gardens*, New York: Hacker Art Books.

Daley, B. (1986) "The 'Closed Garden' and the 'Sealed Fountain': Song of Songs 4:12," in E.B. Macdougall (ed.) *Medieval Gardens*, Washington DC: Dumbarton Oaks.

Esherick, W. (1927) *The Song of Solomon*, Philadelphia PA: Centaur.

Falk, M. (1982) *Love Lyrics from the Bible: A Translation and Literary Study of the Song of Songs*, Sheffield: Almond Press.

—— (1993) *The Song of Songs: A New Translation*, San Francisco: Harper.

Feliks, Y. (1983) *Song of Songs: Nature Epic & Allegory*, Jerusalem: The Israel Society for Biblical Research.

Fell, H.G. (1897) *The Song of Solomon*, London: Chapman and Hall.

Fox, M.V. (1985) *The Song of Songs and the Ancient Egyptian Love Song*, Madison WI: University of Wisconsin.

Gregory, C. (1962) *Cantique des Cantiques*, Paris: Le Club Francais du Livre.

Landy, F. (1979) 'The Song of Songs and the Garden of Eden,' *Journal of Biblical Literature* 98: 513–528.

—— (1983) *Paradoxes of Paradise: Identity & Difference in the Song of Songs*, Sheffield: Almond Press.

—— (1987) "The Song of Songs," in R. Alter and F. Kermode (eds) *The Literary Guide to the Bible*, Cambridge MA: Harvard.

Pope, M. (1977) *Song of Songs. The Anchor Bible*, New York: Doubleday.

Raban, Z. (1923) *The Song of Songs*, Jerusalem: Hasefer.

Ricci, L.B. (1996) "Gardens in Italian Literature during the Thirteenth and Fourteenth Centuries," in J.D. Hunt (ed.) *The Italian garden: Art, design and culture*, New York: Cambridge.

"The Song of Songs" (1982) *The Writings: Kethubim*, Philadelphia PA: Jewish Publication Society of America.

Stokstad, M. and Stannard, J. (1983) *Gardens of the Middle Ages*, Lawrence KS: Spencer Museum of Art University of Kansas.

Thacker, C. (1979) *The History of Gardens*, Berkeley CA: University of California.

Weems, R. J. (1992) "Song of Songs," in C. Newson and S. Ringe (eds) *The Women's Bible Commentary*, London: SPCK.

Wragg, A. (c.1950) *The Song of Songs*, London: Selwyn & Blount.

15 Gendering ghetto and gallery in the graffiti art movement, 1977–1986

Kristina Milnor

In 1984, the *Philadelphia Inquirer* published an extremely hostile review by their regular critic Edward Sozanski of an exhibition of New York graffiti art at the Moore College of Art. Graffiti art, by this time, was a well-established movement in the art community, having been born during the early 1970s in the streets of the south Bronx and subsequently adopted by New York's commercial galleries. The exhibit reviewed by Sozanski, entitled *Femmes Fatales*, was an installation of paintings, some on canvas and some directly on the walls of the gallery, by an artist named Sandra Fabara, known in graffiti circles as Lady Pink. Sozanski's criticisms of the exhibit are numerous, but center around his belief that graffiti cannot be thought of in the same terms which describe "real" cultural production: "while occasionally entertaining, graffiti do not promote retrospection or introspection, only annoyance ... [A graffiti painting] isn't especially interesting, and it certainly isn't art" (Sozanski 1984: 3D). Thus, the review focuses less on the installation itself and more on the trappings of cultural legitimacy that Sozanski sees surrounding it, such as the presence of wine, cheese, and the mayor of the city at the exhibition's opening reception.

In this, Sozanski was following many of his fellow art critics, who objected on principle to the idea of graffiti art, and who found in the movement an excellent opportunity to make statements about what art should do and who artists should be. In this context, there is one particularly striking set of images to be found in Sozanski's review of *Femmes Fatales*. The article opens with a description of Lady Pink as she looked when the reviewer saw her first, in the act of creating the installation:

> She looked like someone who would crawl through a mine field to "tag" the space shuttle. She was dressed androgynously in jeans and a sleeveless top, and her hair was tucked under a maroon beret that she wore with military panache. A bandoleer filled with Magic Markers would have been the perfect accessory.
>
> (Sozanski 1984: 1D)

Sozanski sees Fabara as a soldier engaged in a symbolic war in which graffiti is her weapon and whose violence and danger are revealed by her "androgynous military" dress. The reviewer goes on to comment on the transformation that graffiti has undergone in its transition from street vandalism to gallery art, a process that he compares to the "domestication" of "a formerly feral animal." He supports this vision with his impressions of Fabara as she appeared at the opening of the exhibit: "Lady Pink herself showed the mayor a totally domesticated face and figure. The paint-smeared Sandinista had been transformed into a fragile and demure princess" (Sozanski 1984: 3D). The "domestication" of graffiti art, tamed in its move from street to gallery, thus finds its parallel in the "domesti-cation" of Lady Pink, who no longer appears violent and dangerous but "fragile and demure." The language here and the images that Sozanski presents are not simply racist – they are also highly gendered, as Fabara goes from an "androgynous" soldier smeared with paint in place of blood to the delicate "princess" who appears with the mayor at the exhibition's opening. In Sozanski's imagination, the cultural legitimacy that the gallery has given to graffiti is inscribed on Fabara herself as femininity; as her work moves from vandalism to high art, Lady Pink becomes a woman.

Sozanski's vision of the restoration of Sandra Fabara's womanhood in the space of the art gallery is, I would argue, paradigmatic of the discourses that framed the graffiti art movement from its inception in the late 1970s to the collapse of the "spraycan art" market in the mid-1980s. Like other aspects of the 1970s' hip-hop subculture, graffiti was quickly discovered by the mainstream establishment, which became interested in it not as art per se but as a commodified version of ghetto experience: as one critic remarked, "it is not talent but defiance that is [being] sold" (Kriegel 1993: 434). But because graffiti was an art form defined by its location in the urban environment, the debate over its commercialization – similar in some ways to that surrounding the popularization of rap – became expressed in particularly spatial terms, as critics, collectors, government authorities, and artists sought to articulate what it meant for graffiti literally to move from the street to the studio. Insofar as graffiti was synonymous with the inner city, it was understood as wild and lawless – and incompatible with the rarefied confines of the commercial gallery. Where the one space was black, poor, angry, and primitive, the other was white, wealthy, sophisti-cated, and civilized; in other words, where the one was raw and amoral nature, the other was pure and untainted culture. Thus, the graffiti art movement came to be understood within a highly gendered dichotomy of place, one that opposed the masculine public spaces of the urban jungle with the civilized, pure, and feminine interiors of the professional art world. Sozanski's review, therefore, is drawing on ideas about gender, place, and art that, by 1984, pervaded any discussion of the graffiti art movement and that were the legacy of the social and cultural forces that first gave rise to New York's version of spraycan art.

Historians of the graffiti movement in New York generally trace its origins back to the early 1970s, and a young Greek immigrant who began writing his "tag" or signature – Taki 183 – on buildings all over Manhattan. Until then, graffiti had been strictly a local phenomenon, confined to the particular neighborhood in which the writer lived, but Taki and his imitators inaugurated the tradition that soon became street graffiti's most famous attribute: namely, the competition in the geographical distribution of tags, as each writer sought to have his signature in the most locations around the city (Lachmann 1988: 229–50). This, in turn, led to the appearance of graffiti on subway cars, which had the merit of moving through different neighborhoods and thus displaying the writer's name to many more people than would see it written in a stationary location. Tags became larger, more elaborate, until they developed into huge mural paintings that covered entire cars. Different styles evolved, and, along with break dancing and rap music, graffiti came to symbolize the cultural movement taking shape in urban communities of color in the late 1970s and early 1980s – namely, hip-hop (Rose 1994: 41–47; Atlanta and Alexander 1988: 156–68).

The origins of hip-hop, and more specifically of the kinds of graffiti associated with it, must be understood within the social and economic environment of New York as the 1970s drew to a close. The city was still reeling from the effects of an unprecedented fiscal crisis, in which it came within an ace of declaring bankruptcy, with the result that social and public services in the city had been drastically cut. The building of the Cross-Bronx expressway at the beginning of the decade had necessitated the dissolution of a large number of established communities and the relocation of many people of color, particularly to the area later known as "the cradle of hip-hop," the South Bronx. The disappearance of many social services and a city-wide crisis in low-income housing meant that the plight of the urban poor was increasingly desperate, and the material fabric of the neighborhoods in which they lived was gradually degenerating. Areas like the South Bronx, with a high percentage of poor African-Americans and Latinos, were alternately ignored and deplored by the city government as run-down, crime-ridden disgraces to the economic and physical renewal of the city. Thus, by the late 1970s and early 1980s, the mutual hostility between the well-to-do and the poor had become articulated in the popular imagination in essentially spatial terms, as a conflict between midtown and the ghetto over ownership of the urban environment (Rose 1994: 27–34).

Within this framework, graffiti arose, in the words of Tricia Rose's *Black Noise*, "as a source for youth of alternative identity formation and social status in a community whose older local support institutions had been all but demolished along with large sectors of its built environment" (Rose 1994: 34). As writers collaborated in the creation of paintings, evaluated one another's work, and competed in the beauty and distribution of their graffiti around the subways, a sense of community and identity was established. Additionally, graffiti offered a means for marginalized youth to establish a kind of ownership of public property that had been denied them, a way to insist on their right to participate in the culture of a city that continually sought to efface them. Insofar as graffiti was defined by its transgressive presence in public places, it mirrored the marginalization of its creators, who were themselves "out of place" in a world defined as white and middle class (Stewart 1987: 161–80). Graffiti thus simultaneously established the basis for a community of relationships among writers and provided the means by which that community could powerfully announce its existence to the rest of the city.

On the other side, the proliferation of graffiti in New York, particularly on the subways, became a frequent topic of debate in public fora. For some, graffiti was an important part of the city's cultural fabric: Norman Mailer praised it in *The Faith of Graffiti*, and Claus Oldenburg rather famously compared a graffiti-covered train to "a bouquet of flowers from Latin America" (quoted in Hagopian 1987: 106). For many others, however, including the Mayor's office, graffiti became the visible manifestation of urban decay, a symbol of the chaos into which the city was imagined to be disintegrating. Graffiti was compared in public discourse to dirt, to disease, to anarchy, and, perhaps most tellingly, to theft (Cresswell 1992: 329–44). Graffiti came to represent what the city feared most from its dispossessed, namely that they would take by force what had been denied them, so that graffiti's most frightening role in the public imagination was as, to quote one scholar, "a symbolically violent attack on an equally symbolic category of property" (Cresswell 1992: 337). It was this sense of threat to the dominant culture's ownership of places that Fab Five Feddy, a famous graffiti writer, articulated when he said in 1982, "When those who don't like it think of graffiti, they are really scared, because the first image that comes to their mind is that more and more of their world is going to get written over" (quoted in Moufarrege 1982: 89).

The power of graffiti to excite the anger of authority is revealed in the millions of dollars that New York spent in the 1970s and 1980s to keep the transit system safe from graffiti, as well as in the increasingly harsh legal penalties leveled against writers. Artists, while not denying the illegality of graffiti, insisted that it was a victimless crime. Lee, another famous writer, said of his work at one time, "here is a thing that doesn't hurt you . . . all it does is excite your heart, make your eyes follow it. It doesn't take your wallet" (quoted in Atlanta and Alexander 1988: 166). The city government's opposition to graffiti symbolized for writers the arbitrary nature of urban authority, which sought to punish them for a simple act of artistic creation. Lee, in his painting *Roaring Thunder*, coined the motto that became the graffiti writers' credo, often quoted in conversation with reporters and in mural paintings around the city: "Graffiti is Art, and if Art is a Crime, let God forgive all" (Atlanta and Alexander 1988: 167). In this sense, then, the transgression against the law, the danger of being caught by police or other authorities, became an important part of graffiti's thrill. The risks taken were part of the joy and of the writers' self-conception as creative desperados working against a repressive system.

"Graffiti writer" thus became not just an occupation but an identity. Yet it was one that was strictly gendered. Both the place of graffiti (the midnight alleyways and train yards of the ghetto) and the role of the writer as a cultural outlaw were understood as excluding women from graffiti writing. Richard Lachmann, who interviewed many writers in the 1980s, writes, "they often define the dangerousness of writing on the subways in terms of women's inability to participate," and quotes one writer who said, "You got to get into the yards . . . by going under or over those barbed-wire fences. They have dogs loose. Women get scared and can't keep up" (Lachmann 1988: 235). The homosociality of graffiti culture was articulated by one artist in a *Rolling Stone* interview, when he remarked that "graffiti writers are like bitches: a lot of lying, a lot of talking, a lot of gossip," while at the same time insisting that graffiti must be an exclusively male pursuit: "all [girls] say is (in a whiny voice) 'you're crazy . . . Write my name'" (Heldman 1995: 49). Moreover, women who did attempt to become involved in street graffiti reported that they were often told by male writers that "at three o'clock in the morning a girl should be sleeping," rather than heading out with spray can in hand (Guevara 1987: 165). They also found that their work was often dismissed as "biting" – merely a poor imitation of someone else's style – and that it was more frequently written over by other artists. Thus, women were imagined to be incapable both of the physical daring and of the rebellious creativity that marked the genuine graffiti writer.

Street writing, therefore, was understood as masculine both in place and process, in the streets that were its original home and the heroic rebellion against authority that defined the graffiti artist. The meaning, however, both of street writing and the role of the street writer, became complicated with the advent of the graffiti art movement, as subway writers' work was adopted by commercial galleries and sold to a largely white upper- and middle-class audience. Starting in the late 1970s, New York's art community began to embrace graffiti as the successor to pop art, the newest artistic movement whose origins outside the realm of high art seemed to guarantee its authenticity and raw power. Gallery owners and collectors saw graffiti writers' insistence on continuing to write despite the obstacles thrown in their way as a manifestation of an idealized artistic impulse, a pure and primitive drive to create. As Sidney Janis wrote in the catalogue for a 1983 show: "[the graffiti artist] expresses in his work his own experiences of the street with an urge and drive so obsessive that formerly he broke the law to do so" (Janis 1983: 1). On the other hand, writers were frequently portrayed as in the grip of a passion for graffiti that rendered them safe from the taint of the ghetto: one art dealer said in 1982, "[Our artists] are not into drugs or violence. They walk around, thousands of them, with sketch books, doodling and drawing. They're consumed by art" (quoted in Gablik 1982: 35). The movement gained popularity rapidly, so that, by the early 1980s, the phenomenon of graffiti in the gallery was well established enough that one average citizen could state in the *New Yorker* without conscious irony, "graffiti has its place, *but not on the side of buildings*" ("Taking Action" 1983: 25, emphasis in original).

The irony, of course, is that it was the sense of lawlessness associated with graffiti that gave it its power and appeal; once it had been given not just a new place of display but a new moral and ethical legitimacy, it lost the thing that made it different from simple painting. It is not surprising, moreover, that the graffiti art movement soon gave rise to raging controversy within both the art community and the city as a whole. Conservative art critics in particular were outraged by the affront the movement represented to "real art," refusing to acknowledge that graffiti could occupy the same spaces and be described in the same terms as Picasso or Monet. More liberal critics were also hesitant in the face of dealers' enthusiasm, but for different reasons: they argued that by robbing graffiti of its location, the gallery also destroyed its power; instead of speaking from the street, from outside the realms of artistic production, graffiti had been "tamed" and silenced by its relocation to the space of "high" art. Thus they deplored the process by which the dominant culture achieved by appropriation what it could not achieve by repression and the way in which what was powerful and authentic in the public street became "aestheticized" by the controlled privacy of the gallery setting. "What was formerly a matter of desecration and even violence is now modulated to a matter of domestic consumption – enter *taste*" (Stewart 1987: 173).

The use of "domestic" as it is found in this critique is typical of reviews of graffiti art shows: forms of the word appear again and again to describe the place and process that have robbed graffiti of its power. With its connotations of privacy and hominess, the word may signal to us that, like the earlier debate around graffiti on the subways, what is at stake here is not only the meaning and ownership of art, but also of places. The question of where – the gallery or the subway, midtown Manhattan or the South Bronx? – becomes crucial in determining what graffiti means, whether it is vandalism or art, meaningful or ridiculous, moral or immoral. Place gives meaning to graffiti, but it is also true that graffiti gives meaning to place. Observers' perceptions of the public spaces of the city influence what they think about the graffiti written there, but, correspondingly, the graffiti also influences what they see in the urban environment. Graffiti, moreover, is often seen as a means of claiming a place for a use and by a population to whom it has been denied. Kerry Carrington in her article "Girls and Graffiti" traces the ways in which, through graffiti, working-class girls make railroad bathrooms into spaces of relaxation and self-expression in defiance of the transportation authority (Carrington 1989). Susan Sontag makes the point specifically of New York subway graffiti:

[G]raffiti have to be seen as an assertion of something, a criticism of public reality ... [a] tide of indecipherable signatures of mutinous adolescents which has washed over and bitten into the facades of monuments and the surfaces of public vehicles in the city where I live: graffiti as an assertion of disrespect, yes, but most of all simply an assertion ... the powerless saying: I'm here, too.

(Sontag 1987: 122)

By definition, then, graffiti is out of place, an expression, whether verbal or pictorial, that does not belong where it is found. Just as the dominant culture casts the self-representations of certain people as more appropriate than those of others, it reserves the right to designate appropriate places for the creation of those self-representations. Graffiti challenges these formulations by insisting that anyone can create art, and that art belongs wherever its creator chooses to put it. As one critic astutely observes, "[u]nlike pornography, graffiti is not a crime of content" (Stewart 1987: 174); more act than artifact, it is the transgression that it embodies that gives graffiti meaning. Thus, it does not matter that many writers' signatures are indecipherable to the layperson – it is the fact that they are written in the street and not the content of the writing that matters. Graffiti in its primitive wildness is imagined as the product of the urban jungle, and as such is incompatible with the civilized interiors of the art world. Moreover, as we saw in Edward Sozanski's vision of the "domesticated" Lady Pink quoted earlier, the word has particularly gendered overtones, as the private, controlled world of the gallery is understood as feminized and feminizing.

Where conservative and liberal critics appear to have agreed, however, is in the representation of graffiti's transition from subway car to canvas as profoundly gendered. While the one group lamented the "emasculation" of an artistic form whose strength and authenticity depended on its link with the streets of its birth, the other professed itself appalled by the "rape" of the gallery by the male brutality of ghetto art. Thus, in an *Arts Magazine* article from 1982, Addison Parks writes that graffiti in the gallery is a threat to the very idea of art, "a wolf in the fold," the embodiment of violence, anger, and hatred. Calling graffiti "the skin cancer of our civilization" and "bloody, bloodless brutality," Parks argues that the graffiti art movement is a reaction against the progress that the art world has made in recent years away from the primitive, the savage, the masculine:

[The graffiti artist] likes to do it because he is a he, and has to voice his opinion ... So why now when the world was just getting a taste of the feminine aesthetic and sensibility are we getting crass male blowhard brutality again? ... Why are no women included among the new spirited expressionists? ... Do we fall back on bucks and brutality after all the good-intentioned enlightenment of the previous two decades? Why are big collectors going to clubs with bad boys in fatigues?

(Parks 1982: 73)

For Parks, graffiti is fundamentally a male pursuit, inextricably bound up with a kind of harsh primitiveness and pleasure in transgression understood to be masculine. "Sensibility," "enlightenment," and a "feminine aesthetic" are the markers of real art, which is threatened by the macho stance of male graffiti artists and their work. Parks' "moral landscape" divides the harsh, criminal, and unprincipled world of graffiti writers from that of the art gallery, imbued with a kind of pure and chaste femininity. Parks triumphantly proves his assertion that graffiti represents a male backlash against the civilized femininity of the art world by pointing out that graffiti writers are all men, the "bad boys in fatigues" who are being courted by collectors. "Why are no women included . . .?" he asks rhetorically, with the implicit answer being that there are none to include.

If street writing, then, was coded masculine, "legitimizing" forces external to it that were seen as impinging on its freedom and lawlessness were represented as feminine. A remarkable example of this may be found in a 1983 film entitled *Wild Style*, made as a collaboration between the director Charles Ahearn and the graffiti writers who starred in it. The story of the movie is complicated, but basically traces the transformation of a subway graffiti writer named Ray after he is "discovered" by a reporter who comes to the South Bronx in order to experience the world of graffiti. The film is a peculiar combination of genres: the story is fictionalized, but based on the life of the graffiti writer, Lee, who plays the main character. Other people in the film also play themselves, more or less: the gallery owner Neva, who commissions Ray to paint a canvas for her, was a well-known collector, and the reporter is played by Patti Astor, owner of one of the first galleries in New York to display graffiti art (Jacobson 1983; Stein 1983; Jaehne 1984).

The professional art world's appropriation of graffiti in the film is represented first in the person of Virginia, the *Village Voice* reporter, who assures Ray, "I'm going to make you a millionaire" (Ahearn 1983). She is stereotypically blond, helpless, and clearly slumming in the South Bronx – after she and Ray are attacked by gang members, she remarks, "wait 'till I tell everyone at the party that I almost got killed. They'll love it" (Ahearn 1983). Whiteness, privilege, and femininity are mapped onto one another in the image of Virginia, who embodies the interest that the art world has in graffiti: in her own words, "these people are ready for new thrills" (Ahearn 1983). Later, when Ray, his friend Phase, and Virginia head into Manhattan to attend a party in Neva's penthouse, the journey maps in reverse Virginia's earlier trip into the Bronx – and as the home of graffiti is symbolized by the crowd of young male artists whom the reporter encounters, so does the elegant collector Neva come to signify the professional art world. Moreover, Neva's invitation to Ray to paint her a graffiti canvas is offered in sultry tones as she lounges on a satin covered bed: from Ray's evident discomfort, to Phase's raised eyebrows as he watches his friend disappear into the bedroom, to Neva's question after she hands Ray a check, "how does it feel?" – this is certainly profession-alization as seduction (Ahearn 1983). The graffiti artist is lured into the world of commercial art by money offered in the perfumed boudoir of the collector. Addison Parks' representation, which I quoted earlier, of the opposition between the civilized, feminine world of gallery art and the mas-culine gesture of street graffiti is played out from the artist's perspective in *Wild Style.*

Lee, however, who plays the main character, Ray, was not the only graf-fiti writer to appear in the film. Sandra Fabara – that is, Lady Pink, whose 1984 exhibition *Femmes Fatales* inspired the critical review with which I opened – also plays an important role, as Ray's sometime girlfriend Rose, or (her graffiti name in the film) "Lady Bug." Rose is also a writer, but her approach to graffiti is very different from her boyfriend's: whereas Ray prefers the darkness and anonymity of subway writing, Rose is portrayed from the beginning of the film as already deeply enmeshed in the commer-cialization of graffiti art. In fact, although it is Ray whom she eventually picks up and takes to Manhattan, Virginia originally comes to the South Bronx to interview Rose and her graffiti gang The Union. During the course of that interview, Rose explains that she and her gang have started a business painting advertising murals for local stores: "we've got quite a few jobs lined up . . . we try to do a lot for the community now, liven up a few things, make it a job, make it a living, I guess" (Ahearn 1983). Indeed, Rose's interest in making graffiti "legitimate" is the source of tension between her and Ray, a tension expressed in the film in sexual terms. Early in the film, Ray encounters an acquaintance named Chico, who criticizes Ray for refusing to participate in the mural-painting business. He goes on to say, "I saw your female with them too: what's up with her? I been hearing she's been giving that stuff out to all them graffiti guys" (Ahearn 1983). Ray reacts angrily to the sexual innuendo, but when he later sees Rose painting a mural with another man, he stalks off to draw a large image of "LOVE" crossed out and including a broken heart. The film thus makes an analogy between Rose's "flirtation" with commercialization and the supposed sexual betrayal of her relationship with Ray.

Significantly, however, Rose is also the source of Ray's redemption at the end of the film. After he returns from Manhattan to work on his commission for Neva, Rose appears to criticize both the painting itself and the entire enterprise of putting graffiti on canvas. Ray explodes in frustration:

> All I've been wanting to do, by this (gesturing to the painting) and by all the other things I've been doing is that I really want to be your

man, and I want to continue being your man. I mean, what do you think I've been doing? I want to prove it.

(Ahearn 1983)

Rose responds, "and how're you going to do that?", which Ray answers with a lingering kiss (Ahearn 1983). The commission is then forgotten. The emasculization of Ray and his art by Neva and the world of the art collector are thus redeemed by his return to Rose and to "real" graffiti. Ultimately, it is again Rose who leads Ray to the final resolution of the conflict between his desire to maintain his artistic connection to the streets of the ghetto and the need to make graffiti more than a criminal act. The film concludes with an outdoor rap concert in which the entire community participates and for which Ray is commissioned to paint a huge mural as a stage backdrop. He begins with an image that he describes as:

> the hands of doom ... what I'm trying to draw is the artist in the middle and like he's like painting all by himself in his own world and whatnot; he don't care about nobody around him and that's what the hands are: everybody around him.

(Ahearn 1983)

Unable to complete the painting on the original plan, Ray consults with Rose, who says bluntly:

> I don't like your mural; I don't like the idea ... You're only worried about Zoro [Ray's graffiti identity] ... Concentrate on what the whole thing is about: the jam ... Rappers are coming down; they're going to be the stars of this thing, not you.

(Ahearn 1983)

Ray exclaims, "you did it!" and finishes the painting as a disembodied set of hands sending out bolts of electricity toward the center stage, where there is no figure but only a large multi-colored star (Ahearn 1983). Based on the earlier conversation, the message seems to be that the community ("everybody around him") is not a drain but a source of energy, and not just for the graffiti artist but for the many different creative people who join together to make the "jam" (Ahearn 1983). Again, however, it is Rose who facilitates Ray's passage from criminal to legitimate graffiti artist, from lawless loner to upstanding participant in community life.

It is not coincidental, I would argue, that Sandra Fabara-as-Rose in *Wild Style* is assigned the task of representing a form of legitimacy that Ray can adopt without becoming the art collectors' lap dog, a kind of commercialization that does not involve selling out. Fabara, in fact, had always cut an anomalous figure in the graffiti art community, as a female artist working within a genre that had been defined as paradigmatically masculine. Insofar as graffiti was understood to be the product of the dangerous, violent, masculine street, Fabara's participation was problematic. Yet at the same time, those aspects of her graffiti identity that made her role as street writer difficult to define seem to have facilitated her entry into the professional art scene. Lady Pink and her work were "out of place" in the streets that were the original home of graffiti, but, in part because of this, they were more easily able to find a place in the world of the gallery. Edward Sozanski's vision, then, of Fabara's transformation from an androgynous soldier in graffiti's army to a fragile princess of the art world highlights the discourses that worked against male graffiti artists, but ironically assisted Fabara's own transition from street writer to artist. The civilized and feminized confines of the commercial art establishment that threaten to emasculate the male graffiti artist are safe for Lady Pink: she is "at home" in the domesticity of a gallery setting.

In fact, it is clear that Fabara made a conscious choice to develop a style in her art that might at first be understood to carry out Sozanski's view of her "domestication." Her early work was imbued with a kind of hyper-articulated femininity, beginning with the choice of her graffiti name, Lady Pink. Fabara was always eager to distinguish between her own work and that of her male colleagues: she said once in an interview, "Men have a passion for black! I sometimes exclude black altogether. I work more with light colors. Things that are a little softer, more tender, sensitive" (in Guevara 1987: 165). Thus, in Fabara's early work, both on the street and in the gallery, she deliberately painted subjects that foregrounded her gender: she sometimes accompanied her tag on walls and trains with an image of the Japanese children's cartoon character My Melody, a large-eyed bunny in a pink hat. Again, the painting of hers that was displayed in *Post-Graffiti*, the last group exhibition in New York before the market for graffiti art collapsed in 1984, contrasts remarkably with others shown in the same exhibition. While the canvases by male graffiti artists such as Crash, Daze, A-One, and Bear depicted guns, explosions, and images of war, Pink's contribution was a large painting of a single rose. Indeed the contrast is so dramatic that we may begin to wonder about the very performativity of gender that occurs here both in the work of the male graffiti artists and in that of Lady Pink: the men perform their wildness and Fabara her domesticity in exactly the ways that the gallery expects. Interestingly, the quote from Fabara that accompanied a photograph of this painting in the exhibition's catalogue reads, "My paintings . . . are executed in their original medium; spray paint on canvas. It can no longer be called graffiti but art, and is accepted as such" (catalogue for *Post-Graffiti* 1983: 11).

Turning back to *Femmes Fatales*, however, which was Lady Pink's first solo exhibition, two things about it stand out as rather remarkable. First, it is interesting that this exhibition was put together in 1984, a time when the graffiti art market in New York was in the process of rapidly collapsing. Lady Pink would continue to enjoy a success that eluded her male colleagues, going on to collaborate with Jenny Holzer and to exhibit widely in Europe (Siegel 1993). Second, and perhaps more to the point, we see in *Femmes Fatales* an unmistakable stylistic departure from Fabara's earlier work – whereas before she focused on flowers, landscapes, and "sensitive" subjects, the images in *Femmes Fatales* are much more dramatic. The women here are strong and angry, beautiful and daring, some of them victimized but many of them armed. A central image of the cover for the catalogue of the show, drawn by Pink, is a poster mounted on a graffiti-covered wall that depicts a woman's face and the words, "Wanted: Dead!!" In front of the wall is a masked female figure holding a gun – woman as hunted, yes, but also as hunter. An Italian critic declares in the show's catalogue:

> [T]he female hand, painted up for seduction, stiffens, and the canni-balism of passion is translated into nail polish and ash. The femme fatale who scratches with her filed nails and offers herself up with sweet-smelling creams is, in fact, horrid in her beauty, and is a declaration of Lady Pink's severely critical position.
>
> (Celant 1984)

But critical of what? In a sense, *Femmes Fatales* stands as the answer to the polarization of the "primitive" graffiti artist and "sophisticated" commercial art along the lines of male and female. Lady Pink has taken her earlier stylized femininity and made it dangerous and powerful, so that the resulting images "belong" neither to the sensitive confines of the art world or to the masculine brutality of the graffiti form. The image of the femme fatale becomes for Fabara a way to negotiate the contradictions in her identity as a woman working in a self-articulated masculine medium, and as a "primitive" street artist exhibiting in a "civilized" gallery.

Graffiti came and went on the professional art scene very fast: by 1986, the market had entirely collapsed and New York graffiti artists were left with nowhere to sell their paintings and no one to sell them to (Powers 1996). The criminality of graffiti was not just part of its romance; it was an inextricable aspect of the creative act. Graffiti's relocation to the pure, untainted world of high art made it, in a very real sense, no longer art. A few writers moved to Europe and continued their careers, but most gave up on commercial art as well as subway graffiti. This was heralded by some critics as the triumph of art over vandalism, of civilization over brutality, of the feminine space of the gallery over the masculine savagery of the inner city. But I think that we may also see the story of the graffiti art movement as a window onto the process by which ideologies of gender come to intersect with race and class in creating a moral map of the urban environment – one which situates the ghetto and the gallery on either side of the same gulf that separates nature from culture, black from white, poor from rich, criminal from innocent, and male from female. It is in decon-stucting that map that we may not only critique the ideological dichotomies on which it is based, but also open up space for new definitions of "real" cultural production and the place of the artist in American society.

Bibliography

Ahearn, C., dir. (1983) *Wild Style*, New York: Pow Wow Productions.

Atlanta and Alexander (1988) "Wild Style: Graffiti Painting" in Angela McRobbie (ed.) *Zoot Suits and Second-Hand Dresses: An Anthology of Fashion and Music*, Boston MA: Unwin Hyman.

Carrington, K. (1989) "Girls and Graffiti," *Cultural Studies* 3: 89–100.

Celant, G. (1984) "Lady Pink," trans. Meg Shore, in the catalogue for *Femmes Fatales*, Philadelphia PA: Moore College of Art.

Cresswell, T. (1992) "The Crucial 'Where' of Graffiti: A Geographical Analysis of Reactions to Graffiti in New York," *Environment and Planning D: Society and Space* 10: 329–44.

Gablik, S. (1982) "Report from New York: The Graffiti Question," *Art in America* 70: 33–39.

Guevara, N. (1987) "Women Writin' Rappin' Breakin'" in Mike Davis, Manning Marable, Fred Pfeil, and Michael Sprinker (eds), *The Year Left 2: An American Socialist Yearbook*, London: Verso.

Hagopian, P. (1987) "Reading the Indecipherable: Graffiti and Hegemony," *Polygraph* 1: 105–11.

Heldman, K. (1995) "Mean Streaks," *Rolling Stone* February 9: 38–42.

Jacobson, H. (1983) "Wild Style: Charles Ahearn interviewed," *Film Comment* 19: 64–66.

Jaehne, K. (1984) "Charles Ahearn: Wild Style," *Film Quarterly* summer: 2–5.

Janis, S. (1983) *Catalogue for Post-Graffiti*, New York: Sidney Janis Gallery.

Kriegel, L. (1993) "Graffiti: Tunnel Notes of a New Yorker," *The American Scholar* 62: 431–36.

Lachmann, R. (1988) "Graffiti as Career and Ideology," *American Journal of Sociology* 94(2): 229–50.

Moufarrege, N. (1982) "Lightning Strikes (Not Once but Twice): An Interview with Graffiti Artists," *Arts Magazine* November: 87–93.

Parks, A. (1982) "One Graffito, Two Graffito . . . ," *Arts Magazine* September: 73.

Powers, L.A. (1996) "Whatever Happened to the Graffiti Art Movement?" *Journal of Popular Culture* 29(4): 137–42.

Rose, T. (1994) *Black Noise: Rap Music and Black Culture in Contemporary America*, Hanover NH: Wesleyan University Press.

Siegel, F. (1993) "Lady Pink: Graffiti with Feminist Intent," *Ms.* March/April: 66–68.

Sontag, S. (1987) "The Pleasure of the Image," *Art in America* 75: 122–31.

Sozanski, E. (1984) "Art: Bringing Graffiti Inside," *Philadelphia Inquirer* April 27: 1–3D.

Stein, E. (1983) "Wild Style," *American Film* 9: 48–50.

Stewart, S. (1987) "Ceci Tuera Cela: Graffiti as Crime and Art" in John Fekete (ed.) *Life After Postmodernism: Essays on Value and Culture*, New York: St Martin's Press.

"Taking Action" (1983) *The New Yorker* August 15: 25–26.

Index

Page references in *italic* indicate illustrations

Abbot, Britton 38, 39, 41, 45, 48, 51n18
abjection 27
Academy of Women 158–9
Account of the Produce of a Cottager's Garden in Shropshire (1806) 46
Adams, Henry 198nn9,11
Adams, J.H. 99
Adirondacks, photographed by Seneca Ray Stoddard 9, 124–42, *132, 133, 135, 136, 139*
Adler, Jeanne 141n10
Adolph, Robert 223
aestheticization 274
aesthetic theory, Picturesque 55, 57, 64–5
African-Americans 125, 130–1
Agyeman, J. 43
Ahearn, Charles 276, 277, 278
airline travel 111, 112–13, *114*, 115, 121
Aksakov, Sergey 242
Alan of Lille 263
alchemy 229
Alcock, Sir Rutherford 61–2, *63*
alcohol, masculinity and colonialism 22, 24, 25, 26, 29
Alexander 272
Alexander, Catherine 35
Algoma Sketches and Pictures Exhibition (Toronto, 1919) 203
Allingham, Helen 48
allotments 39–42, 46, 47
Alter, R. 259
alternative science 229, 231–3
alternative social space 7
Alvarez, E. 90n12
Amazon, the *see* Barney, Natalie Clifford
American Airlines 110
amitié, Barney's literary salon 145–61

anachronistic space 216–17, 218
Anderson, Benedict 24, 188, 207, 211
Anderson, K. 7
Antheuil, George 152
anthropology 168, 178
antimodernism 215–16
Antoinette 169, 175
A-One (graffiti artist) 279
Apollinaire, Guillaume 153
Arabic poetry 264–5
architecture 7, 168, 178; and female body in Columbian Exposition (1893) 10–11, 182–202
Ardener, Shirley 179n6
Arendt, H. 5
Arnold, D. 21
Aron, Cindy 104nn3,9
art *see* book illustration; graffiti art movement; painting
Arts and Letters Club 217
Arts Magazine 275
"Art V" 178
Ashcroft, Bill 208
Astor, Patti 276
astrology 229, 233
astronomy 232
Atchison, Topeka and Santa Fe Railroad 87
Atlanta 272
Atlantic City, middle-class beach culture 9, 94–108
Aurel (Antoinette Gabrielle Mortier de Faucamberge) 160n8
Austen, Jane 69n13
authenticity 208, 216
authority 208
authorship, Russian women writers 11, 240–53

Automobile Club of Southern California 110, 116
automobile travel 113–20

Babylonian fertility cults 258
Bachelder, J.B. 83
Bacon, Francis 223, 225, 226, 229–30
Bacon, Roger 229
Baele, Nancy 205–7
Banks, Joseph 57–8
Barnes, Djuna 152, 160n8
Barney, Natalie Clifford, literary salon 10, 145–61, *147, 148, 150, 155*
Barthes, Roland 236
Battle of Tippecanoe (1812) 164
beach: family travel 110–12, 121; public landscape 9, 94–108, *101, 103*
Bear (graffiti artist) 279
Bederman, Gail 90n9, 140n1, 191, 193, 195
Belinsky, Vissarion 243, 244, 245, 247, 249, 250
Bellamy, Francis J. 197n1
Bender, T. 89n4
Benhabib, S. 5
Benjamin, Walter 212
Bennett, Edward H. 197n8, 198n12
Bennett, Tony 104n3
Berenson, Bernard 151
Berger, R. 51n6
Bernard, Sir Thomas 38, 39, 45
Bernard of Clairvaux, St 255
Bible, illustrative history of "Song of Songs" 11–12, 254–68, *257, 260, 261, 262, 266*
Blackpool 96, 104n3
Blathwayt, William 234
Bloch, A. 258, 263, 265
Bloch, C. 258, 263, 265
Bloom, Harold 240, 241
Blunt, Alison 171
Bly, Robert 267
body: and masculinity in Ceylon 22; middle-class beach culture 9, 94–108; *see also* female body
Bolotin, N. 186, 200n22
book illustration, "Song of Songs" 11–12, 254–68, *257, 260, 261, 262, 266*
Bordo, Jonathan 214
Borland, Hal 115
boundaries: cottage garden 46–50; crossing by Russian women writers 249, 251
Boyd, William 19–20, 24, 26, 28
Boyer, M. Christine 198n10

Boyle, Robert 236
Bremer, Frederica 96–7
Brennan, Timothy 207
British East India Company 31n10
Britishness, colonial masculinity and 26–8, 29–30; *see also* Englishness
Brooks, Romaine 152, 154–6, 160n8
Brown 22
Brown, A. 25
Brown, Eric 203, 204
Brown, Gillian 199n20
Brown, John 131
Brown, Lancelot "Capability" 71nn22,29
built environment, as means to implement control 10
Burchardt, J. 45, 51n7
Burg, David F. 190, 199n17
Burke, Edmund 67–8
Burnham, Daniel Hudson 184, 197n8, 198n12
Busro *see* West Union

Callahan, L. 118
camping 115, 118–21; in Adirondacks 9, 124–42
Canaanite fertility cults 258
Canada: representation in art 11, 203–20, *206, 215*; Vancouver's Chinatown constructed as immoral landscape 7
Canadian Pacific Railway 216
Canadian Railway 110
Carmichael, Franklin 219n4
Carolus-Duran 154
Carpen 26
Carrington, Kerry 274
cartography 11, 223–39, *227, 228, 234*; Shaker sites *162*, 164, 165–8, *166, 167*
Casson, A.J. 219n4
Cavendish, Margaret, Duchess of Newcastle 231–2
Celant, G. 280
center/periphery, Russian women writers 11, 240–53
Ceylon, colonialism and British home 8, 19–33
Chambers's Information for the People (1842) 40–2
character, cult of 22
Chartist Land Plan 51n18
Chase, M. 51nn7,18
Chicago: Great Fire (1871) 183; sex and architecture in the Columbian Exposition (1893) 10–11, 182–202, *185, 192, 194, 196*
Chicago, Milwaukee & St. Paul Railway 77

"Christian Distinction – No.2" 176
Christian interpretation of "Song of
   Songs" 11, 255–7
cinema 102–3, 104
citizenship 4; Canada 209–11; Columbian
   Exposition (1893) 187, 193
civilization 20, 22, 24, 28, 43, 79, 83–4,
   125, 126, 184, 280
Clark, Clifford 79
Clarke, T.J. 105n14
class 3; Adirondacks 130; American
   railroad and public domesticity 79, 80,
   85, 86, 89; built environment 10;
   Canadian national identity 210–11,
   216; colonialism and masculinity 8, 24,
   30; Columbian Exposition
   (1893) 191; cottage garden 36–50;
   graffiti art 12, 280; literature 243;
   middle-class beach culture 9,
   94–108; and scientific knowledge 230,
   231
Claude Lorrain 55, 59, 60, 62, 70nn16,18
climate, morality of, in Ceylon 8, 20–3,
   24, 29
Cockburn, Edwin 48
Codman, Henry 184
cognitive cartography 235
Colette 152, 160nn6,8
colonialism: cartography 235; Ceylon 8,
   19–33; and gardening 43; national
   identity and narratives of virgin
   landscape 208, 213, 218; Picturesque 8,
   56, 60–1
color *see* race
Columbian Exposition (1893) 10–11,
   182–202, *185, 192, 196*; map *194*
comfort: Adirondacks 134, *135*; airline
   travel 112–13, *114*; American railroad
   as home 79–89
Coney Island 96, 104n7
conflict 6
conservation, Adirondacks 137–40
Constantine, Stephen 34–5
consumption and consumerism 79–80,
   85–6, 109
Cook, M. 130
Coolidge, Calvin 118
corporate office, domestication 89n3
Cosgrove, Denis E. 212
Cosmopolitan 99
cottage garden 8, 34–54
Cottage Gardener 44
courtship 49
Crandell, Gena 55
Crash (graffiti artist) 279
Crayon, The 140n4

cultural codes 11
cultural geography 178
cultural landscape, West Union Shaker
   village 10, 162–81 168
cultural nationalism 11, 203–20
cultural production, graffiti art movement
   12, 269–81
cultural studies 1, 168
Culture/Nature 50
Cunniff, M.G. 88
Currie, G. 5
Curt Teich Company 95, 101–2

Dante Alighieri 265
Darley, Gillian 37
Darrow, Ruth 172
Daze (graffiti artist) 279
Delarue-Mardrus, Lucie 160n8
Deloria, Philip 141n9
democracy 6, 90n12
Denison 46
Descartes, René 229
Detroit Publishing Company 95, 101–2
Deutsche, Rosalyn 6, 211, 212–14
development 20; *see also* industrialization
Digby, W. 28
Ditchfield, P.H. 34, 52n19
division of labor: American family
   vacation 120–1; Shaker 167–8, 169–71,
   170, 173–4
Doctor and Patient 141n8
domestication, graffiti art 269–81
domesticity 241
domestic landscape 8
Dostoevsky, Fyodor Miklailovich 243
Doty, Mrs D. 84
Dowler, L. 5
dress, American beach 98–101
Driver, F. 7
Droege, J.A. 91n15
Dryden, John 232
dualism, cartography 230–1, 236
DuBois 2
Du Bois, W.E.B. 2, 88
Duncan, Carol 204
Duncan, Isadora 145
Duncan, J. 6
Dunning, Alvah 127

Eckert, Max 234
Egyptian poetry 258
elitism, Adirondacks 125, 130–1,
   132–4
Ely, Christopher 242
embodiedness *see* body; female body
Emerson, Ralph Waldo 141n6

emotion, knowledge and cartography 11, 226, 235, 236
Encyclopaedia Britannica 41
England: cartography 11, 223–39; country garden 8, 34–54; home and domestic landscape 8; home reconfigured in colonial Ceylon 8, 19–33; moral landscape 7; Picturesque 8, 55–74; seaside resorts 96, 104n3
English, Barbara 52n28
English Empiricism 224–7, 229–30
English literature, literary identity of women writers 240
Englishness 42, 50; *see also* Britishness
environmental determinism 20–1
environmentalism 168
equality/inequality: nation-state and gender difference 210; Shaker 169, 178; "Song of Songs" 264
Esherick, Wharton 259, *260*
Evangelicalism 22, 25, 30n1, 31n11
Evans, F.W. 176
Everett, N. 71n30
exclusion, Adirondacks 130–1

Fabara, Sandra 269–70, 275, 277, 278–80
Fab Five Feddy (graffiti artist) 272
Fadeeva, N.A. 246, 247
Falk, Marcia 260, 261, 264
Farrar, Virginia 234–5, *234*
Fëderovna, Maria (formerly Princess Sophia-Dorothea of Württemberg) 241
Fell, H. Granville 265–6, *266*
female body: Adirondack wilderness as 125, 137, 138; as Picturesque 56–60, 65–8; representation in Columbian Exposition (1893) 10–11, 182–202; representation in "Song of Songs" 254–68, *257, 260, 261, 262, 266*
femininity: cartography 11, 223, 226, 228–9, 231–7, *234*; Columbian Exposition (1893) 190–3, 195; forms of travel 110; graffiti art 270, 275–6, 277, 279, 280; of home realm 30; of wildreness 2, 9
feminism: approach to gender and landscape 1–12 (*see also individual topics*); critique of colonial discourse 218; critique of public/private 4–6, 35; literary history 240; visual pleasure 212–13
Femmes Fatales Exhibition (1984) 269, 277, 279–80
Ferguson, J. 29
Fielding, Henry 71n21
Fildes, Sir Samuel Luke 49

First Nations *see* indigenous peoples
First World War, Canadian participation 203, 209–10
Fishbein, Morris 118
Flanner, Janet 149, 152
Flint, Timothy 177
Forrest, M. 51n6
Fox, M.V. 258
France, Anatole 153
France, seaside resorts 96
Frank Leslie's Illustrated Newspaper 91n14
freedom 116, 122
French, Daniel C. 192
French literature 240
French Revolution 71n29
Freud, Sigmund 66, 240
Fried, Michael 66
friendship, Barney's literary salon 145–61
Fuller, Thomas 205

Gablik, S. 274
Galileo 229
Gan, Elena 242, 243, 244, 245–7, 250
garden(s): Barney's *fêtes pastorales* 157; country garden in England 8, 34–54; kitchen garden in Columbian Exposition (1893) 196; metaphor for home 8; National Gallery of Canada 205–7; Shaker 172, 173, 174–5; "Song of Songs" 12, 254–68, *257, 260, 261, 262, 266*
Garden: An Illustrated Weekly Journal of Horticulture in All Its Branches, The 51nn13,14
Gardener's Magazine of Rural and Domestic Improvement, The 44, 51n12
Garden of Eden 175, 254, 256, 257, 263, 265
Gatton, J.S. 159n1
Gay, P. 31n7
Gellner, Ernest 219n6
gender: relation to landscape 1–12; *see also individual topics*
Gendered Landscapes conference (Pennsylvania State University, 1999) 2
gender stereotypes 1
gender studies 1
generalization, cartography 236–7
Genesis 254, 256, 257
Genlis, Madame de 241
geography 232
Geological Survey of Canada 212
Gide, André 145
Gilbert, S. 240, 241
Gilman, Charlotte Perkins 141n8

Gilpin, William 64, 69n13
Glanvill, J. 224, 226
Gobineau, Joseph-Arthur, comte
    de 27
Godey's Lady's Book 83, 87, 90n11
Goethe, Johann Wolfgang von 56, 57
Gogol, Nikolai 242, 248
Goncharov, Ivan Aleksandrovich 242
Good Housekeeping 120
Gourmont, Remy de 145, 153
government policy 2–3
Graefer, John Andrew 58, 59
graffiti art movement 12, 269–81
Graham, F. 140n5
Gramont, Elizabeth de 152, 160n8
grand tour 56, 58–9, 66, 70–1n21
Graves, J.T. 82
Great Britain: British home reconfigured
    in colonial Ceylon 8, 19–33; *see also*
    England
Great Exhibition (1862) 70n17
Griffiths, Gareth 208
group identity 10; Barney's literary salon
    156, 157
Group of Seven 203–20
Group of Seven: Art for a Nation Exhibition
    (1995) 207
Gruffudd, P. 7
Gubar, S. 240, 241
Guevara, N. 279

Hackert, Jakob Philipp *60*
Hagopian, P. 272
hairstyles, American beach 100
Hall, C. 31n11
Hall, Radclyffe 145, 152
Halttunen, Karen 79
Hamilton, Emma 56–60, *57*, 65–8, *67*
Hamilton, Geoff 37, 44
Hamilton, Sir William 56–60, 65–8
Haraway, Donna 236
Harley, Brian 235
Harpers Weekly 105n20
Harris, J. 233
Harris, Lawren 219nn2,4
Hayden 4
Hayden, Sophia G. 190
health: camping in Adirondacks 128–9;
    cottage gardening 44–5; Picturesque as
    healing 56, 58, 65–8, 69
Heldman, K. 273
Hepworth, M. 30
hermeneutic knowledge 224–7
hermetic movement 231
Herzen, Aleksandr 243
Heston, Alfred 97–8

heterogeneity 59–69
Hine, Al 115, 117
hip-hop subculture 271
Hippocrates 20
hitchhikers 117–18
Holiday 9, 110–12, 115–17, 118, 121
Holt, F. 104n9
Holzer, Jenny 280
home 5, 8; on American railroad 9,
    77–93; dual meaning in colonial
    Ceylon 8, 19–33; graffiti art 274;
    redefinition as public space 10;
    transplantation to non-western world 8,
    56–69
homoeroticism 24, 157, 158
homophobia 27
Hooke, Robert 230
hooks, bell 5
Horticultural Register 44
hortus conclusus, garden of "Song of Songs"
    12, 255–67
Howe, Julia Warde 195
Howells, William Dean 85–6, 186, 187
Howkins, A. 51nn7,18
How to Make Home Happy (1884) 90n7
Hoyles, Martin 37, 43, 44, 45, 46, 50,
    51nn5,19
Huckins, P.L. 98
Hulme, Peter 218
humanism 235
Huntington, Emily 196
Hyams, Edward 36, 43
hybridity, racial 27

Iddings, L.M. 86
identity 6; cottage garden 36; graffiti and
    271–2; West Union Shaker village
    162–81; *see also* literary identity
I Love Lucy (TV program) 117–18
immigration *see* migration
imperialism 43, 200n20; *see also*
    colonialism
India, colonialism 31n10
indigenous peoples 2; and settler-invader
    subject 208, 216, 218, 219n3; *see also*
    Native Americans
industrialization 136–40, 209
Ingram, V. 51n6
institutions, geography 7
interdisciplinarity 1, 2, 4
interpretation 3
Israel, landscape and nation 207, 208
Italy, landscape gardening 57–60

Jackson, A.Y. 205–6, *206*, 219n4
Jackson, J.B. 3

Jacobs, J.M. 51nn9,10
Jacobs, Jane 182
Janis, Sidney 273
Japan, Picturesque 8, 61–3, *63*
Jekyll, Gertrude 36–7, 48
Jenkins, I. 65
Jenkins, Richard Wade 26
Jewish interpretation of "Song of Songs"
    11, 255
Johnson 45
Johnston, Anna 208, 219n3
Johnston, Frank 219nn2,4
Jones, Chillian 205
Jones, K. 232

Karamzin, Nikolai 249
Kasson, John 104n7
Kelly, C. 249
Kent, William 69n9
Khvoshchinskaia, Nadezhda 242, 244,
    250
Khvoshchinskaia, Praskovia 250
Khvoshchinskaia, Sof'ia 242, 244, 249–50
Kilian, T. 6
King, Ronald 37
Kireevsky, Ivan 244
Knight, Richard Payne 57
knowledge, women's access to 5
Kokhanovskaia, Nadezhda *see*
    Sokhanskaia, Nadezhda
Kolodny, A. 3, 172
Kristeva, J. 27
Kul'man, Elizaveta 246

Labourers' Friendly Society 40
Lachmann, Richard 273
Ladies' Home Journal 99
Lady Pink *see* Fabara, Sandra
Laing, C. 186, 200n22
Landes, J. 5
Landowska, Wanda 152
landscape: private and public language
    4–6; relation to gender 1–15; *see also
    individual topics*
Landscape Architecture 205
landscape architecture 55; Canadian
    National Gallery 204–7
landscape design and gardening 7, 36–7,
    55, 57–60, *60*
landscape history 178
landscape painting, Canada 203–20, *206,
    215*
landscape studies, creating new language
    2–4
Landy, Francis 256–7, 259, 263, 264
Lang, K. 51n5

language, of landscape 2–4
Lawson, Alan 208, 218, 219n3
Lears, T.J.Jackson 215, 216
Lee (graffiti artist) 272, 276, 277
Lee, Ann 163, 164, 168
Lefort, Claude 211
legitimacy, graffiti art 270, 274, 276–81
leisure: cottage gardening 42, 44, 48;
    discipline of 109, 121
Lenéru, Marie 160n8
Lerner 2
Leslie, M.F. (Mrs Frank) 77, 85, 91n14
Le Vaillant, François 60–1, *61*
Lewis, F. 29–30
Lewis, Peirce 3
Lewis, R.E. 25
Lippincott's Magazine 128
Lismer, Arthur 219n4
literary identity: Barney's literary salon 10,
    145–61, *147, 148, 150*; Russian women
    writers 11, 240–53
literature: gender and landscape 2;
    Russian 11, 240–53
Livingstone, D.N. 20
Locke, John 225–6, 230, 236
Look Inside a Shaker Kitchen, A 171
Lotman, Iuri M. 241, 251
Loudon, J.C. 36, 38, 43, 44, 45
Louÿs, Pierre 151, 153, 160n6
love, illustrative history of "Song of
    Songs" 11–12, 254–68
Lowentahl, David 3
Loy, Mina 152, 160n8
Lutyens, Edwin 51n14

M. 175
MacCannell, Dean 139
McClintock, Anne 210, 212, 213, 214,
    216–17
McCracken, D.P. 52n24
MacDonald, J.E.H. 219nn2,4
McGibbon, J. 5
McKay, Ian 215
Mailer, Norman 272
Mannerism 229
Marie-Antoinette, Queen of France 149
Markovich, Mariia *see* Vovchock, Marko
marriage, illustrative history of "Song of
    Songs" 11–12, 254–68
Marsh, George 137
Martineau, Harriet 177
Mary, Virgin, "Song of Songs" 255–6,
    257
masculinity: camping in Adirondacks
    124–42; Canadian national identity and
    landscape painting 214; cartography

223–39; Columbian Exposition (1893) 184–5, *185*, 198n9; country gardens 44–5; forms of travel 110, 113–20; graffiti 273, 275–6, 277, 279; and landscape 2; and morality in colonial Ceylon 8, 19–33
Mason, William 58–9
Massey, Doreen 43
Mata Hari 160n6
Matisse, Henri 254
May, Elaine Tyler 109
Meacham, Joseph 179n4
Medcalf, S. 224
Meinig, Donald 3
Menken, A. 81
Mercure de France 145
Meredith, A. 51n6
metaphors 2; garden of the "Song of Songs" 254–68; knowledge and cartography 224, 226, 228–9
Metcalf, T.R. 28
middle class: beach culture 9, 94–108; gardening 42–3, 48–9, 50
migration: American railroad and public domesticity 87; Canadian attitude to immigrants 210–11
Miller, N. 240
Miller, Page Putnam 179n1
Miller, R. 193
Millie, P.D. 20, 21, 22–3, 28
Mitchell, Silas Weir 128–9
Mitchell, Tom 209–10
Mitchell, W.J.T. 207–8
mobility 8–9
modernization theory 219n6
Moll, Herman *228*, 235
Monk, J. 2, 3, 4
Montesquiou, Robert de 145
Montpensier, Catherine-Marie de Lorraine, Duchesse de 159n4
morality: of climate 8, 20–3, 24, 29; and public/private distinctions 6;
moral landscape 1, 2–3, 6–7; Adirondacks 125, 127, 128, 132, 138; American family travel 109–23; American home and railroad 79–80, 83–5, 86–7, 89; Barney's literary salon 157–9; Canadian national identity 210, 211; colonial Ceylon 8, 19–33; Columbian Exposition (1893) 195; cottage gardening 8, 34, 36, 44–5, 46, 48, 50; garden in "Song of Songs" 12; graffiti 274, 276, 280; knowledge and cartography 226, 229; middle-class beach culture 94–105; redefinition of public space as home 10–11; Russian

women writers 251; Shaker 174–6, 177; transplantation of British home abroad 8, 28–30
moral reform movement, Canada 210
Morden, Robert 228
Morena, Marguerite 152
Morgan, Richard 28
Moser, Barry 264
Mosse, G. 29
mountains, association with masculinity 9, 110, 118–21, *119*, 124–42
Muccigrosso, R. 183, 186, 189
Mullins, Tom 41
Murray, William H.H. 126
myth 2

Nairn, Tom 211
narcissism, and Picturesque 65–8
National Gallery of Canada 11, 203–20, *205*
national identity: in Columbian Exposition (1893) 10–11, 182–202; representation of Canada in art 11, 203–20; Russian 242
nationalism 7, 8; cultural 11, 203–20
National Museum of Canada 212
Native Americans 125, 130–1, 133, 164
Natsume, Sôseki 70n19
natural scenery 55
nature: perfected, of garden of "Song of Songs" 263; Picturesque 56–7, 64–8; Plain Style domination of 229, 230; Shaker view 174; tamed in cottage garden 34, 50; women's access to 9–10
Nead, Lynda 37, 48
Nekrasova, E.S. 246, 247
Nelson, Horatio, Viscount 56, 65
Neva 276–7, *278*
New, W.H. 203
New-England Magazine 177
Newton, Isaac 229
New York, ghetto and graffiti art 12, 269–81
New Yorker, The 1, 274
New York Times, The 83, 84, 87, 104n8, 136, 190, 195, 199n15
New Zealand, landscape and nation 207–8
Nicholas I, Tsar of Russia 251
North, Lord (Governor of Ceylon) 31n10
North American Review 176
Northwestern Railroader 86
Norton, Charles Ledyard 97
Nye, Bill 127

Oberlander, Cornelia Hahn 204, 205–7
Oblenis, Charles 135
Official Guide to the World's Columbian Exposition 186, 187–8, 189, 190, 191, 192, 193
Oldenburg, Claus 272
Olmstead, Frederick Law 184
Osborne, Duffield 97, 98
Other, Picturesque and 8, 59–69
Outing 137, 138

Pacific Tourist and Guide of Travel Across the Continent, The 77
painting: Canada 203–20, *206, 215*; images of rurality 48–9
Palestine, landscape and nation 207
Palmer, Bertha Honore 190
Palmer, I.C. 98
Pan American 113
Paris, Barney's literary salon 10, 145–61, *147, 148, 150, 155*
Paris Exposition (1878) 189
Parks, Addison 275–6, 277
patriarchy 213, 218
Pavlova, Karolina 245
Paxton, Joseph 44
Peirce, Charles 236
Pennsylvania Railroad 78, 82
Penny Cyclopaedia of the Society for the Diffusion of Useful Knowledge, The 39–40, 41, 42
Pepys, Samuel 232
performing art, attitudes of Emma Hamilton 56–60, 66, *67*
Phelps, Orson "Old Mountain" 127
Philadelphia Inquirer 105n19, 269
Philadelphia Photographer, The 135
Phillips, R. 20
Phillips, Wendell 140n5
photography: Adirondacks 9, 124–42; American cities 198n10; beach culture 94, 95, 101, 102, 104; *see also* postcards
Picturesque 8, 55–74
Picturesque Chicago 190
pioneering, Canada 217
Piroli, Tommaso *57*
Pitt, Moses 230
place(s) 3, 4
Plain Style knowledge and cartography 11, 223–31, *227, 228*, 235–7
Pleasant Hill Shaker village 167, 171, 177
Pletnev, Petr 245
politics: cartography 229, 231; cottage gardening as diversion from 44, 45, 48; literature 243

Poovey, M. 246
pop art 273
Pope, Marvin 264
Popenoe, Lucille 121
popular magazines: American family travel 9, 109–23; beach culture 95, 99
Porter, Edwin S. 105n26
positionality 3, 8, 213
positivism 224, 236
postcards: American seaside 95, 98, 100, 101–2, *101, 103*; Columbian Exposition (1893) 198n10
postcolonialism 218
post colonial studies 1
Post-Graffiti Exhibition 279
postmodernism, cartography 235–6
Pound, Ezra 145, 152
power 1, 2, 3, 4–6, 7, 12; Canadian national identity and landscape painting 214; cottage garden 36; scientific knowledge and cartography 232–3, 236; Shaker communities 168–9; "Song of Songs" 264; transplantation abroad of British home 8
Price, Sir Uvedale 57, 64–5, 66
privacy, and domesticity 199n20, 274, 275
private/public space 4, 10; American family vacations 9, 109–23; Barney's literary salon 10; cartography 224; colonialism in Ceylon 8, 28–30; Columbian Exposition (1893) 191, 193, 195, 199n20; English garden 8, 34–54; language of landscape 4–6; middle-class beach culture 9, 94–108; mobility 8–9; reconstruction of West Union Shaker site 10, 171–2; Russian women writers 11
Proctor-Smith, Marjorie 168
progress 20, 28, 215, 216
Protestant work ethic 24
Proust, Marcel 153
provinciality, Russian women writers 11, 240–53
public art 168
public domesticity, railroad travel 9, 77–93
public identity, Russian women writers 11, 240–53
public landscape, beach as 9
public/private space *see* private/public space
public schools 25
Public Vision Exhibition (1982) 213–14
Pullman, George Mortimer 83–4
Pullman, George Morton 183

Pullman Company 82
Pushkin, Alexander 242, 246, 250–1
Putnam, Samuel 154

Raban, Ze'ev 260–1, *261*
race 3; Adirondacks 130–1; American
    railroad and public domesticity 80, 86,
    87, 88–9; built environment 10;
    Canadian nationalism 210, 213, 218;
    colonialism and masculinity in Ceylon
    8, 19–30; Columbian Exposition (1893)
    185, 189, 191, 195; graffiti art 12, 276,
    280
Rachilde (Marguerite Eymery Valette)
    160n8
racial degeneration 20, 26–8, 29
racism 20, 27; Vancouver's Chinatown
    constructed as immoral landscape 7
Railway Age 86, 90n10
Railway Age Gazette 90n11
railway(s) 121; Adirondacks 136–7;
    American family travel 111, 112, 113,
    115, 121; graffiti 274; public
    domesticity 9, 77–93
realism, Russian literature 242, 243, 244,
    248, 249
"Recent Travelers in America" 177
Reed, Michael 35
reflexivity 3, 68
regional differences 6
religion: and landscape 3; Plain Style
    knowledge and cartography 229, 231;
    see also Shakers
*Reports of the Special Assistant Poor Law
    Commissioners on the Employment of Women
    and Children in Agriculture* (1843) 46
Rhode, Eleanor 49–50
Rich, Adrienne 203
Richter, Amy 122n1
Rilke, Rainer Maria 153
road travel 113–20
Robinson, William 36, 43, 51n12
Rolling Stone 273
Romanticism 55, 70n20, 214
Root, John Wellborn 184
Rosa, Salvator 55
Rose, Gillian 3, 4, 5, 171
Rose, Tricia 271
Rothenberg, C. 5
Rousseau, Jean-Jacques 71n22, 243
Rowlandson, Thomas *67*, 68
Royal Society of London 223, 225–7, 228,
    229, 230, 231, 232, 233, 236
rurality: images in art 48–9; Russian
    women writers 11, 240–53
Ruskin, John 63

Russia, literary identity of women writers
    11, 240–53
Russian Revolution 209
Rutherford, J. 25
Ryan, Mary P. 200n23
Rybczynski, Witold 204–5
Rydell, Robert W. 197n1, 199n17

Sabattis, Mitchell 127, 131
sacred 11–12, 254–68
Safdie, Moshe 204, 219n5
Sale, Ewell 111–12
Sand, George 241, 243, 251
Sappho 157–8
Sarratoga Springs 129–30
Saturday Evening Post, The 9
Savannah, Florida and Western Railway
    Company 82
Savkina, Irina 241–2, 248
Schaffer, S. 227
Schuyler, Montgomery 187, 188, 189
scientific philosophy 11, 223, 224–7;
    cartography 228–31
Scribners magazine 97
Seashore Frolics (film) 102–3
sea travel 111, 112, 115
Second World War 49
Sedgwick, C.M. 176
self-discipline: American home and
    railroad 79–80, 84, 85, 87; masculinity
    in Ceylon 19–33
self-expressive home 79–80, 85–9
self-knowledge, visiting West Union
    Shaker village 163
self-representation: graffiti 275; Russian
    women writers 11, 240–53
Seller, John *227*
semiotics 230, 236
Senex, John 230, 235
Senkovsky, Osip 246
Sentimentalism 247
settler-invader subject, representation of
    Canada 11, 203–20
Settles, Mary 177
sexism 27, 58
sexuality: American beach culture 95, 98,
    99, 100, 103; and architecture in
    Columbian Exposition (1893) 10–11,
    182–202; Canada 210; and colonialism
    22, 24, 25–8, 29; overtones in
    Picturesque 65, 66–8; "Song of Songs"
    257, 264
sexuality and gender studies 1
Shakers 10, 162–81, *162, 166, 167, 170*;
    outsiders' views 176–8
Shapin, S. 227

Sharp, Joanne P. 204, 218
Shaw, Marian 200n25
Shearer, F.E. 77
Shearer, Lloyd 115–16, 117
Shelgunov, Nikolai 243–4
Shiga, Shigetaka 63
Silvester, Susan 39, 46
Sloan, K. 65
social history 168
social interaction 4
social purity movement 25, 210
Society for Bettering the Conditions and
    Increasing the Comforts of the Poor 38
Sokhanskaia, Nadezhda 242, 244, 245
"Song of Songs", illustrative history
    11–12, 254–68, *257, 260, 261, 262,
    266*
Sontag, Susan 274–5
Sörlin, S. 7
South Africa, Picturesque 8, 60–1, *61*
Southern Pacific 112
Sozanski, Edward 269–70, 275, 279
space 1, 4
Speake, Eliza 177
spectacle: Columbian Exposition (1893)
    188, 196–7; middle-class beach culture
    9, 94–108
Speed, John 227
spiritual place: Canada 216; West Union
    Shaker village 10, 162–81
Spooner, R. 43
Sprat, T. 225, 226
squatting 41
Starchevsky, A.V. 246
statuesque object, Emma Hamilton as
    69n7
Stein, Gertrude 145, 152, 160n8
Stendhal 55
Stevenson, Louise 79
Stewart, Susan 173
Stillman, William James 140n4
Stoddard, Seneca Ray 9, 124–42, *132,
    133, 135, 136, 139*
Stoler, L.A. 19, 22, 24, 26, 30
Strachey, J.E. 152
Strasman, John 219n5
Street, Alfred 141n11
subaltern studies 1
subjectivity: knowledge and cartography
    11, 226, 229, 235, 236–7; settler-
    invader 11, 203–20
Sullivan, Louis 186, 199n13
Sullivan Union 177
Sund, Judy 195, 200n21
Sunday Magazine 49
Sunset 110, 111, 112, 113, *114*, 121

Sunset Ideas For Family Camping 119–20
surveillance 9–10
Suzzallo, Henry 210–11
Sweet, C.S. 81
symbolic knowledge and cartography
    224–7, 231
symbolic landscapes 10, 145–61

tableaux vivants 56
Taki 271
Taylor, Benjamin Franklin 77
Taylor, James 23, 26–8
Teitelbaum, Matthew 214
television, representation of automobile
    travel 117–18
Temple de l'Amitié (Temple of Frienship)
    146–7, 149–51, *150*, 153–4, 157, 158–9
Temple of Love, Versailles 149
Terrell, Mary Church 88
Terrie, Philip 140n3
textual mapping 11–12
Thacker, Christopher 260
Thompson, Flora 47–8
Thomson, Tom 214, 219n4
Thorpe, Thomas Bangs 129
Tiffin, Helen 208
Till, Karen 4
tobacco cards 99
Toklas, Alice B. 152
Tolstoy, Leo 242, 243, 244
Tomashevsky, Boris 241, 244, 246
Tosh, J. 24–5
totalitarianism 211
tourism: Adirondacks 9, 124–42; Canada
    216; family travel 9; middle-class beach
    culture 9, 94–108; Shaker cultural
    landscape 163, 171, 173–4, 178–9
Trans World Airlines 110
Travel 112
travel landscapes 8–9; American family
    vacations 9, 109–23; Picturesque 61–3,
    61, 63
Troubridge, Lady Una 152
Trudeau, Edward L. 141n7
Tsuji, N. 70n19
Turgenev, Ivan 242, 243, 244
Turner, Frederick Jackson 198n9

United Airlines 113
United States: Adirondacks photographed
    by Seneca Ray Stoddard 9, 124–42,
    132, 133, 135, 136, 139; architecture
    and female body in Columbian
    Exposition (1893) 10–11, 182–202,
    185, 192, 194, 196; cartography
    234–5, *234*; family travel 9, 109–23;

gender identity and spiritual place
in West Union Shaker village 10,
162–81, *162, 166, 167, 170*; ghetto
and graffiti art 12, 269–81; middle-class
beach culture 9, 94–108, *101, 103*;
railroad and public domesticity 9,
77–93
universalism, architecture of the
Columbian Exposition (1893) 189, 190
University of Toronto, Hart House 217
urban environment: gardens 35; graffiti
12, 269–81
urban/rural division: identity of Russian
women writers 11, 240–53; moral
landscape 7
utopianism, of reconstructions of Shaker
sites 168–71

Valéry, Paul 152
value formation 6
Valverde, Mariana 210
Van Brunt, Henry 183–5, 186–7, 189,
190
Van Vechten, Carl 152
Varley, F.H. *215*, 219n4
Verevkin, N.V. 244–5
Vincennes Commercial 177
Virginia, cartography 234–5, *234*
virgin landscape: conservation of
Adirondacks 137–40; representation of
Canada 11, 203–20, *206, 215*
vision 8; distinguished from visuality 3;
feminist theories 212–13; Picturesque as
mode of 8, 55
visuality 3
Vivien, Renée 153, 157, 160n8
Vovchok, Marko (Mariia Markovich) 242,
243–4
voyeurism 94

Walker, M. 6–7
Wallace, Edwin 141n5
Walpole, Horace 56–7, 71n28
Wasf literary device 264–5
Waters, M. 36, 51nn5,11
Wauthier, Magdeleine 159n4
Weimann, J.M. 190, 197
Wells, Ida B. 88, 199n14
Westling, L.H. 2–3
West Union Shaker site 10, 162–81, *162,
166, 167, 170*

Westways 110, 111, 116, 117, 118, *119,*
121
Wickham, Anna 160n8
Wilde, Dolly 145
wilderness: Adirondacks photographed by
Seneca Ray Stoddard 9, 124–42;
equated with nation in Canadian art
11, 203–20, *206, 215*; family travel 8,
118–21; viewed as female 2, 9
Wild Style (film) 276–8
Wilkins, Augusta 120–1
Williams, Raymond 38
Williams, William Carlos 153
Willmott, Robert A. 43, 44
women: American railroad travel 9,
77–93; Adirondacks 128–9; and
cartography 223; colonial Ceylon
23, 26–30; colonial discourse 217–18;
country gardening 39, 46–50;
graffiti artists 269–70, 273; *see also*
femininity
women's emancipation 241
women's issues 223
women's movement, Canada 209, 216,
217
Woodruff 21
work: female, and family travel 110, 114,
116, 120–1; and leisure in cottage
gardens 42, 44, 46–50; *see also* division
of labor
working class: Canada 216; development
of cinema 104; English country gardens
34–54
Wragg, Arthur *262*
Wright, G. 199n12
Wright, Gwendolyn 89n5
Wright, Lucy 179n4

Yaeger, P. 173
Young, Arthur 38–9
Young, Iris Marion 27
Young, Isaac 166–8
Young, R. 27

Zeller, Suzanne 212
Zemans, Joyce 203–4, 218
Zeneida R-va *see* Gan, Elena
Zhelikhovskaia, V.P. 246, 247
Zhukova, Maria 241–2, 243, 244, 247–9,
250
Zlotnick, S. 52n22